THE ROLE OF SCIENTIFIC AND TECHNICAL DATA AND INFORMATION IN THE PUBLIC DOMAIN

PROCEEDINGS OF A SYMPOSIUM

Julie M. Esanu and Paul F. Uhlir, Editors

Steering Committee on the Role of Scientific and Technical
Data and Information in the Public Domain

Office of International Scientific and Technical Information Programs

Board on International Scientific Organizations

Policy and Global Affairs Division

NATIONAL RESEARCH COUNCIL
OF THE NATIONAL ACADEMIES

THE NATIONAL ACADEMIES PRESS
Washington, D.C.
www.nap.edu

THE NATIONAL ACADEMIES PRESS   500 Fifth Street, N.W.   Washington, DC 20001

NOTICE: The project that is the subject of this report was approved by the Governing Board of the National Research Council, whose members are drawn from the councils of the National Academy of Sciences, the National Academy of Engineering, and the Institute of Medicine. The members of the committee responsible for the report were chosen for their special competences and with regard for appropriate balance.

Support for this project was provided by the Center for Public Domain (under an unnumbered grant), the James D. and Catherine T. MacArthur Foundation (under grant no. 02-73708-GEN), the National Library of Medicine (under purchase order no. 467-MZ-200750), the National Oceanic and Atmospheric Administration/National Weather Service (under an unnumbered purchase order), and the National Science Foundation (under grant no. GEO-0223581). Any opinions, findings, conclusions, or recommendations expressed in this publication are those of the author(s) and do not necessarily reflect the views of the organizations or agencies that provided support for the project.

International Standard Book Number 0-309-08850-X (Book)
International Standard Book Number 0-309-52545-4 (PDF)

Copies of this report are available from Board on International Scientific Organizations, 500 Fifth Street, NW, Washington, DC 20001; 202-334-2807; Internet, http://www7.nationalacademies.org/biso/.

Additional copies of this report are available from the National Academies Press, 500 Fifth Street, N.W., Lockbox 285, Washington, DC 20055; (800) 624-6242 or (202) 334-3313 (in the Washington metropolitan area); Internet, http://www.nap.edu.

Copyright 2003 by the National Academy of Sciences. All rights reserved.

Printed in the United States of America

THE NATIONAL ACADEMIES
Advisers to the Nation on Science, Engineering, and Medicine

The **National Academy of Sciences** is a private, nonprofit, self-perpetuating society of distinguished scholars engaged in scientific and engineering research, dedicated to the furtherance of science and technology and to their use for the general welfare. Upon the authority of the charter granted to it by the Congress in 1863, the Academy has a mandate that requires it to advise the federal government on scientific and technical matters. Dr. Bruce M. Alberts is president of the National Academy of Sciences.

The **National Academy of Engineering** was established in 1964, under the charter of the National Academy of Sciences, as a parallel organization of outstanding engineers. It is autonomous in its administration and in the selection of its members, sharing with the National Academy of Sciences the responsibility for advising the federal government. The National Academy of Engineering also sponsors engineering programs aimed at meeting national needs, encourages education and research, and recognizes the superior achievements of engineers. Dr. Wm. A. Wulf is president of the National Academy of Engineering.

The **Institute of Medicine** was established in 1970 by the National Academy of Sciences to secure the services of eminent members of appropriate professions in the examination of policy matters pertaining to the health of the public. The Institute acts under the responsibility given to the National Academy of Sciences by its congressional charter to be an adviser to the federal government and, upon its own initiative, to identify issues of medical care, research, and education. Dr. Harvey V. Fineberg is president of the Institute of Medicine.

The **National Research Council** was organized by the National Academy of Sciences in 1916 to associate the broad community of science and technology with the Academy's purposes of furthering knowledge and advising the federal government. Functioning in accordance with general policies determined by the Academy, the Council has become the principal operating agency of both the National Academy of Sciences and the National Academy of Engineering in providing services to the government, the public, and the scientific and engineering communities. The Council is administered jointly by both Academies and the Institute of Medicine. Dr. Bruce M. Alberts and Dr. Wm. A. Wulf are chair and vice chair, respectively, of the National Research Council.

www.national-academies.org

Preface

The body of scientific and technical data and information (STI)[1] in the public domain in the United States is massive and has contributed broadly to the economic, social, and intellectual vibrancy of our nation. The "public domain" can be defined in legal terms as sources and types of data and information whose uses are not restricted by statutory intellectual property laws or by other legal regimes, and that are accordingly available to the public for use without authorization. In recent years, however, there have been growing legal, economic, and technological pressures that restrict the creation and availability of public-domain information—scientific and otherwise. It is therefore important to review the role, value, and limits on public-domain STI.

New and revised laws have broadened, deepened, and lengthened the scope of intellectual property and neighboring rights in data and information, while at the same time redefining and limiting the availability of specific types of information in the public domain. National security concerns also constrain the scope of information that can be made publicly available. Economic pressures on both government and university producers of scientific data similarly have narrowed the scope of such information placed in the public domain, thus introducing access and use restrictions on resources that were previously openly available to researchers, educators, and others. Finally, advances in digital rights management technologies for enforcing proprietary rights in various information products pose some of the greatest potential restrictions on the types of STI that should be accessible in the public domain.

Nevertheless, various well-established mechanisms for preserving public-domain access to STI—such as public archives, libraries, data centers, and ever-increasing numbers of open Web sites—exist in the government, university, and not-for-profit sectors. In addition, innovative institutional and legal models for making available digital information resources in the public domain or through "open access initiatives" are now being developed by different groups in the scientific, educational, library, and legal communities.

In light of these rapid and far-reaching developments, the Office of International Scientific and Technical Programs organized the *Symposium on the Role of Scientific and Technical Data and Information in the Public Domain*. The symposium was held on September 5-6, 2002, at the National Academies in Washington, D.C. The meeting brought together leading experts and managers from the public and private sectors who are involved in the creation, dissemination, and use of STI to discuss (1) the role, value, and limits of making STI available in the public domain for research and education; (2) the various legal, economic, and technological pressures on the

---

[1] STI has been used for several decades to refer to scientific and technical information, generally limited to scientific and technical (S&T) literature. For purposes of this symposium, it was used in the broader sense to refer also to scientific data so as to be comprehensive.

producers of public-domain STI and the potential effects of these pressures on research and education; (3) the existing and proposed approaches for preserving the STI in the public domain or for providing "open access" in the United States; and (4) other important issues in this area that may benefit from further analysis.

The main question that was addressed by the symposium participants was, What can the S&T community itself do to address these issues within the context of managing its own data and information activities? Enlightened new approaches to managing public-domain STI may be found to be desirable or necessary, and these need to be thoroughly discussed and evaluated. The primary goal of this symposium, therefore, was to contribute to that discussion, which will be certain to continue in many other fora and contexts.

The symposium was organized into four sessions, each introduced by an initial framework discussion and then followed by several invited presentations. The first session focused on the role, value, and limits of scientific and technical data and information in the public domain. This was followed in the second session by an overview of the pressures on the public domain. Session three explored the potential effects on research of a diminishing public domain, and the final session focused on responses by the research and education communities for preserving the public domain and promoting open access to various types of STI.

Different aspects of the issues discussed in this symposium have been addressed in some detail already in several reports previously published by the National Academies. In 1997, the report, *Bits of Power: Issues in Global Access to Scientific Data,* examined the scientific, technical, economic, and legal issues of scientific data exchange at the international level.[2] In 1999, another report, *A Question of Balance: Private Rights and the Public Interest in Scientific and Technical Databases*, looked at the competing public and private interests in scientific data and analyzed several different potential legislative models for database protection in the United States from the perspective of the scientific community.[3] In 2000, the study, *The Digital Dilemma: Intellectual Property Rights in the Information Age*, discussed the conundrums of protecting intellectual property rights in information on digital networks.[4] And, in 2002, the National Academies released a report that investigated the resolving of conflicts in the privatization of public-sector environmental data.[5]

Most recently, the National Academies' U.S. National Committee for CODATA collaborated with the International Council for Science (ICSU), the United Nations Educational, Scientific, and Cultural Organization (UNESCO), the International Committee on Data for Science and Technology (CODATA), and the International Council for Scientific and Technical Information (ICSTI) to convene a related event, the International Symposium on Open Access and the Public Domain in Digital Data and Information for Science.[6] This meeting, held in March 2003 at UNESCO headquarters in Paris, focused on the same categories of issues as the September 2002 Public Domain symposium, except that it reviewed the existing and proposed approaches for preserving and promoting the public domain and open access to S&T data and information on a global basis, with particular attention to the needs of developing countries, and identified and analyzed important issues for follow up by ICSU in preparation for the World Summit on the Information Society.

This publication presents the proceedings of the September symposium. The speakers' remarks were taped and transcribed, and in most cases subsequently edited. However, in several instances, the speakers opted to provide a formal paper. The statements made in these Proceedings are those of the individual authors and do not necessarily represent the positions of the steering committee or the National Academies. Finally, the committee also commissioned a background paper by Stephen Maurer, "Promoting and Disseminating Knowledge: The Public Private Interface," which is available on the National Academies' Web site only.[7]

---

[2] See National Research Council (NRC). 1997. *Bits of Power: Issues in Global Access to Scientific Data,* National Academy Press, Washington, D.C.

[3] NRC. 1999. *A Question of Balance: Private Rights and the Public Interest in Scientific and Technical Databases,* National Academy Press, Washington, D.C.

[4] NRC. 2000. *The Digital Dilemma: Intellectual Property Rights in the Information Age*, National Academy Press, Washington, D.C.

[5] NRC. 2001. *Resolving Conflicts Arising from the Privatization of Environmental Data*, National Academy Press, Washington, D.C.

[6] The Proceedings of the March open access symposium will be published in print and online by the National Academies Press in 2004. For additional information on the symposium, see the CODATA Web site at http://www.codata.org.

[7] Copies of this paper are available on the National Academies Web site at http://www7.nationalacademies.org/biso/Maurer_background_paper.html or from the National Academies' Public Access Records Office.

Acknowledgments

The Office of International Scientific and Technical (S&T) Information Programs and the Board on International Scientific Organizations of the National Research Council of the National Academies wish to express their sincere thanks to the many individuals who played significant roles in planning the Symposium on the Role of Scientific and Technical Data and Information in the Public Domain. The Symposium Steering Committee was chaired by David Lide, Jr., formerly of the National Institute of Standards and Technology. Additional members of the Steering Committee included Hal Abelson, Massachusetts Institute of Technology; Mostafa El-Sayed, Georgia Institute of Technology; Mark Frankel, American Association for the Advancement of Science; Maureen Kelly, independent consultant; Pamela Samuelson, University of California, Berkeley; and Martha Williams, University of Illinois at Urbana-Champaign. R. Stephen Berry of the University of Chicago served as chair of the symposium.

The Office of International S&T Information Programs also would like to thank the following individuals (in order of appearance) who made presentations during the workshop (see Appendix A for final symposium agenda): William A. Wulf, President, National Academy of Engineering; R. Stephen Berry, University of Chicago; James Boyle, Duke University School of Law; Suzanne Scotchmer, University of California, Berkeley; Paul David, Stanford University; Dana Dalrymple, U.S. Agency for International Development; Rudolph Potenzone, LION Bioscience; Bertram Bruce, University of Illinois, Urbana-Champaign; Francis Bretherton, University of Wisconsin; Sherry Brandt-Rauf, Columbia University; Jerome Reichman, Duke University School of Law; Robert Cook-Deegan, Duke University; Justin Hughes, Yeshiva University Cardozo School of Law; Susan Poulter, University of Utah School of Law; David Heyman, Center for Strategic and International Studies; Julie Cohen, Georgetown University School of Law; Peter Weiss, National Weather Service; Stephen Hilgartner, Cornell University; Daniel Drell, U.S. Department of Energy; Tracy Lewis, University of Florida; Jonathan Zittrain, Harvard University School of Law; Stephen Maurer, Esq.; Harlan Onsrud, University of Maine; Ann Wolpert, Massachusetts Institute of Technology; Bruce Perens, formerly with Hewlett Packard; Shirley Dutton, University of Texas at Austin; and Michael Morgan, Wellcome Trust.

This volume has been reviewed in draft form by individuals chosen for their technical expertise, in accordance with procedures approved by the National Academies' Report Review Committee. The purpose of this independent review is to provide candid and critical comments that will assist the institution in making its published report as sound as possible and to ensure that the report meets institutional standards for quality. The review comments and draft manuscript remain confidential to protect the integrity of the process.

We wish to thank the following individuals for their review of selected papers: Hal Abelson, Massachusetts Institute of Technology; Bonnie Carroll, Information International Associates; Robert Chen, Columbia University; Joseph Farrell, University of California, Berkeley; Mark Frankel, American Association for the Advancement of Science; Oscar Garcia, Wright State University; Gary King, Harvard University; Monroe Price, Yeshiva University; Pamela Samuelson, University of California, Berkeley; and Carol Tenopir, University of Tennessee.

Although the reviewers listed above have provided constructive comments and suggestions, they were not asked to endorse the content of the individual papers. Responsibility for the final content of the papers rests with the individual authors.

Finally, the Office of International S&T Information Programs would like to recognize the contributions of the following National Research Council (NRC) staff and consultants: Paul Uhlir, director of International S&T Information Programs, was project director of the symposium and coauthored the discussion frameworks with Jerome Reichman, professor at Duke University School of Law and NRC consultant; Stephen Maurer served as a consultant on this project and prepared a background paper on "Promoting and Disseminating Knowledge: The Public Private Interface"; Subhash Kuvelker of the Kuvelker Law Firm also served as a consultant and moderated the pre-symposium online discussion forum; Julie Esanu helped to organize the symposium and served as the primary editor of the proceedings; and Amy Franklin and Valerie Theberge organized and coordinated the logistical arrangements and assisted with the production of the manuscript.

Contents

Session 1: The Role, Value, and Limits of Scientific and Technical (S&T) Data and Information in the Public Domain

1 Discussion Framework — 3
   Paul Uhlir, National Research Council

The Role, Value, and Limits of S&T Data and Information in the Public Domain in Society

2 The Genius of Intellectual Property and the Need for the Public Domain — 10
   James Boyle and Jennifer Jenkins, Duke University School of Law

The Role, Value, and Limits of S&T Data and Information in the Public Domain for Innovation and the Economy

3 Intellectual Property—When Is It the Best Incentive Mechanism for S&T Data and Information? — 15
   Suzanne Scotchmer, University of California, Berkeley

4 The Economic Logic of "Open Science" and the Balance between Private Property Rights and the Public Domain in Scientific Data and Information: A Primer — 19
   Paul David, Stanford University

5 Scientific Knowledge as a Global Public Good: Contributions to Innovation and the Economy — 35
   Dana Dalrymple, U.S. Agency for International Development

6 Opportunities for Commercial Exploitation of Networked Science and Technology Public-Domain Information Resources — 52
   Rudolph Potenzone, LION Bioscience

The Role, Value, and Limits of S&T Data and Information in the Public Domain for Education and Research

   7 Education   56
      Bertram Bruce, University of Illinois, Champaign-Urbana

   8 Earth and Environmental Sciences   60
      Francis Bretherton, University of Wisconsin

   9 Biomedical Research   65
      Sherry Brandt-Rauf, Columbia University

Session 2: Pressures on the Public Domain

10 Discussion Framework   73
      Jerome Reichman, Duke University School of Law

11 The Urge to Commercialize: Interactions between Public and Private Research and Development   87
      Robert Cook-Deegan, Duke University

12 Legal Pressures in Intellectual Property Law   95
      Justin Hughes, Yeshiva University Cardozo School of Law

13 Legal Pressures on the Public Domain: Licensing Practices   99
      Susan Poulter, University of Utah School of Law

14 Legal Pressures in National Security Restrictions   104
      David Heyman, Center for Strategic and International Studies

15 The Challenge of Digital Rights Management Technologies   109
      Julie Cohen, Georgetown University School of Law

Session 3: Potential Effects of a Diminishing Public Domain

16 Discussion Framework   119
      Paul Uhlir, National Research Council

17 Fundamental Research and Education   125
      R. Stephen Berry, University of Chicago

18 Conflicting International Public Sector Information Policies and their Effects on the Public Domain and the Economy   129
      Peter Weiss, National Weather Service

19 Potential Effects of a Diminishing Public Domain in Biomedical Research Data   133
      Stephen Hilgartner, Cornell University

Session 4: Responses by the Research and Education Communities in Preserving the Public Domain and Promoting Open Access

20 Discussion Framework 141
    Jerome Reichman, Duke University School of Law

21 Strengthening Public-Domain Mechanisms in the Federal Government: A Perspective From Biological and Environmental Research 161
    Ari Patrinos and Daniel Drell, U.S. Department of Energy

22 Academics as a Natural Haven for Open Science and Public-Domain Resources: How Far Can We Stray? 165
    Tracy Lewis, University of Florida

23 New Legal Approaches in the Private Sector 169
    Jonathan Zittrain, Harvard University School of Law

24 Designing Public–Private Transactions that Foster Innovation 175
    Stephen Maurer, Esq.

New Paradigms in Academia

25 Emerging Models for Maintaining Scientific Data in the Public Domain 180
    Harlan Onsrud, University of Maine

26 The Role of the Research University in Strengthening the Intellectual Commons: the OpenCourseWare and DSpace Initiatives at Massachusetts Institute of Technology 187
    Ann Wolpert, MIT

New Paradigms in Industry

27 Corporate Donations of Geophysical Data 191
    Shirley Dutton, University of Texas at Austin

28 The Single Nucleotide Polymorphism Consortium 194
    Michael Morgan, Wellcome Trust

29 Closing Remarks 198
    R. Stephen Berry, University of Chicago

Appendixes 201

A Final Symposium Agenda 203

B Biographical Information on Speakers and Steering Committee Members 206

C Symposium Attendees 215

D Acronyms and Initialisms 225

SESSION 1: THE ROLE, VALUE, AND, LIMITS OF SCIENTIFIC AND TECHNICAL DATA AND INFORMATION IN THE PUBLIC DOMAIN

1

Discussion Framework

Paul Uhlir[1]

Factual data are fundamental to the progress of science and to our preeminent system of innovation. Freedom of inquiry, the open availability of scientific data, and full disclosure of results through publication are the cornerstones of public research long upheld by U.S. law and the norms of science. The rapid advances in digital technologies and networks over the past two decades have radically altered and improved the ways that data can be produced, disseminated, managed, and used, both in science and in all other spheres of human endeavor. New sensors and experimental instruments produce exponentially increasing amounts and types of raw data. This has created unprecedented opportunities for accelerating research and creating wealth based on the exploitation of data as such. Every aspect of the natural world, from the subatomic to the cosmic, all human activities, and indeed every life form can now be observed and stored in electronic databases. There are whole areas of science, such as bioinformatics in molecular biology and the observational environmental sciences, that are now primarily data driven. New software tools help to interpret and transform the raw data into unlimited configurations of information and knowledge. And the most important and pervasive research tool of all, the Internet, has collapsed the space and time in which data and information can be shared and made available, leading to entirely new and promising modes of research collaboration and production.

Much of the success of this revolution in scientific data activities, apart from the obvious technological advances that have made them possible, has been the U.S. legal and policy regime supporting the open availability and unfettered use of data. This regime, which has been among the most open in the world, has placed a premium on the broadest possible dissemination and use of scientific data and information produced by government or government-funded sources. This has been implemented in several complementary ways: by expressly prohibiting intellectual property protection of all information produced by the federal government and in many state governments; by contractually reinforcing the traditional sharing norms of science through open data terms and conditions in federal research grants; and by carving out a very large and robust public domain for noncopyrightable data, as well as other immunities and exceptions favoring science and education, from proprietary information otherwise subject to intellectual property protection.

---

[1] Note: This presentation is based on an article by J. H. Reichman and Paul F. Uhlir, *A Contractually Reconstructed Research Commons for Scientific Data in a Highly Protectionist Intellectual Property Environment*, 66 Law and Contemporary Problems 315 (Winter-Spring 2003), and is reprinted with the permission of the authors.

PUBLIC-DOMAIN INFORMATION DEFINED

It is worthwhile at the outset to clearly define what we mean by "public-domain information." Jerome Reichman and I characterize it as "sources and types of data and information whose uses are not restricted by intellectual property or other statutory regimes and that are accordingly available to the public for use without authorization or restriction." For analytical purposes, data and information in the public domain can be divided into two major categories:

1. information that is not subject to protection under exclusive intellectual property rights or other statutory restriction; and

2. information that qualifies as protectible subject matter under some intellectual property regime, but that is contractually designated as unprotected.

The first major category of public-domain information can be further divided into three subcategories: (i) information that intellectual property rights cannot protect because of the nature of the source that produced it; (ii) otherwise protectible information that has lapsed into the public domain because its statutory term of protection has expired; and (iii) ineligible or unprotectible components of otherwise protectible subject matter.

This presentation focuses primarily on categories 1(i) and 2, so I will just say a few words about categories 1(ii) and (iii) here. Information that has lapsed into the public domain because it has exceeded the statutory term of protection under copyright is currently the life of the author plus 70 years in the United States. This category constitutes an enormous body of creative works with great cultural and historical significance. Because of the long lag time in entering the public domain, however, it is not of great importance to most types of research.

The other subcategory of information that is not subject to protection under exclusive intellectual property rights consists of ineligible or unprotectible components of otherwise protectible subject matter, such as an idea, fact, procedure, system, method of operation, concept, principle, or discovery, all of which are expressly excluded from statutory copyright protection. Thus, all ideas or facts contained in an otherwise copyrighted work have been—at least in the past—unprotected and could be used freely. Although this public-domain material is highly distributed among all types of works, it is of particular concern to research and education and will be discussed in some detail later in the symposium.

Finally, a third, related category is information that becomes available under statutorily created immunities and exceptions from proprietary rights in otherwise protected material. Instead of being in the public domain because it is unprotectible subject matter, it is otherwise protected content that is allowed to be subject to certain unprotected uses under limited circumstances, subject to case-by-case interpretation. Such immunities and exceptions allow for the use of proprietary information for purposes such as scholarship, research, critical works, commentaries, and news reporting, but their specific nature and extent varies greatly among different jurisdictions. Known as "fair uses" in the United States, they also tend to be quite controversial and are frequently in dispute by rights holders. Although copyright law has not formally treated these immunities and exceptions as public domain *per se*, a number of them may be construed as functional equivalents of "public-domain uses." Because many immunities and exceptions are allowed only in the context of not-for-profit research or education, this category of public-domain uses, although relatively small, is especially important.

The main goal in this opening presentation is to coarsely map out the scope and nature of the public domain that has served the American scientific enterprise so well for many years. We divide the producers of scientific data into three sectors, namely, government agencies (primarily but not exclusively federal), academic and other nonprofit research institutions, and private-sector commercial enterprises. Figure 1-1 presents a conceptual framework for analysis of this information regime, as well as a useful summary of the relative rights across the spectrum of information producers in all three sectors. Because of time constraints, I will focus only on the scope of the public domain in government and academia.

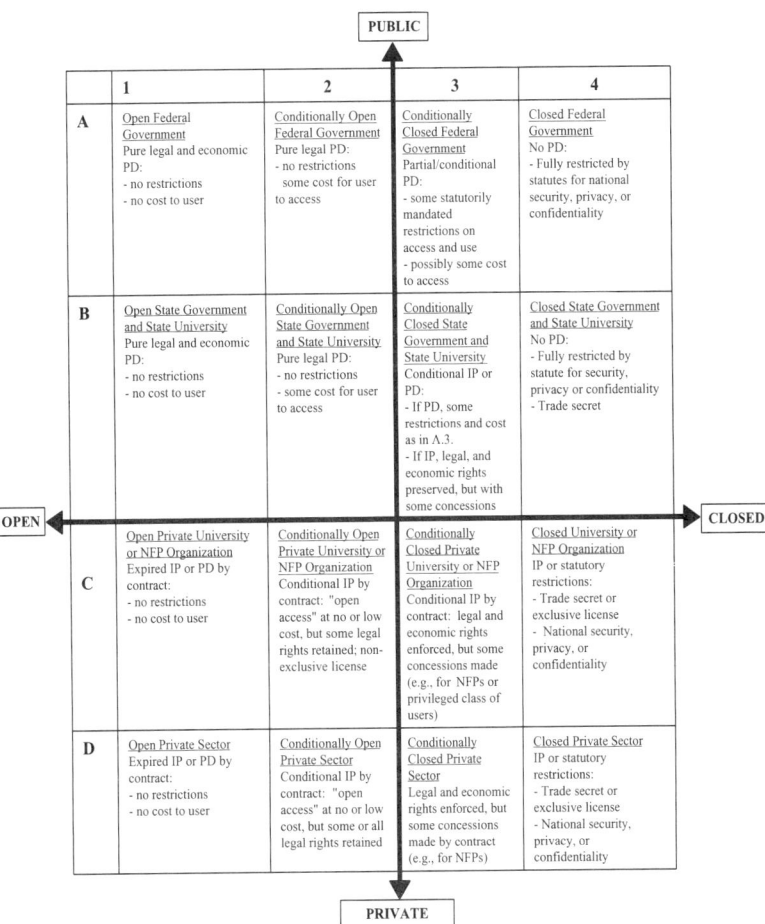

FIGURE 1-1 Conceptual framework for analysis of the S&T information regime. Copyright 2000 by Paul F. Uhlir. Notes: "PD"-public domain; "IP"-intellectual property; "NFP"-not-for-profit.

GOVERNMENT INFORMATION IN THE PUBLIC DOMAIN

The role of the government in supporting scientific progress in general, and in the creation and maintenance of the research commons in particular, cannot be overstated. The U.S. federal government produces the largest body of public-domain data and information used in scientific research and education. For example, the federal government alone now spends more than $45 billion per year on its research (as opposed to "development") programs, both intramural and extramural, with a significant percentage of that invested in the production of primary data sources; in higher-level processed data products, statistics, and models; and in scientific and technical information, such as government reports, technical papers, research articles, memoranda, and other such analytical material.

At the same time, the United States—unlike most other countries—overrides the canons of intellectual property law that could otherwise endow it with exclusive rights in government-generated collections of data or other information. To this end, Section 105 of the 1976 Copyright Act prohibits the federal government from claiming protection in its publications. A large portion of the data and information thus produced in government programs automatically enters the public domain, year after year, with no proprietary restrictions, although the sources are not always easy to find. Much of the material that is not made available directly to the public can be obtained by citizens through requests under the Freedom of Information Act.

The federal government is also the largest funder of the production of scientific data by the nongovernmental research community, as a major part of that $45 billion investment in public research, and many of the government-funded data also become available to the scientific community in the research commons. However, government-funded data follow a different trajectory from that of data produced by the government itself, as I will discuss later on. The following are a number of well-established reasons for placing government-generated data and information in the public domain:

• A government entity needs no legal incentives from exclusive property rights that are conferred by intellectual property laws to create information, unlike individual authors or private-sector investors and publishers. Both the activities that the government undertakes and the information produced by the government in the course of those activities are a public good.

• The taxpayer has already paid for the production of the information. One can argue that the moral rights in that information reside with the citizens who paid for it, and not with the state entity that produced it on behalf of the citizens.

• Transparency of governance and democratic values are undermined by restricting citizens from access to and use of public data and information. As a corollary, citizens' First Amendment rights of freedom of expression are compromised by restrictions on redissemination of public information, and particularly of factual data. It is no coincidence that the most repressive political regimes have the lowest levels of available information and the greatest restrictions on expression.

• Finally, there are numerous positive externalities—particularly through network effects—that can be realized on an exponential basis through the open dissemination of public-domain data and information on the Internet. Many such benefits are not quantifiable and extend well beyond the economic sphere to include social welfare, educational, cultural, and good governance values.

The federal government's policies relating to scientific data activities date back to the advent of the era of "big science" following World War II, which established a framework for the planning and management of large-scale basic and applied research programs. The hallmark of big science, or "megascience," has been the use of large research facilities or research centers and of experimental or observational "facility-class" instruments. The data from many of these government research projects, particularly in the past two decades, have been openly shared and archived in public repositories. Hundreds of specialized data centers have been established by the federal science agencies or at universities under government contract.

Scientific and technical articles, reports, and other information products generated by the federal government are also not copyrightable and are available in the public domain. Most research agencies have well-organized and extensive dissemination activities for such information, typically referred to as "scientific and technical information."

Limitations on Government Information in the Public Domain

Without delving at this point into the growing pressures on the public domain, we need to outline a number of countervailing policies and practices that limit the free and unrestricted access to and use of government-generated data and information. First, there are important statutory exemptions to public-domain access based on national security concerns, the need to protect personal privacy of human subjects in research, and to respect confidential information. These limitations on the public-domain accessibility of federal government information, although often justified, must nonetheless be balanced against the rights and needs of citizens to access and use it.

Another limitation derives from the fact that government-generated data are not necessarily provided cost free. The federal policy on information dissemination, the Office of Management and Budget's (OMB) Circular A-130, stipulates that such data should be made available at the marginal cost of dissemination (i.e., the cost of fulfilling a user request). In practice, the prices actually charged vary between marginal and incremental cost pricing, and these fees can create substantial barriers to access, particularly for academic research.

Another major barrier to accessing government-generated data and information in practice, which I have already alluded to, arises from the failure of agencies to disseminate them or to preserve them for long-term availability. As a result, large amounts of public-domain information are either hidden from public view or have been irretrievably lost.

Yet another limitation derives from OMB Circular A-76, which bars the government from directly competing with the private sector in providing information products and services. This policy substantially narrows the amount and types of information that the government can undertake to produce and disseminate. As regards science, the traditional view has been that basic research, together with its supporting data, are a public good that properly fall in the domain of government activity, although the boundary between what is considered an appropriate public and private function continues to shift in this and other respects.

Finally, the government is required to respect the proprietary rights in data and information originating from the private sector that are made available for government use or, more generally, for regulatory and other purposes, unless expressly exempted. To the extent that more of the production and dissemination functions of research data are shifted from the public to the private sector, this limitation becomes more potent.

ACADEMIC INFORMATION IN THE PUBLIC DOMAIN

The second major source of public-domain data and information for scientific research is that which is produced in academic or other not-for-profit institutions with government and philanthropic funding. Databases and other information goods produced in these settings become presumptively protectible under any relevant intellectual property laws unless affirmative steps are taken to place such material in the public domain. In this case, the public domain must be actively created, rather than passively conferred. This component of the public domain results from the contractual requirements of the granting agencies in combination with longstanding norms of science that aspire to implement "full and open access" to scientific data and the sharing of research results to promote new endeavors.

The policy of full and open access or exchange has been defined in various U.S. government policy documents and in National Research Council reports as "data and information derived from publicly funded research are [to be] made available with as few restrictions as possible, on a nondiscriminatory basis, for no more than the cost of reproduction and distribution" (that is, the marginal cost, which on the Internet is zero).[2] This policy is promoted by the U.S. government (with varying degrees of success) and for most government-funded or cooperative research, particularly in large, institutionalized research programs (such as "global change" studies or the Human Genome Project) and even in smaller-scale collaborations involving individual investigators who are not otherwise affiliated with private-sector partners. It is necessary, however, to look beyond the stated policies and rules built around government-funded data to gain some deeper insights into how they are actually implemented in academic science. In this regard, it is useful to further subdivide the producers of government-funded data into two distinct, but partially overlapping, categories.

The Zone of Formal Data Exchanges

In the first category, the research takes place in a highly structured framework, with relatively clear rules set by government funding agencies on the rights of researchers in their production, dissemination, and use of data. Publication of the research results constitutes the primary organizing principle. Traditionally, the rules and norms that apply in this sphere—which we call "the zone of formal data exchanges"—aim to achieve a bright line of demarcation between public and private rights to the data being generated, with varying degrees of success.

For example, government research contracts and grants with universities and academic researchers seek to ensure that the data collected or generated by the beneficiaries will be openly shared with other researchers, at least following some specified period of exclusive use—typically limited to 6 or 12 months—or until the time of publication of the research results based on those data. This relatively brief period is intended to give the researcher sufficient time to produce, organize, document, verify, and analyze the data being used in preparation of a research article or report for scholarly publication. Upon publication, or at the expiration of the specified period

---

[2] NRC. 1997. *Bits of Power: Issues in Global Access to Scientific Data*, National Academy Press, Washington, D.C.

of exclusive use, the data in many cases (particularly from large research facilities) are placed in a public archive where they are expressly designated as free from legal protection, or they are expected to be made available directly by the researcher to anyone who requests access.

In most cases, publication of the research results marks the point at which data produced by government-funded investigators should become generally available. The standard research grant requirement or norm has been that, once publication occurs, it will trigger public disclosure of the supporting data. To the extent that this requirement is implemented in practice, it represents the culmination of the scientific norms of sharing. From here on, access to the investigator's results will depend on the method chosen to make the underlying data publicly available and on the traditional legal norms—especially those of copyright law—that govern publications.

These organizing principles derive historically from the premise that academic researchers typically are not driven by the same motivations as their counterparts in industry and publishing. Public-interest research is not dependent on the maximization of profits and value to shareholders through the protection of proprietary rights in information; rather, the motivations of academic and not-for-profit scientists are predominantly rooted in intellectual curiosity, the desire to create new knowledge and to influence the thinking of others, peer recognition and career advancement, and the promotion of the public interest.

Science policy in the United States has long taken it for granted that these values and goals are best served by the maximum availability and distribution of the research results, at the lowest possible cost, with the fewest restrictions on use, and the promotion of the reuse and integration of the fruits of existing results in new research. The placement of scientific and technical data and databases in the public domain, and the long-established policy of full and open access to such resources in the government and academic sectors, reflects these values and serves these goals.

The Zone of Informal Data Exchanges

In the second category, individual scientists establish their own interpersonal relationships and networks with other colleagues, largely within their specialized research communities. In this category, which we call "the zone of informal data exchanges," based on the work of Stephen Hilgartner and Sherry Brandt-Rauf, the scientists will more likely be working autonomously in what we refer to as "small science" research. They will generate and hold their data subject to their personal interests and competitive strategies that may deviate considerably from the practices established within the zone of formal exchanges and institutionalized by structured federal research programs. In this informal zone there will be other (nonfederal) sources of funding used, or the federal support that is available will be less prescriptive about open terms of data availability. Moreover, for research involving human subjects, strong regulations protecting personal privacy adds an additional cloak of secrecy.

The "small science," independent investigator approach traditionally has characterized a large area of experimental laboratory sciences, such as chemistry or biomedical research, and field work and studies, such as biodiversity, ecology, microbiology, soil science, and anthropology. The data or samples are collected and analyzed independently, and the resulting data sets from such studies generally are heterogeneous and unstandardized, with few of the individual data holdings deposited in public data repositories or openly shared. The data exist in various twilight states of accessibility, depending on the extent to which they are published, discussed in papers but not revealed, or just known about because of reputation or ongoing work, but kept under absolute or relative secrecy. The data are thus disaggregated components of an incipient network that is only as effective as the individual transactions that put it together. Openness and sharing are not ignored, but they are not necessarily dominant either. These values must compete with strategic considerations of self-interest, secrecy, and the logic of mutually beneficial exchange, particularly in areas of research in which commercial applications are more readily identifiable.

What occurs is a delicate process of negotiation, in which data are traded as the result of informal compromises between private and public interests and that are worked out on an ad hoc and continued basis. Small science thus depends on the flow of disaggregated data through many different hands, all of whom collectively construct a fragile chain of semicontractual relations in which secrecy and disclosure are played out against a common need for access and use of these resources.

The picture that we paint of "big" and "small" science, and of the formal and informal zones of data exchange is, of course, overstated to clarify the basic concepts. The big science system is not free of individual strategic behavior that is typical of the small sciences, and the latter often benefit from access to public repositories where data are freely and openly available, rather than through the ad hoc transactional process. However, the "brokered networks" typical of small science are endemic to all sciences, and access to data is everywhere becoming more dependent on negotiated transacting between private stakeholders.

For the purposes of this symposium, we have chosen big science geophysical and environmental research that uses large observational facilities and small science biomedical experimental research that involves individual or small independent teams of investigators as emblematic of these two types of research and related cultures. We use them to provide real-world examples of the opportunities and challenges now inherent in the role of scientific and technical data and information in the public domain.

THE IMPETUS FOR THIS SYMPOSIUM

Although the scope and role of public-domain data and information are well established in our system of research and innovation, for reasons that subsequent speakers in this symposium will elaborate, there is another trend that is currently under way, which may be characterized as the progressive privatization and commercialization of scientific data and by the attendant pressures to hoard and trade them like other private commodities. This trend is being fueled by the creation of new legal rights and protectionist mechanisms largely from outside the scientific enterprise, but increasingly adopted by it. These new rights and mechanisms include greatly enhanced copyright protection in digital information, the ways in which access to and use of digital data are being contractually restricted and technologically enforced, and the adoption of unprecedented intellectual property rights in collections of data as such.

This countervailing trend is based on perceived economic opportunities for the private exploitation of new data resources and on a legal response to some loss of control over certain proprietary information products in the digital environment. At the same time, it is disrupting the normative customs and foundation of science, especially the traditional cooperative and sharing ethos, and producing both the pressures and the means to enclose the scientific commons and to greatly reduce the scope of data and information in the public domain.

Viewed dispassionately, the need to appropriately reconcile these two competing interests in a socially productive framework is imperative and the goal of such a reconciliation seems clear. A positive outcome would maximize the potential of private investment in data collection and in the creation of new information and knowledge, while preserving the needs of public research for continued access to and unfettered use of data and other public-domain inputs. These pressures, their potential impact on science, and the potential means for reconciling them in a win-win approach are the subjects of the next three sessions, respectively, of this symposium.

2

The Genius of Intellectual Property and the Need for the Public Domain

James Boyle & Jennifer Jenkins[1]

I will start by talking about the genius of the intellectual property system and about what this system is supposed to do. Then I will turn to a broad, overall vision of recent expansions of intellectual property protection and talk a bit about the available conceptual tools, both economic and noneconomic, for thinking about appropriate limits on intellectual property protections. I will conclude with some suggestions about an appropriate larger framework for conducting the inquiry around which this symposium is built.

Although I am going to say a number of negative things about the current state of intellectual property protection, I want to start by stressing what a wonderful idea intellectual property protection is. The economists tell us that intellectual property allows us to escape the problems caused by goods that are nonrival and nonexcludable by conferring a limited monopoly (although some would argue about that nomenclature) and allowing producers of information or innovation to recover the investment that they put into it. But the real genius of intellectual property is that it tries to do something that my colleague Jerry Reichman calls "making markets": that is, it tries to make markets where markets would otherwise be impossible.

For many of you, making markets might seem a cold and unattractive phrase. But you should view it with the romance with which economists view the possibility of making markets. When I say markets, you should get excited: You should think about a spontaneous and decentralized process where all of us have the possibility to innovate and to recover something of the costs of our innovation.

If you are a radical feminist and want to write a book condemning the contemporary workplace, if you have a unique image of what can be done with a slack key guitar, if your pen can draw the horrors of cubicle life in a way that would amuse millions over their breakfast newspaper, the idea is that you could venture out there into the world; and if the market works (if the intermediaries do not stifle your expression, which they often do), you would be able to offer your creation to the public. Similarly, innovations in other areas—the drug, the invention, the Post-it note—would be put to the test of a decentralized set of needs and desires. The market would in turn allow us to say, "this is what we want."

---

[1] The oral presentation was given by James Boyle, William Neal Reynolds Professor, Duke University School of Law. The prose version was cowritten with Jennifer Jenkins, Director of the Duke Center for the Study of the Public Domain. For further discussion of the issues in this presentation, see James Boyle, "The Second Enclosure Movement and the Construction of the Public Domain," 66 *Law & Contemp. Probs.* (Winter/Spring 2003), draft available at http://james-boyle.com/; James Boyle, "The Second Enclosure Movement?" Duke Law School Conference on the Public Domain Webcast Archive (Nov. 9, 2001), available at http://www.law.duke.edu/pd/realcast.htm; and "Tensions between Free Software and Architectures of Control" (Apr. 5, 2001), available at http://ukcdr.org/issues/eucd/boyle_talk_text.shtml.

At present it seems that we want Post-it notes and Viagra and Britney Spears: The point is that this particular construction of culture and innovation is made possible by intellectual property. It is that transaction between the successful innovator or cultural artist and the public that intellectual property tries to make possible, together of course with all the failed innovations, for which there are no audiences, or at least no interested audiences.

The genius of intellectual property, then, is to allow the possibility of a decentralized system of innovation and expression, a system in which I do not have to ask for polka or house music to be subsidized by the state, where I do not need to go to some central authority and plead the case for the funding of a particular type of innovation. Instead, through the miracle either of my own initiative (and the capital markets) I can say "I have faith in this, and we will see whether I am right or wrong." I will make a bet, or my investors will make a bet. That is the genius of intellectual property.

Now I am going to say a lot of very negative things, but do remember that initial positive idea. I want to start with an image of the old intellectual property system. (This is going to sound something like a golden age, which is not what I intend, but given the time constraints, I have to simplify somewhat drastically.)

The first thing to realize is that intellectual property rights used to be quite difficult to violate, because the technologies and activities necessary to do so were largely industrial. You could tell a friend about a book, lend it to him, or sell it to him, but you would need a printing press readily to violate its intellectual property protection. So intellectual property was an upstream regulation that operated primarily between large-scale industrial producers competing against each other. Horizontally it operated to stop them from simply taking the book du jour and ripping that off, or taking the drug formula and ripping that off. But it had relatively little impact on individual users or consumers and on everyday creative acts, whether in the sciences or in the arts.

Second, intellectual property protection was extremely limited in time. Prior to 1978, 95 percent of everything that was written, as my colleague Jessica Litman has pointed out, went immediately into the public domain, because a work was not protected by copyright unless you affixed a copyright notice onto it. Even if you did affix a copyright notice, the copyright would expire after some period of time. We started off in the United States at 14 years of copyright protection. Even up until 1978, this initial term had been extended only to 28 years, at the end of which you had to renew your copyright to maintain protection. A considerable percentage of the things under copyright went into the public domain at that point, because they were not renewed, because there was no salient market. They were then available for all to use. This has changed completely. Now the default is that everything is automatically protected. Every note written in this auditorium will not pass into the public domain until after your lifetime plus 70 years.

Third, intellectual property was tightly limited in scope. If you look at the areas that intellectual property now covers, you can see a sort of steroidlike bloating in every dimension. I look back at some of the confident phrases in my old syllabi: "No one would ever claim that a business method could be patented." Wrong. "No one would ever claim that you could have intellectual property rights over a simple sequence of genomic data, like CGATTGAC." Wrong. And of course, "copyright protection only lasts for life plus 50 years." Wrong.

Intellectual property rights now cover more subject matter, for a longer time, with greater penalties, and at a finer level of granularity. These rights are now backed by legally protected technical measures, which try and make intellectual property protection a part of your use of a work, so you do not have the choice to violate an intellectual property right. They may also come with an associated contract of adhesion, which tells you how, when, and why you may make use of the work.

We have, in other words, a bloating or expansion on every substantive level—breadth, length, depth, and scope. But also a kind of expansion that, because of the changes produced by digital technology and other copying technologies, has changed fundamentally from being industrial regulation that affected upstream producers—the publisher, the manufacturer of the drug—to personal regulation that applies literally on the "desktop."

In a networked world, copying is implicated in everyday acts of communication and transmission, even reading. It is hard for us to engage in the basic kinds of acts that make up the essence of academia or of the sciences without at least potentially triggering intellectual property liability. It is in fact the paradox of the contemporary period that at the moment when the idea of every person with his or her own printing press becomes a reality, so does his or her own copyright police; and for every person's digital lab, the potential of a patent police nearby.

So we have this enormous expansion of intellectual property rights. Although I used the negative terms "bloating" and "expansion," in theory this could all be perfect. It might be that our legislators, responding doubtless to carefully phrased economic arguments rather than to straightforward economic payments, had wonderfully hit the economically optimal balance between underprotection and overprotection. That could have happened.

But can we at least be skeptical? Do we really want, for example, to conduct the following experiment with the American system of science and innovation? "We never really knew how the American system of science worked in the first place, but, by and large, it did. Now let's change its fundamental assumptions and see if it still works." This is an interesting experimental protocol. And, if human subjects are not involved, it might even be a useful one. But it would seem that if you only have one system of science and innovation, then breaking it might have negative consequences. So fundamental tinkering in the absence of strong evidence that the tinkering is needed would not seem like the optimal strategy.

If, then, we are slightly skeptical and are engaged in a process of trying to reflect on what the correct balance is between intellectual property and the material that is not protected by intellectual property, what are the conceptual, practical, political, and economic tools that we have at our disposal?

Well, by simplifying 300 years of history into a minute and a half, there is a longstanding tradition of hostility to and skepticism of intellectual property. Thomas Jefferson in the United States and Thomas Macaulay in Britain expressed this very nicely. It is an antimonopolistic sentiment coming from the Scottish Enlightenment. It is a sentiment that sees the biggest danger as being that monopoly makes things scarce, makes them dear, and makes them bad, as Macaulay said, and that intellectual property is a monopoly like any other. It may be a necessary evil, but it is an evil, and we should therefore carefully circumscribe it. We should put careful limits around it—constitutional limits, limits of time, and so forth.

The antimonopolistic critique of intellectual property offers, of course, a useful set of correctives to our current steroid-fueled expansion. But what we really do not see for nearly 150 or 200 years is any mention of something called "the public domain." Instead, there is an assumption that what is not covered by intellectual property will be "free," a term that is tossed around just as loosely by the philosophers I have mentioned as it is by the contemporary philosophers of the Internet. What do we mean by "free"? No cost? Easily available? Uncontrolled by a single entity? Available with a flat fee on a nondiscriminatory basis? "Free" as in "free speech" or "free" as in "free beer," as Richard Stallman famously observed? Although there are a lot of definitions, it is not very clear what norm of freedom we are trying to instantiate.

Now, over the past 25 to 30 years, one sees in debates about intellectual property an affirmative mention of the public domain. The Supreme Court said that it would be unconstitutional to withdraw material from the public domain, or to restrict access to that which is already available. Academics, including my colleagues David Lange and Jerry Reichman, have written very interestingly about the role and function of the public domain, and I have added a little bit to that discussion myself. Paul David, Suzanne Scotchmer, and others have also contributed to these discussions in fascinating ways. But I think they would all agree that when we talk about the "public domain," it is not quite clear how that term is defined.

What is the public domain? Is material made available through the fair-use privilege under copyright law part of the public domain or not? One could debate that. Is material that is in the public domain only that material which is entirely uncovered by intellectual property, such as a book whose copyright term has expired? Or can it include the little chunk of unprotected material within a work that is otherwise covered by intellectual property, such as the ideas or facts underlying protected expression? We do not have a very good notion.

How does the public domain function? We really do not know that very well either. The intellectual property system we have inherited had a strategy of braiding a thin layer of intellectual property rights around a public domain of ideas and facts, which could never be owned. But one could own the expression or the invention made out of those ideas and facts, leaving the ideas above and facts below in the public domain for the next generation to build on. That actually sounds like an interesting strategy of a mixed public domain and property system.

However, this system is one that we are busily demolishing through expanding intellectual property protections, without, so far as I can tell, either good empirical evidence that it is necessary to change its fundamental premises or good economic models that the changes will work. There is remarkably little empirical study of the actual effects of intellectual property, or of the actual need for particular intellectual property rights rather than the concept itself. I think that this is what we should demand. As one of my colleagues has said: We should not succumb either to merely good data or to good economic models, we need a little bit of both.

Where does this leave us? We are in a situation now that is akin to where the movement to protect the environment was in the 1940s and 1950s. At that time, there was no environment per se. This way of quilting together a series of problems and issues and ecosystems—problems of pollution, ecosystem functions, our relationship to the biosphere—was literally invented in the 1940s, 1950s, and 1960s by a series of brilliant scientific, popular, and other works, which told us that we should see these problems as a whole.

We are in the process of creating our "environment" in the intellectual world. That environment, in my view, is the public domain, and we are still looking for our writers of *Silent Spring* or *A Sand County Almanac*. We need the writers who would manage both to conjure up what it is the public domain does for us and to delineate the ways in which it functions. We need the writers who would give us the equivalent of the notion of ecological interconnections—the unexpected reciprocal connections that turn out to be fundamental. We need the writers who would give us the sense of modesty that environmentalists gave us about interventions in living systems, the sense that perhaps we do not understand terribly well how the public domain operates; so if we tinker with it radically, things will not automatically go well. We seem to lack that sense of modesty; indeed, we seem to be possessed of an extraordinary hubris.

We also lack the associated scholarly analysis, from economic, legal, historical, and other perspectives. There is a need to look at the optimal balance and interconnection between that which is protected and that which is not; to look at the types of production systems that the Internet makes possible, like peer-to-peer production; to follow the kinds of fascinating studies like the one produced by Jerry Reichman and Paul Uhlir on the possibility of establishing well-functioning scientific commons in the new world of scientific inquiry.

As we look at the contemporary mixture of that which is protected and that which is not, it is apparent that we cannot simply say that everything should be free. As Jerry Reichman and Paul Uhlir point out, there are multiple subtle types of limitations and restrictions that may actually be necessary to allow for the free flow of data. Nor can we romanticize the past of big science under the Cold War model, under which so much research was actually paid for by the state but material was not automatically available to everyone. Big science, as we know, has had its own monopolies and blockages.

Instead, we need new cognitive orientations from which to think about these issues. We need, for example, to take a leaf out of the books by the theorists of the commons, such as Elinor Ostrom, who have studied how well-functioning commons continue. We have to analyze carefully how commons are not in fact automatically tragedies, but, as Carol Rose has described, can be comedies: collective resources that are better managed collectively than they would be individually. (The environmentalists have taught us that rhetoric is no substitute for careful science, but we also need a process of actually getting people to understand the stakes.)

So how are we to proceed? Our job in the National Academies is to produce studies, to produce expert opinions, and to concentrate on this field. I would suggest here too that the environmental movement offers us some hints. One of the fascinating things about the environmental movement is the way that elite scientific, economic, and other forms of academic inquiry are put together with a popular movement that is much broader based. A similar approach is needed here to understand the impact of intellectual property expansions. We cannot simply focus on the old questions of the impact of the Bayh-Dole Act, or whether there should be a research exemption to patent law, although these are indeed important issues. The reason that we cannot do that is because there is an urgency to the task, which is going to require speed greater than that produced by the occasionally glacial process of convincing Congress or other bodies, including private bodies, that they have taken a wrong step.

The urgency of our task is underscored by a number of developments accompanying the recent expansions in intellectual property. First, there is a generational shift occurring. Many of the people in the audience were brought up in an era in which the Mertonian ideals of science were taken for granted. They start from the assumption that data should be free and need to be convinced that this assumption should be changed in particular cases. (I will not say they cannot be convinced, but they need to be convinced; and the burden of proof lies with the people who would advocate restriction.) I would argue that, for many younger scientists, this ideal is no longer true. The idea of paying for data is no stranger than paying for reagents. They start from the assumption that data are a costly input like all others, rather from the belief that some economies actually work better when fundamental, upstream research is available freely, even if property rights are introduced further downstream. I do not think one can assume that scientists will automatically continue to have the same kind of skepticism that they currently express toward intellectual property expansions like those enacted in the European Directive on the Legal Protection of Databases and in recently proposed database protection legislation in the United States.

A second development is that the universities and the science establishment, traditionally the "public defenders for the public domain," are changing. When new expansions of intellectual property were proposed, this community, together with librarians and a few others, would suggest a little restraint, would advocate limitations on expansions that went a little far. This was an extraordinarily valuable role for them to play in the legislative process. But it is a role that we can no longer take for granted. Universities, in particular, are now major beneficiaries of intellectual property expansion. Every university has a chief technology officer or an intellectual property licensing officer—a person whose job it is to maximize revenue. That person does not get promoted by giving away a lot of data. It would be crazy for us to think that, under that kind of impetus, the universities will always play the same role in the policy-making process that they have done before, unless activity is taken through both the National Academies and other learned societies to awaken us to the danger posed.

The types of actions that need to be taken span the traditional boundaries between the arts and sciences, or between the realms of patent and copyright. We need to explore a wide range of options, from the equivalent of the Nature Conservancy—a private initiative to preserve a particular resource that is under threat—to the large-scale scientific study, to the economic modeling that tells us the limits of the ecosystem, to the coalition building that is exemplified by organizations like Greenpeace. Happily, organizations paralleling these types of functions are currently moving forward, such as Creative Commons, the Electronic Frontier Foundation, Duke Law School's newly created Center for the Study of the Public Domain, and Public Knowledge. We need to support these organizations and tie together the interests of these and other groups engaged in individual struggles by articulating our common interest in protecting a valuable, shared resource.

So what should we take away from all of this discussion? I would offer the following central points to guide the discussion that follows. Debates about the desirable limits of intellectual property are as old as the field itself. Similarly, debates about the role of public and private financing of scientific investigation have a long and distinguished history. In striking respects, however, contemporary events have shown the limits of both the "antimonopoly" perspective on intellectual property and of the assumption that "public" automatically equals "free" or that "private" automatically means "controlled." The hypertrophy of intellectual property protections over the past 20 years has exacerbated these problems, while the increasing reliance of universities on patent funding and licensing income promises to destabilize an already unstable political situation. As I pointed out earlier, universities traditionally played the role of "public defender for the public domain" in the legislative and policy process. That role can no longer be relied on, and much work needs to be done to make sure it does not altogether disappear. This is a point of fundamental importance to our intellectual property policy. Finally, the traditional alternative to propertized scientific research—government-funded, Cold War-style science—has both practical and theoretical problems.

What is to be done? The answer is a complex one. Theoretically we need a much better understanding of the role of the public domain and of the commons—two terms that are not equivalent—in innovation policy. Practically, we need a series of concrete initiatives, aimed at both public and private actors, to stave off the ill effects of "the second enclosure movement." Some of those initiatives are already under way, but much more effort is needed. On the positive side, this symposium represents, if not the solution, at least an important recognition of the magnitude of the problem. Given the role of the National Academies, it is hard to think of a more pressing issue for it to confront over the next 10 years.

3

Intellectual Property—When Is It the Best Incentive Mechanism for S&T Data and Information?

Suzanne Scotchmer[1]

Intellectual property has invaded the academy, especially in the sciences. Along with others, economists have been thinking about this invasion, and the proper roles of the public and private sectors in the conduct of science. In particular, the focus has been on the question of whether the invasion of intellectual property is either inevitable or appropriate. As I tried to focus my thoughts on how to summarize economic thinking on this topic, and reviewed conversations with colleagues, it seemed to me that instead of trying to explain, as a positive matter, how economists view the relationship between the public and private domain, I should comment on the justifications I hear for intellectual property that I view as wrong. So I have unofficially retitled this talk, "Why Intellectual Property: Four Wrong Answers."

The first wrong answer is "otherwise, we won't get inventions." The second is "intellectual property has benefits without costs"; it rewards inventors and, if licensed, does not hurt users. The third wrong answer is that "the private sector is more efficient than the public sector." The fourth is that, "without intellectual property, even public domain inventions will not be used efficiently"—the Bayh-Dole argument.

Consider the first wrong answer, that otherwise we won't get inventions. This is wrong as an historical matter, as a factual matter about the modern world, and as a logical matter. As an historical matter, it has never been true that the only way we get inventions is by inciting private actors to make them in return for profit. Many alternative mechanisms have been used. Among the most prolific users of alternative mechanisms have been the French, but also Europeans more generally.

Some examples may be well known to you. In the 1840s, the French simply bought the invention of photography and made it freely available, thus engendering a great many improvements to it. The English, in contrast, made it proprietary, thus inhibiting further research on photography in England. Napoleon needed to feed his vast armies marching on Russia and elsewhere. To find a way to do this, the French offered a prize for a way to preserve food. This led to canning and heat sterilization, still in use today. The French also offered a prize for water turbines. The Germans offered a prize for demonstrating Maxwell's equations, which led to the inventions of Hertz. And, of course, a very important and well known prize was that given for discovering a way to calculate longitude at sea. That was an English prize.

---

[1] See N. Gallini and S. Scotchmer, 2002. "Intellectual Property: When is it the best incentive system," in *Innovation, Policy and the Economy*, Vol 2, A. Jaffe, J. Lerner, and S. Stern, eds., MIT Press, Massachusetts. Also available at http://socrates.berkeley.edu/~scotch/G_and_S.pdf.

In short, history informs us that there are many incentive systems other than intellectual property. Some of these are embodied in our own public institutions, such as the National Science Foundation and the National Institutes of Health — a fact that we often forget, and certainly do not emphasize in the current climate where even publicly funded discoveries are subject to intellectual property.

Something that everyone should know is that about half of research and development is funded by public sponsors. If you pull up basic raw data from the Organisation for Economic Cooperation and Development Web site,[2] you will find that about one-third of U.S. research and development is federally funded. In other countries, which devote less money to military expenditures, a higher fraction of research and development is funded by the government. In Latin America, it is well over half. In most European countries it is a little under half. The fraction funded by taxpayers is not trivial anywhere in the world. However, the question of public/private domain is not just who funds the research, but whether the discoveries are put in the public domain. Increasingly, even the discoveries that are publicly funded are restricted by intellectual property, e.g., the many discoveries made in national laboratories funded by the Department of Energy.

Wrong answer number two is that "intellectual property rewards inventors, and doesn't hurt users provided it is licensed." It is true that intellectual property rewards inventors, and may even be an important engine of growth and of invention. But, as I have just pointed out, there are plenty of alternatives. This is not a definitive answer. And the second part of this statement, that it doesn't hurt users provided it is licensed, is misleading. It is important to notice that, even if licensed, intellectual property still creates deadweight loss, including for scientific research tools.

A license price is a price like any other. The licensor makes profit by licensing at a proprietary price. Such a price excludes users; else the licensor could probably increase profit by increasing the price. At a higher price, some users will be retained, and the additional profit received from the retained users will outweigh the profit lost from those that refuse the license. Thus, distributing the use of a research tool by licensing instead of putting it in the public domain will reduce its use just as any other monopoly price reduces use. This may seem obvious to you, but wrong answer number two is a common justification. While licensing certainly puts the intellectual property in use, it does not solve the problem of deadweight loss.

The argument about deadweight loss is illustrated by Figure 3-1, which shows the proprietary price. On the horizontal axis, we have arrayed all the users by their willingness to pay. If the proprietor charges a single price, then all the users arrayed to the left, for whom willingness to pay is higher than the price, will take a license. Each is getting some surplus despite the proprietary price, because each is willing to pay more than he actually has to pay. On the other hand, the proprietary license price is excluding all the other users to the right. That is precisely the social burden of intellectual property.

There is one small caveat to this story, namely, that price discrimination can cure the deadweight loss that arises from charging a single price to all users or potential users. If the proprietor could charge each user exactly his willingness to pay, then there is no reason to exclude, because the way to maximize profit is to embrace all the users within a system of discriminatory pricing. Although no user is excluded, the proprietor gets the entire surplus.

I mention the price discrimination caveat because it is a deeply subversive argument. It implies that there is nothing wrong with intellectual property. The argument has appeared in at least one court opinion, *ProCD*, a Seventh Circuit opinion.[3] If you believed that perfect price discrimination was possible or likely in most cases, it would remove the basic reason that intellectual property is a burden on society, namely, deadweight loss, and removes any argument for reining it in. I believe the problem of excluding users is fundamental and basic, so I would not want the argument to be subverted in that way. Price discrimination can be very difficult, at least the perfect price discrimination that underlies this argument. In my view, we should take seriously the deadweight loss due to intellectual property, and take steps to preserve the public domain.

Wrong answer number three is that "the private sector is more efficient than the public sector." First of all, what does this mean? The public sector does not make inventions. Scientists make inventions. Most public sector

---

[2] See www.oecd.org/pdf/M00026000/M00026476.pdf for additional information.
[3] *ProCD v. Zeidenberg*, 8 F.3d. 1447 (7th Cir. 1996).

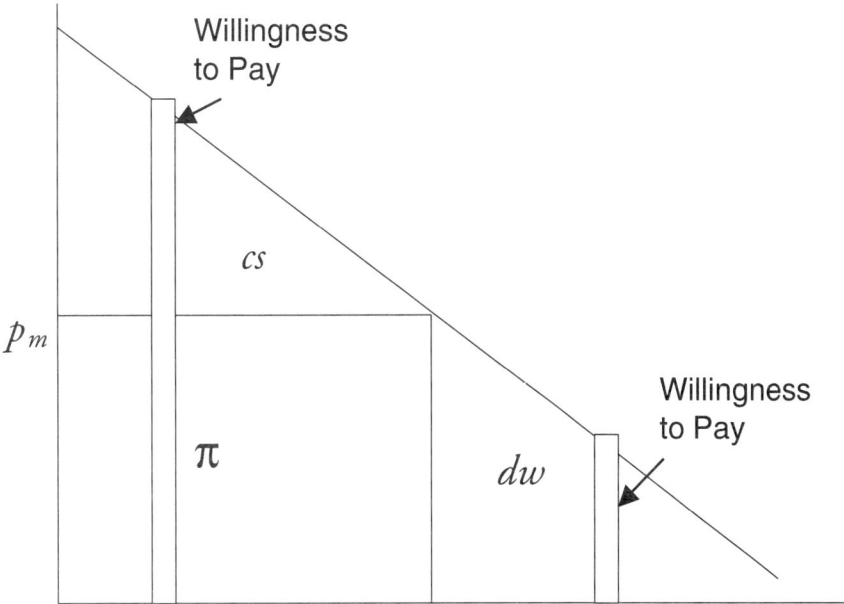

FIGURE 3-1 Deadweight loss illustration, which shows the proprietary price.

research is done by the private sector, that is, by scientists either in industry, the academy, or in national laboratories. The same people do inventions whether they are in the private sector or the public sector. Only about 25 percent of publicly funded research is performed by employees of the government. Most of the rest is contracted out to industry and universities.[4]

The question is not whether the public sector is more efficient than the private sector. The question is, how do you give incentives not to waste money? How do you give incentives to invest in the right things? There are plenty of incentive mechanisms that mimic the incentives for effort that intellectual property creates, while avoiding deadweight loss.

It is often said in favor of intellectual property that the large reward of an intellectual property right will incite effort, and such effort is of great benefit to society. That is true, but, as I have already pointed out, prizes serve that same function. Every academic knows that under the funding systems of, for example, the National Institutes of Health and the National Science Foundation, you will not get funded the second time if you did not produce the first time. These are prize systems, and they are very good at inciting effort. Whether there is a disparity between the effort that such funding incites, and the effort incited under intellectual property rights, is a nuanced question. The incentives for effort are not different in kind, but only in degree, and the incentives created by both systems can, in fact, be modified.

But wrong answer number three, that the private sector is more efficient, misses the main point, even if true. The main question in my view is not who does the research, but rather the terms for using the output. The issue is proprietary pricing versus public domain. There may well be some reason for concern on the efficiency question, but I do not think that gives us license to ignore the question of access.

Finally we come to wrong answer number four, that "without intellectual property, basic inventions will not be used even if they are in the public domain." That is part of the argument for the 1980 Bayh-Dole Act, which authorized the patenting of publicly sponsored inventions. The argument makes no sense to many economists. To many economists, if someone makes a useful invention and puts it in the public domain, people will use it. Even

---

[4] See National Science Board. 2002. *Science and Engineering Indicators 2002*, National Science Foundation, Arlington, VA. Available at http://www.nsf.gov/sbe/srs/seind02/start.htm.

if potential users do not know what inventions are available, they have every incentive to find out. In any other realm of property, we would naturally assume that would happen. Why should we ascribe less rationality to potential users of knowledge?

One rejoinder is that basic inventions need collateral investments in order to become commercially viable. In order to get the private sector to make the collateral investments, they need exclusive licenses on an underlying property right.

However, that is not an answer. If improvements need protection not available under the law, then the Bayh-Dole Act is trying to overcome a deficit of intellectual property law. That is a completely different spin on the story. My view is that we should fix the thing that is broken, rather than something else. If collateral improvements need protection, then they should be protected. Protecting collateral improvements or applications is not a reason to allow restrictions on use of the underlying basic research, which was performed at public expense. If the thing that is broken is that there is not sufficient protection for improvements and commercialization, then that is the thing to fix. If fixed, the underlying basic research could be preserved in the public domain to stimulate improvements and commercialization in many parallel paths simultaneously, with no restrictions on use.

Let me conclude by coming back to the question, why intellectual property? I have given four answers that I regard as wrong. Are there any "right" answers? There are, in my view, some right answers, but they are very limited in scope, and they do not apply very convincingly to scientific knowledge or to scientific data. I believe there are two right answers.

The first is that the private sector may have better knowledge of what the private sector needs than a public sponsor would. There is a set of issues not about the efficiency of how research and development gets done, and not about monopoly pricing versus open access, but rather about the right things to invest in. There are many realms of scientific inquiry, especially those that are closely tied to industry and industrial development, that we would not want to put under the jurisdiction of a public sponsor. Under intellectual property rights, the private sector will apply at least a weak test of whether an R&D investment is worth the cost, namely, whether it is likely to turn a profit. In many realms, it is useful to apply that test.

However, this perspective also seems to be a clear rationale for treating basic research differently from follow-on applications. Nobody disagrees that we need genomic data or weather data. Nobody disagrees that we need certain kinds of environmental data. If there is no disagreement, the argument for intellectual property, in my view, is weak. We do not need the commercial sector to use their superior private knowledge or to apply the profit test as to whether the investment in R&D is a good one.

The second justification for intellectual property is that it concentrates costs among the users. There is a clear rationale, for example, for funding computer games through intellectual property, because the people who use them are a narrow population of 12 year olds. But there is no such justification, in my view, for using intellectual property to fund vaccines for infectious diseases, basic scientific knowledge, Census data or weather data that benefit all of us. Whether we know it or not, we are all users. The test of concentrating the costs among the users simply does not apply with very much force for those kinds of scientific discoveries.

In summary, I believe there are two valid tests for whether a particular subject matter should be supported under private incentives and an intellectual property regime rather than supported by public sponsors to be put in the public domain. It seems to me that most scientific research done in universities and national laboratories does not pass these tests. The other four justifications that I have heard for the expansion of intellectual property rights are not, in my view, sound arguments.

4

The Economic Logic of "Open Science" and the Balance between Private Property Rights and the Public Domain in Scientific Data and Information: A Primer

Paul David

The progress of scientific and technological knowledge is a cumulative process, one that depends in the long-run on the rapid and widespread disclosure of new findings, so that they may be rapidly discarded if unreliable, or confirmed and brought into fruitful conjunction with other bodies of reliable knowledge. "Open science" institutions provide an alternative to the intellectual property approach to dealing with difficult problems in the allocation of resources for the production and distribution of information. As a mode of generating reliable knowledge, open science depends upon a specific non-market reward system to solve a number of resource allocation problems that have their origins in the particular characteristics of information as an economic good. There are features of the collegiate reputational reward system—conventionally associated with open science practice in the academy and public research institutes—that create conflicts between the ostensible norms of "cooperation" and the incentives for non-cooperative, rivalrous behavior on the part of individuals and research units who race to establish "priority." These sources of inefficiency notwithstanding, open science is properly regarded as uniquely well suited to the goal of maximizing the rate of growth of the stock of reliable knowledge.

High access charges imposed by holders of monopoly rights in intellectual property have overall consequences for the conduct of science that are particularly damaging to programs of exploratory research that are recognized to be vital for the long-term progress of knowledge-driven economies. Like non-cooperative behaviors among researchers in regard to the sharing of access to raw data-steams and information, and systematic underprovision of the documentation and annotation required to create reliably accurate and up-to-date public database resources, lack of restraint in privatizing the public domain in data and information can significantly degrade the effectiveness of the entire research system. Considered at the macro-level, open science and commercially oriented research and development (R&D) based on proprietary information constitute complementary sub-systems. The public policy problem, consequently, is to keep the two sub-systems in proper balance by public funding of open science research, and by checking excessive incursions of claims to private property rights over material that would otherwise remain in the public domain of scientific data and information.

INFORMATION AND THE PUBLIC GOODS PROBLEM IN RESEARCH

Acknowledging the peculiar character of information as an economic commodity is a necessary point of departure in grasping the basic propositions that have been established by modern economic analyses of knowledge-producing (research) activities of all kinds.

Data, Information, and Knowledge as Economic Goods

An idea is a thing of remarkable expansiveness, being capable of spreading rapidly from mind to mind without lessening its meaning and significance for those into whose possession it comes. In that quality, ideas are more akin to fire than to coal. Thomas Jefferson, writing in 1813, remarked upon this attribute, which permits the same knowledge to be jointly used by many individuals at once: "He who receives an idea from me, receives instruction himself without lessening mine; as he who lights his taper at mine receives light without darkening me. . . ." (see David, 1993, on Jefferson's observations in this connection).

Modern economists have followed Nelson (1959) and Arrow (1962) in recasting this insight and pointing out that the potential value of an idea to any individual buyer generally would not match its value to the social multitude, since the latter would be the sum of the incremental benefits that the members of society derived from their use of the idea. The individual benefits conveyed, however, will not readily be revealed in a willingness to pay on the part of everyone who would gain thereby; once a new bit of knowledge is revealed by its discoverer(s), some benefits will instantly "spill over" to others who are therefore able to share in its possession—at little incremental cost. Why should they then offer to bear any of the initial costs that were incurred in bringing the original thought to fruition?

Commodities that have the property of "expansibility," permitting them to be used simultaneously for the benefit of a number of agents, are sometimes described as being "non-rival" in use. This characteristic is a form of non-convexity, or an extreme form of decreasing marginal costs as the scale of use is increased: although the cost of the first instance of use of new knowledge may be large, in that it includes the cost of its generation, further instances of its use impose at most a negligibly small incremental cost. It sometimes is noticed that this formulation ignores the cost of training potential users to be able to grasp the information and know what to do with it. But while it is correct to point out that there can be fixed costs of access to the information, these do not vitiate the proposition that re-use of the information will neither deplete it nor impose further costs. It may well be costly to teach someone how to read the table of the elements or the rules of the differential calculus. Nevertheless, any number of individuals thus instructed can go on using that knowledge without imposing further costs either on themselves or upon others.

A second peculiar property of ideas, which has to be underscored here, is that it is difficult, and generally costly, to retain exclusive possession of them whilst putting them to use. Of course, it is possible to keep a piece of information or a new idea secret. The production of results not achievable otherwise, however, discloses something about the existence of a method for doing so. Quite understandably, scientific and technical results obtained by methods that cannot or will not be disclosed are felt to be less dependable on that account; their production is deemed to be more in the nature of magical performances than as contributions to the body of reliable knowledge. Even the offer of a general explanation of the basis for achieving a particular, observable result may be sufficient to jeopardize the exclusivity of its possession, because the knowledge that something can be done is itself an important step toward discovering how it may be done.

Public Goods and the "Appropriability Problem"

The dual properties of non-rival usage and costly exclusion of others from possession define what a "pure public good." The term "public good" does not imply that such commodities cannot be privately supplied, nor does it mean that a government agency should or must produce it. Nevertheless, it follows from the nature of pure public goods that competitive market processes will not do an efficient job of allocating resources for their production and distribution, simply because where such markets work well they do so because the incremental costs and benefits of using a commodity are assigned to the users. In the case of public goods, such assignments are not automatic and they are especially difficult to arrange under conditions of competition.

One may see the essence of the problem posed by the public goods characteristics of knowledge by asking: How can ideas be traded in markets of the kind envisaged by disciples of Adam Smith, except by having aspects of their nature and significance disclosed before the transactions were consummated? Rational buyers of ideas, no less than buyers of coal, and of fish and chips, first would want to know something about what it is that they will be getting for their money. Even were the prospective deal to fall through, it is to be expected that the potential purchaser would enjoy (without paying) some benefits from what economists refer to as "transactional spill-overs." The latter are not without considerable value, because there may be significant commercial advantages from the acquisition of even rather general information about the nature of a discovery, or an invention—especially one that a reputable seller has thought it worthwhile to bring to the attention of people engaged in a particular line of business.

This leads to the conclusion that the findings of scientific research, being new knowledge, would be seriously undervalued were they sold directly through perfectly competitive markets. Some degree of exclusivity of possession of the economic benefits derived from ideas is therefore necessary if the creators of new knowledge are to derive any profit from their activities under a capitalist market system. Intellectual property rights serve this end in the form of patent and copyright monopolies. But imposing restrictions on the uses to which ideas may be put also saddles society with the inefficiencies that arise when monopolies are tolerated; a point harped upon by economists ever since Adam Smith. In addition, as will be seen below, access charges that holders of intellectual property are free to impose on its users, and the secrecy practices that business companies embrace to protect their investments in the R&D process when it aims at securing proprietary control of new information, are a further source of inefficiencies in society's utilization of existing bodies of knowledge.

THE ECONOMIC LOGIC OF OPEN SCIENCE

For the purposes of this basic exposition, the institutional conditions of primary interest are those that sharply distinguish the sphere of open science supported by public funding and the patronage of private foundations, on the one hand, from both the organized conduct of scientific research under for commercial profit under "proprietary rules," and the production and procurement of defense-related scientific and engineering knowledge under conditions of restricted access to information about basic findings and their actual and potential applications.

Ethos, Norms, and Institutions of "the Republic of Science"

The formal institutions of modern science are ones with which academic economists already are thoroughly familiar, for, it is a striking phenomenon in the sociology of science that there is high degree of mimetic professional organization across the various fields of academic endeavor. Whether in the social sciences, the natural sciences, or the humanities, most fields have their professional academies and learned societies, journal refereeing procedures, public and private foundation grant programs, peer-panels for merit review of funding applications, organized competitions, and prizes and public awards. Within the sciences proper, however, there are recognized norms and conventions that constitute a clearly delineated ethos to which members of the academic research community generally are disposed to publicly subscribe, whether or not their individual behaviors conform literally to its strictures. The norms of "the Republic of Science" that were famously articulated by Robert Merton (1973, especially Chapter 13) sometimes are summarized under the mnemonic CUDOS: communalism, universalism, disinterestedness, originality, skepticism. (See Ziman, 1994, p. 177.)

The "communal" ethos emphasizes the cooperative character of inquiry, stressing that the accumulation of reliable knowledge is a social, rather than an individual program; however much individuals may strive to contribute to it, the production of knowledge which is "reliable" is fundamentally a social process. Therefore, the precise nature and import of the new knowledge ought not to be of such immediate personal interest to the researcher as to impede its availability or detract from its reliability in the hands of co-workers in the field. Research agendas, as well as findings, ought therefore to be under the control of personally (or corporately) disinterested agents. The force of the universalist norm is to allow entry into scientific work and discourse to be open to all persons of "competence," regardless of their personal and ascriptive attributes. A second aspect of openness concerns the disposition of knowledge: the full disclosure of findings and methods form a key aspect of the cooperative, communal program of inquiry. Disclosure serves the ethos legitimating and, indeed, prescribing skepticism, for it is that which creates an expectation that all claims to have contributed to the stock of reliable knowledge will be subjected to trials of verification, without insult to the claimant. The "originality" of such intellectual contributions is the touchstone for the acknowledgment of individual claims, upon which collegiate reputations and the material and non-pecuniary rewards attached to such peer evaluations are based.

An essential, differentiating feature of the institutional complex associated with modern science thus is found in its public, collective character, and its commitment to the ethos of cooperative inquiry and to free sharing of knowledge. Like other social norms, of course, these are not perfectly descriptive of the behaviors of individuals who identify themselves with the institutions; indeed, not even of those who would publicly espouse their commitment to "openness" and scientific cooperation. It is evident that, as is the case with public goods more generally, the temptation to "free-ride" on the cooperative actions of others creates tensions within the system. Consequentially, the fact that it has withstood those strains points to the existence of some positive, functional value that it conveys to a substantial number of the scientists working in the "public/academic" research sector, and to the particular benefits that society as a whole has been able to derive through public patronage of this mode of pursuing knowledge.

A Functionalist Rationale for the Norms and Institutions of Open Science

While for most of us the idea of science as the pursuit of "public knowledge" may seem a natural, indeed a "primitive" conceptualization, it is actually a social contrivance; and by historical standards, a comparatively recent innovation at that, having taken form only as recently as the sixteenth and seventeenth centuries (See David, 1998a, 1998b for further discussion of the historical origins of open science). Although economists have only lately begun to analyze the workings of the institutional complex that characterizes "modern science," it has not been difficult to reveal a functionalist rationale: it centers on the proposition that greater economic and social utility derives from behaviors reinforced by the ideology of the open pursuit of knowledge and the norms of cooperation among scientists (see Dasgupta and David, 1987, 1994; David, 2001). That analysis also demonstrates the "incentive compatibility" between the norm of disclosure and the existence of a collegiate reputation-based reward system grounded upon validated claims to priority in discovery or invention.

The core of this rationale is the greater efficacy of open inquiry and complete disclosure as a basis for the cooperative, cumulative generation of predictably reliable additions to the stock of knowledge. In brief, openness abets rapid validation of findings, and reduces excess duplication of research efforts. Wide sharing of information puts knowledge into the hands of those who can put it to uses requiring expertise, imagination, and material facilities not possessed by the original discoverers and inventors. This enlarges the domain of complementarity among additions to the stock of reliable knowledge, and promotes beneficial spill-overs among distinct research programs.

As important as it is to emphasize these societal benefits of the cooperative mode for the advancement of science, it is equally significant to appreciate that the norms of information disclosure—applied both to findings and to the methods whereby research results have been obtained—have a second functional role. They are not an extraneous ethical precept of the open science system, but figure in the mechanism that induces individual scientific effort. The open science reward system typically is a two-part affair, in which researchers are offered both a fixed form of compensation and the possibility of some variable, results-related rewards. The fixed part of the compensation package, e.g., a basic stipend or salary, tied to teaching in the case of academic research institutions, provides individual researchers a measure of protection against the large inherent uncertainties surrounding the process of exploratory science. The variable component of the reward is based upon expert evaluation of the scientific significance of the individual's contribution(s) to the stock of knowledge. This involves both the appraisals by peers of the meaning and significance of the putative "contribution," and acceptance of the individual's claim to "moral possession" of the finding, on the grounds of having priority in its discovery or invention. The latter is particularly important, as it connects the incentives for disclosure of findings under the collegiate reputational reward system of open science with the social efficiency of sharing new information.

One can fairly reasonably monitor minimal levels of input of effort into research activities (turning up in the lab, attending scientific meetings, etc.), or the performance of associated duties such as giving lectures, supervising students, and so on. But beyond that, the assessment of individual performance on the basis of inputs is quite inadequate, and incentives offered for new discoveries and inventions must perforce be based upon observable "outputs"—results. The core of the problem here is that the outputs that one wants to reward are intangible and unique in the sense of being novel. The repetition of someone else's finding is not entirely devoid of social utility, for it may contribute to assessment of the reliability of the scientific assertions in question. But, although you would want to pay a second miner for the ton of coal he brought to the surface more-or-less the same amount that the preceding ton of coal earned for its producer, this is not the case for the second research paper that repeated exactly what was reported in its predecessor. To be rewarded for a novel idea, research technique, or experimental result, one must be seen to have been in the vanguard of its creators. It is the prospect of gaining such rewards—whether in reputational standing and the esteem of colleagues, enhanced access to research resources, formal organizational recognition through promotions accompanied by higher salary, accession to positions of authority and influence within professional bodies and public institutions, and the award of prizes and honors—that serves to induce races to establish "priority," and hence to secure rapid disclosure of "significant" findings.

Because the evaluation of significance will rest largely with the researchers scientific peers, this incentive mechanism performs the further function of directing his or her research efforts toward goals that are more likely to win wider appreciation precisely because it will advance the work of others in the scientific community. What this means is that the incentives created for disclosure under the regime of open science are linked with the orientation of problem choice and the formation of collective research portfolios that are biased toward "research spillovers," rather than "product-design spillovers" in the sphere of commercial innovation.

Open Science and Proprietary Research: Regime Complementarities and System Balance

The advantages to society of treating new findings as public goods in order to promote the faster growth of the stock of knowledge are to be contrasted with the requirements of secrecy for the purposes of extracting material benefits from possession of information that can be privately held as intellectual property. This point can be formulated in the following overly stark, unqualified way. Science (qua social organization) is about maximizing the rate of growth of the stock of knowledge, for which purposes public knowledge and hence patronage or public subsidization of scientists is required, because citizens of the Republic of Science cannot capture any of the social surplus value their work will yield if they freely share all the knowledge they obtain. By contrast, the Regime of Technology (qua social organization) is geared to maximizing wealth stocks, corresponding to the current and future flows of pure "economic rents" (private profits) from existing data, information and knowledge, and therefore requires the control of knowledge through secrecy or exclusive possession of the right to its commercial exploitation. The conditions that are well suited to appropriate value from possession of the rights to exploit new knowledge, however, are not those that favor rapid growth in the stocks of reliable knowledge.

This functional juxtaposition suggests a logical argument for the existence and perpetuation of institutional and cultural separations between the communities researchers forming the Republic of Science and those engaged in proprietary scientific pursuits within the Realm of Technology: the two distinctive organizational regimes serve different and potentially complementary societal purposes. Indeed, neither system alone can perform effectively over the long run as the two of them coupled together, since their special capabilities and limitations are complementary. Maintaining them in a productive balance, therefore, is the central task towards which informed science and technology policies must be directed.

It follows from this logic that such policies need to attend to the delineation of intellectual property rights and protections of the public domain in scientific data and information as the respective codified knowledge infrastructures of the dual regimes, and to the maintenance of balance between them. To do so will be easier if there is a clearer understanding of the ways in which public expenditures for the support of open science serve to enhance the value of commercially oriented R&D as a socially productive and privately profitable form of investment. Such an understanding on the part of representatives of the academic science communities is just as important as it is for government policy-makers, intellectual property lawyers, and business managers.

Application-oriented Payoffs from Open, Exploratory Research: the Uncertain Lure

When scientists are asked to demonstrate the usefulness of research that is exploratory in character and undertaken to discover new phenomena, or explain fundamental properties of physical systems, the first line of response often is to point to discoveries and inventions generated by such research projects that turned out to be of more or less immediate economic value. Indeed, many important advances in instrumentation, and generic techniques, such as polymerase chain reaction and the use of restriction enzymes in "gene-splicing" may be offered in illustration of this claim. These by-products of the open-ended search for basic scientific understanding also might be viewed as contributing to the "knowledge infrastructure" required for efficient R&D that is deliberately directed towards results that would be exploitable as commercial innovations.

The experience of the 20th century also testifies to the many contributions of practical value that trace their origins to large, government-funded research projects that were focused upon the development of new enabling technologies for public-mission agencies. Consider just a few recent examples from the enormous and diverse range that could be instanced in this connection: airline reservation systems, packet switching for high-speed telephone traffic, the Internet communication protocols, the Global Positioning System, and computer simulation methods for visualization of molecular structures—which has been transforming the business of designing new pharmaceutical products, and much else besides.

Occasionally, such new additions to the stock of scientific knowledge are immediately commercializable and yield major economic payoffs that, even though few and far between, are potent enough to raise the average social rate of return on basic, academic research well above the corresponding private rate of return earned on industrial R&D investment. Inasmuch as the high incremental social rate of return in cases like those just mentioned derives from the wide and rapid diffusion of knowledge that provide new "research tools," the open dissemination, validation, and unobstructed elaboration of these discoveries and inventions are directly responsible for the magnitude of the "spillover" benefits they convey.

The skeptical economist's response to recitations of this kind, however, is to ask whether a more directed search for the solutions to the applied problems in question would not have been less costly. Would that approach not be more expedient than waiting for serendipitous, commercially valuable results to emerge from costly research programs that were conducted by scientists with quite different purposes in mind? This is a telling point, at least rhetorically, simply because the theme of such "spin-off" stories is their utter unpredictability. To argue that these "gifts from Athena" are in some sense "free," requires that the research program to which they were incidental was worth undertaking for its own sake, so that whatever else might be yielded as by-products was a net addition to the benefits derived. Yet, the reason those examples are being cited is the existence of skepticism as to whether the knowledge that was being sought by exploratory, basic science was worth the cost of the public support it required. Perhaps this is why the many examples of this kind that scientists have brought forward seem never enough to satisfy the questioners.

The discovery and invention of commercially valuable products and processes are seen from the viewpoint of "the new economics of science" to be rarer among the predictably "useful" results that flow from the conduct of exploratory, open science. Without denying that "pure" research sometimes yields immediate applications around which profitable businesses spring up, it can be argued that those direct fruits of knowledge are not where the quantitatively important economic payoffs from the open conduct of basic scientific inquiry are to be found.

Micro-level Complementarities between Exploratory Science and Proprietary R&D

Much more critical over the long run than spin-offs from exploratory science programs are their cumulative indirect effects in raising the rate of return on private investment proprietary R&D performed by business firms. Among those indirect consequences, attention should be directed not only to "informational spillovers" that lend themselves readily to commercialization, but to a range of complementary "externalities" that are generated for the private sector by publicly funded activities in the sphere of open science, where research and training are tightly coupled.

Resources are limited, to be sure, and in that sense research conducted in one field and in one organizational mode is being performed at the expense of other kinds of R&D. But what is missed by attending exclusively to the competition forced by budget constraints, is an appreciation of the ways in which exploratory science and academic engineering research activities support commercially oriented and mission-directed research that generates new production technologies and products.

First among the sources of this complementary relationship is the intellectual assistance that fundamental scientific knowledge (even that deriving from contributions made long ago) provides to applied researchers, whether in the public or in the private sector. From the expanding knowledge base it is possible to derive time- and cost-saving guidance as to how best to proceed in searching for ways to achieve some pre-specified technical objectives. Sometimes this takes the form of reasonably reliable guidance as to where to look first, and much of the time the knowledge base provides valuable instructions as to where it will be useless to look. How else does the venture capitalist know not to spend time talking with the inventor who has a wonderful new idea for a perpetual motion machine?

One effect this has is to raise the expected rates of return, and reduce the riskiness of investing in applied R&D. Gerald Holton (1996), a physicist at Harvard and historian of science, has remarked that if intellectual property laws required all photoelectric devices to display a label describing their origins, "it would list prominently: 'Einstein, Annalen der Physik 17(1905), pp.132-148.'" Such credits to Einstein also would have to be placed on many other practical devices, including all lasers.

The central point that must be emphasized here is that, over the long-run, the fundamental knowledge and practical techniques developed in the pursuit of basic science serves to keep applied R&D as profitable an investment for the firms in many industries as it has proved to be, especially, during the past half-century (see David, Mowery, and Steinmueller, 1992). In this role, modern science continues in the tradition of the precious if sometimes imprecise maps that guided parties of exploration in earlier eras of discovery, and in that of the geological surveys that are still of such value to prospectors searching for buried mineral wealth.

That is not the end of the matter, for a second and no less important source of the complementary relationship between basic and applied research is the nexus between university research and training on the one hand, and on the other the linkage of the profitability of corporate R&D to the quality of the young researchers that are available for employment. Seen from this angle, government funding of open exploratory science in the universities today is subsidizing the R&D performed by the private business sector. Properly equipped research universities have turned out to be the sites of choice for training the most creative and most competent young scientists and engineers, as many corporate directors of research well know. This is why graduates and postdoctoral students in those fields are sent or find their own way to university labs in the United States, and still to some in the United Kingdom. It explains why businesses participate (and sponsor) "industrial affiliates" programs at research universities.

The "Tacit Dimension" and Knowledge Transfers via the Circulation of Researchers

A further point deserving emphasis in this connection is that a great deal of the scientific expertise available to a society at any point in time remains tacit, rather than being fully available in codified form and accessible in archival publications. It is embodied in the craft-knowledge of the researchers, about such things as the procedures for culturing specific cell lines, or building a new kind of laser that has yet to become a standard part of laboratory repertoire. This is research knowledge, much of it very "technological" in nature in that it pertains to how phenomena have been generated and observed in particular, localized experimental contexts, which is embodied in people. Under sufficiently strong incentives it would be possible to express more of this knowledge in forms that would make it easier to transmit, and eventually that is likely to happen. But, being possessed by individuals who have an interest in capturing some of the value of the expertise they have acquired, this tacit knowledge is transmitted typically through personal consultations, demonstrations, and the actual transfer of people (see Cowan, David, and Foray, 2000, for further treatment of the economics of tacitness and codification of knowledge).

The circulation of post-doctoral students among university research laboratories, between universities and specialized research institutes, and no less importantly, the movement of newly trained researchers from the academy into industrial research organizations, is therefore an important aspect of "technology transfer"—diffusing the latest techniques of science and engineering research. The incentive structure in the case of this mode of transfer is a very powerful one for assuring that the knowledge will be successfully translated into practice in the new location, for the individuals involved are unlikely to be rewarded if they are not able to enhance the research capabilities of the organization into which they move.

A similarly potent incentive structure may exist also when a fundamental research project sends its personnel to work with an industrial supplier from whom critical components for an experimental apparatus are being procured. Ensuring that the vendor acquires the technical competence to produce reliable equipment within the budget specifications is directly aligned with the interests of both the research project and the business enterprise. Quite obviously, the effectiveness of this particular form of user-supplier interaction is likely to vary directly with the commercial value of the procurement contracts and the expected duration and continuity of the research program.

For this reason, big science projects or long-running public research programs may offer particular advantages for the collaborative mode of technology transfers, just as major industrial producers—such as the large automotive companies in Japan—are seen to be able to set manufacturing standards and provide the necessary technical expertise to enable their suppliers to meet them. By contrast, the transfer of technology through the vehicle of licensing intellectual property is, in the case of process technologies, far more subject to tensions and deficiencies arising from the absence of complete alignment in the interest of the involved individuals and organizations. But, as has been seen, the latter is only one among the economic drawbacks of depending upon the use of intellectual property to transfer knowledge from non-profit research organizations to firms in the private sector.

Won't the Private Sector Sponsor Academic, "Open Science" Style Research?

Business firms do indeed cooperate with academic research units and, in some cases, they undertake on their own to support "basic research" laboratories whose employed scientists and engineers are permitted, and even encouraged, to rapidly submit results for publication in scientific and technical journals. Large R&D-intensive companies also adopt strategies of liberally licensing the use of some their patented discoveries and inventions. Their motives in adopting such policies include the desire to develop a capability to monitor progress at the frontiers of science, the hope of being able to pick up early information concerning emerging potential lines of research with commercial innovation potential, being better positioned to penetrate the secrets of their rivals' technological practices, and recruiting talented young scientists whose career-goals are likely to be advanced by the exposure that can be afforded to them by publication and active participation in an open science community. Even some small- and medium-size firms sometimes opt to freely disclose inventions that are patentable. (This may be done as a means of establishing the new knowledge within the body of "prior art," and thereby, without incurring the expenses of obtaining and defending a patent, effectively precluding others from securing property rights that would inhibit the firm from exploiting the knowledge for its own production operations.)

Yet, in regard to the issue of whether market incentives fail to elicit private sector funding of exploratory R&D conducted in the open science mode, the relevant question for society is one that is quantitative, not qualitative. One cannot adequately answer the question "Will there be enough business-motivated support for open science?" merely by saying, "There will be some." Economists do not claim that without public patronage scientific and technical research would entirely cease to be conducted under conditions approximating those of the open science system. Rather, their analysis holds that there will not be enough investment in that style of inquiry, that is to say, not as much as would be carried out were individual businesses able to anticipate capturing all the benefits of such knowledge-producing investments—as is the case for society as a whole.

Moreover, business support for academic-style R&D—as distinguished from industrial contracting for university-based, applications-oriented research with intellectual property rights assigned to the sponsoring firms—is more likely to commend itself as a long-term strategy. Consequently, it is likely to be sensitive to commercial pressures to shift research resources towards advancing existing product development, and improving existing processes, rather than searching for future technological options. This implies that exploratory lines of inquiry funded in this way are vulnerable to wasteful disruptions. Large commercial organizations that are less prone to becoming asset-constrained and, of course, the public sector therefore are better able to take on the job of monitoring what is happening on the international science and technology frontiers. Considerations of these kinds are important in addressing the issue of how to find the optimal balance for the national research effort between secrecy and disclosure of scientific and engineering information, as well in trying to adjust the mix in the national research portfolio of exploratory open science style programs and proprietary, applications-driven R&D projects.

PROPERTY AND THE PURSUIT OF KNOWLEDGE

Intellectual Property Rights and the Norms of Disclosure

The past two decades have witnessed growing efforts to assert and enforce intellectual property rights over scientific and technological knowledge through the use of patents, copyrights, and other, more novel forms of legal protection. (The latter include the special legislation introduced in the United States in 1980 to extend copyright protection to the "mask work" for photo-lithographic reproduction of very large microelectronic circuits on silicon wafer, and the European Union's protection of databases by new national statutes implementing an EC Directive issued in 1996.) These developments have coincided with two other trends that similarly have tended to expand the sphere of private control over access to knowledge, at the expense of the public knowledge domain.

One trend has been the rising tide of patenting activity by universities, especially in the areas of biotechnology, pharmaceuticals, medical devices, and software. This movement started in the United States, where it received impetus under the 1980 Bayh-Dole Act that permitted patent applications to be filed for discoveries and inventions issuing from research projects that were funded by the federal government. It has since spread internationally, being reinforced by the efforts in other countries to foster closer research collaboration between universities and public research institutes on the one hand, and private industry on the other. The other trend has seen a concerted effort by all parties to secure copyright protection for the electronic reproduction and distribution of information, in part to exploit the opportunities created by electronic publishing and in part to protect existing copyright assets from the competition that would be posed by very cheap reproduction of information in digital form over electronic networks.

During the past two decades a renewal of enthusiasm for expanding and strengthening private property rights over information has given rise to a rather paradoxical situation. Technological conditions resulting from the convergent progress in digital computing and computer-mediated telecommunications have greatly reduced the costs of data capture and transmission, as well as of information processing, storage, and retrieval. These developments are working to give individuals, and especially researchers, unprecedentedly rapid and unfettered access to new knowledge. At the same time, and for reasons that are not entirely independent, the proliferation of intellectual property rights and measures to protect these is tending to inhibit access to such information in areas (basic research in general, and most strikingly in the life sciences and computer software) where new knowledge had remained largely in the public domain (see David, 2000, on the connection between the technological and the institutional trends).

Thus, it may be said that a good bit of intellectual ingenuity and entrepreneurial energy is being directed towards the goal of neutralizing the achievements of information scientists and engineers by creating new legally sanctioned monopolies of the use of information, thereby creating artificial scarcities in fields where abundance naturally prevails and access to that abundance is becoming increasingly ubiquitous. The consequent imposition of data and information access charges above the marginal costs of producing and distributing information results in a double burden of economic inefficiency when it falls upon researchers who use those information-goods as inputs in the production of more information and knowledge. The first-order effect is the curtailment of the use of the information, or the increased cost of using it to produce conventional commodities and services, and hence the loss of utility derived from such products by consumers. A second round of inefficiency is incurred by the inhibition of further research, which otherwise would be the source of more public goods in the form of new knowledge.

To understand the full irony of this situation, one has simply to return to the point of departure in this discussion: the observation that knowledge is not like any other kind of good, and certainly does not resemble conventional commodities of the sort that are widely traded in markets. Intellectual property cannot be placed on an equal footing with physical property, for the simple reason that knowledge and information possess a specific characteristic that economists refer to as "non-rival in use"—the same idea and its expression may be used repeatedly and concurrently by many people, without being thereby "depleted." This contrasts with the properties of ordinary "commodities" that are consumed: if Marie eats the last slice of cake in the kitchen, that piece cannot also be consumed by Camille; whereas, both girls may read the same novel either simultaneously or sequentially, and in so doing they will not have rendered the story any the less available for others to enjoy.

The allocation of property rights in the case of information-goods does not attempt to confer a right of exclusive possession, as do property laws governing tangible goods such as land. Indeed, to claim a right of possession one must be able to describe the thing that is owned, but no sooner do you describe your idea to another person than their mind comes into (non-exclusive) possession of it; only by keeping the information secret can you possess it exclusively.

What the creation and assigning intellectual property rights does then is to convey a monopoly right to the beneficial economic exploitation of an idea (in the case of patent rights) or of a particular expression of an idea (in the case of copyright) that has been disclosed, rather than being kept secret. This device allows the organization of market exchanges of "exploitation rights," which, by assigning pecuniary value to commercially exploitable ideas, creates economic incentives for people to go on creating new ones, as well as finding new applications for old ones. By tending to allocate these rights to those who are prepared to pay the most for them, the workings of intellectual property markets also tend to prevent ideas from remaining in the exclusive (secret) possession of discoverers and inventors who might be quite uninterested in seeing their creations used to satisfy the wants and needs of other members of society.

Another potential economic problem that is addressed by instituting a system of intellectual property rights is the threat of unfair competition—particularly the misappropriation of the benefits of someone else's expenditure of effort—which might otherwise destroy the provision of information goods as a commercially viable activity. The nub of the problem here is that the cost of making a particular information good available to a second, third, or thousandth user are not significantly greater than those of making it available to the first one. When Théo listens to a piece of music, modern reproduction and transmission technologies will permit Quentin, Manon, and millions of others to listen to the same piece without generating significant additional costs. The costs of the first copy of a CD are very great, compared to the cost of "burning" a second, third, or millionth copy of that CD. Ever since the Gutenberg revolution, the technical advances that have lowered the costs of reproducing "encoded" material (text, images, sounds) also has permitted "pirates" to appropriate the contents of the first copy without bearing the expense of its development. Unchecked, this form of unfair condition could render unprofitable the investment entailed in obtaining that critical first copy.

Producers of ideas, texts, and other creative works (including graphic images and music) are subject to economic constraints, even when they do not invariably respond to variation in the incentives offered by the market. If they had no rights enabling them to derive income from the publication of their works, they might create less, and quite possibly be compelled to spend their time doing something entirely different but more lucrative. So there is an important economic rationale for establishing intellectual property rights. A strong case also can be made for protecting such rights by the grant of patents and copyrights, especially as that way of providing market incentives for certain kinds of creative effort leaves the valuation of the intellectual production to be determined ex post, by the willingness of users to pay; it thereby avoids having society try to place a value on the creative work ex ante, as would be required under alternative incentive schemes, such as offering prospective authors and inventors prizes or awarding individual procurement contracts for specified works.

Property Institutions, Transactions Costs, and Access to Information

The solution of establishing a monopoly right to exploit that "first copy" (the idea protected by the patent or the text protected by copyright), alas, turns out not to be a perfect one. The monopolist will raise the price of every copy above the negligible costs of its reproduction, and, as a result, there will be some potential users of the information good who will be excluded from enjoying it. The latter represents a waste of resources, referred to by economists as the "dead-weight burden of monopoly": some people's desires will remain unsatisfied even though they could have been fulfilled at virtually no additional cost. Economists as a rule abhor "waste," or "economic inefficiency," but they believe in and rather like the power of market incentives. Not surprisingly, then, the subject of intellectual property policies has proved vexatious for the economics profession, as it presents numerous situations in which the effort to limit unfair competition and preserve incentives for innovation demonstrably results in a socially inefficient allocation of resources.

There is not much empirical evidence as to how altering the legal conditions and terms of intellectual property rights translates into change in the overall strength of economic incentives for the producers, or about the effectiveness of bigger incentives in eliciting creative results; nor is it a straightforward matter to determine the way in which holders of a particular form of intellectual property right would choose to exploit it, and the consequent magnitude of the resultant social losses in economic welfare (the dead-weight burden). Without reliable quantitative evidence of that kind, obviously, it is hard to decide in which direction to alter the prevailing policy regime in order to move toward the notional optimum for any particular market.

The difficulties of arriving at "scientific closure" on such matters, combined with conflicts of economic interests over the distribution of the benefits of new knowledge, quite understandably, have sustained a long history of intense debate in this area. In each era of history new developments affecting the generation or the distribution of knowledge, give rise to a revival of these fundamental questions in new guises. Today, the hot issues arise from questions concerning the desirability of (a) curtailing patent monopolists' rights by letting governments impose compulsory licensing of the local manufacture of certain pharmaceutical products, or of some medical devices; (b) providing those engaged in non-commercial scientific research and teaching with automatic "fair use" exemptions from the force of intellectual law; and (c) permitting purchasers of copyright protected CDs to freely share music tracks with others by means of peer-to-peer distribution over the Internet.

There is no easy general solution to this class of economic problems, and useful answers to the basic questions raised (e.g., are new rights that would better address the new circumstances required, and, if so, what form should they take?) will vary from one case, area, or situation to the next. Most economic and legal analysis favors protecting broad classes of intellectual works, rather than very specific forms that are more likely to be rendered economically obsolete. But having flexible legal concepts that are meant to be applied in novel situations creates added uncertainties for innovators. There is likely to be a protracted period of waiting and struggling to have the courts settle upon an interpretation of the law that is sufficiently predictable in its specific applications to provide a reliable guide for commercial decision making.

Another general principle that finds widely expressed approval is that of harmonizing intellectual property rights institutions internationally, so that arbitrary, inherited legal differences among national entities do not interpose barriers to the utilization of the global knowledge base in science and technology. The catch in this, however, is that harmonization rarely is a neutral procedure. Representatives of polities usually are loathe to cede property rights that their constituents already possess, and, consequently, programs of harmonization turn out to impart an unwarranted global bias towards expanding the range of property rights that will be recognized and raising the strength of the protections afforded.

A more tenable broad policy position on this contested terrain may be derived from the recognition that the generation of further knowledge is among the major important uses of new knowledge, and, at the same time, there are enormous uncertainties surrounding the nature and timing of the subsequent advances that will stem from any particular breakthrough. This is especially true of fields where new discoveries and inventions tend more readily to recombine in a multiplicity of ways that generate further novelties. A reasonably clear policy implication follows from this, and from the additional observation that although we will seldom be able to predict the details and future social value attaching to the consequences of a specific advance in knowledge, it is far more certain that there will be a greater flow of entailed discoveries if the knowledge upon which they rest remains more accessible and widely distributed. Therefore, rather than concentrating on raising the inducements to make "hard-to-predict" fundamental breakthroughs, it will be better to design intellectual property regimes in ways that permit non-collusive pooling and cross licensing. As a practical matter, this consideration generally would call for raising the novelty requirements for patents, awarding protection for narrower claims, requiring renewals with increasing fees, and other, related measures. All of these steps would encourage entry into the process of generating further knowledge by utilizing the breakthroughs that have occurred and been adequately disclosed (see David and Foray, 1995).

The import of this is to strictly limit the scope of grants of monopoly rights over research tools and techniques, curtailing the freedom of the rights-holders to levy whatever "tax" they wished upon others who might use such inventions and discoveries in order to generate still further additions to the knowledge base. Collective knowledge enhancement is thwarted when discoveries cannot be freely commented upon, tested by replication, elaborated upon, and recombined by others. In other words, intellectual property regimes designed to make it easier for many to "see farther by standing on the shoulders of giants" would appear likely to be more fruitful than a strategy that renders those shoulders less easily mounted by others, in the hope that this would stimulate the growth of more, and taller "giants."

The extension of monopoly rights over the application of particular research tools in the life sciences—techniques such as polymerase chain reaction and monoclonal antibodies, new bioinformatic databases and search engines, as well as generic information about the structure of genetic material and the way that these govern the production of proteins—is coming to be seen as especially problematic. The issuing of such patents may indeed be responsible for stimulating more commercially-oriented R&D investment by pharmaceutical companies, and others who look forward to selling them access to new information. Yet, intellectual property protection in this sphere is likely to impose heavy dynamic welfare losses on society. It will do so by impeding access to existing information, or by increasing the wastage of resources in functionally duplicative research aimed at avoiding patent licensing charges. This raises the cost not simply of research directed toward producing a specific new product (e.g., diagnostic test kits for a particular class of genetically-transmitted conditions), but also of exploratory research that may enable the future creation of many applications, including those that still are undreamed of. To use the evocative phrasing of a leading European scientist, cooperatively assembled bioinformatic databases are permitting researchers to make important discoveries in the course of "unplanned journeys through information space." If that space becomes filled by a thicket of property rights, then those voyages of discovery will become more troublesome and more expensive to undertake, unanticipated discoveries will become less frequent, and the rate of expansion of the knowledge base is likely to slow.

Popular wisdom maintains that "good fences make good neighbors." This may apply in the case of two farmers with adjacent fields—one growing crops and the other grazing cattle—or gold diggers excavating neighboring concessions. But unlike land, forage, or other kinds of exhaustible resources, knowledge is not depleted by use for consumption; data sets are not subject to being "over-grazed," but instead are likely to be enriched and rendered more accurate as more researchers are allowed to comb through them (David, 2001).

The issues just examined are entangled with other, and include difficult problems concerning the institutional (as distinct from the technological) determinants of human beings ability to enhance their "capabilities" by finding and making use of existing repositories of knowledge and sources of information (Foray and Kazancigil, 1999). They involve special problems of access to scientific and technological knowledge relevant to developing countries, and raise complex issues of what it implies for resource allocation to insist that every individual in a society has a right to benefit from the collective advance of human knowledge affecting such fundamental, capability enhancing conditions as health and education.

A delicate attempt at regaining a better balance between protection of the public domain of knowledge from further encroachments by the domain of private property rights, is needed at least in regard to some sectors where services are recognized to profoundly affect human well-being (e.g., health, education). The notion of a universal right to health appears to have the strength to countervail against the national and international campaigns led by pharmaceutical companies to secure intellectual property owners the right to unregulated exploitation of their patents. But one must not be deluded into supposing that appeals to principles of equity alone will be sufficient in deciding such contests in the area of political economy.

Some Subtleties in Gauging the Effects of IPR Protections on the Conduct of Science

Apart from a few anecdotes of the sort that already have been introduced in this discussion, is there any really systematic body of empirical material on which to base the warnings that are being sounded about the possible impact of increased international property rights on exploratory research, and academic science more generally? Should we not insist on policy action being evidence-based? More specifically, how can we evaluate the potential for the enforcement of database rights to impede the exploitation of bioinformatic techniques, and scientific discoveries more generally? Heller and Eisenberg (1998) popularized the phrase "tragedy of the anticommons" in an article published in *Science*, suggesting that the great increase of patenting in the biomedical and in biotechnology area more generally might actually inhibit innovation, especially where inventions and discoveries required the assembling of an assortment of complementary bits of knowledge and research tools, each of which might be owned by distinct parties. The argument, basically, is that fragmented ownership rights invite opportunistic bargaining strategies (hold-outs, over-pricing), and so raise negotiation costs that projects which otherwise would be privately profitable, and possibly of great social value, would not be undertaken. But Heller and Eisenberg did not substantiate the existence of a potent "anti-commons effect" by producing concrete instances of such losses.

Not surprisingly, the claim that granting greater protection to IPRs might adversely affect investment in the production of knowledge, by increasing the costs of collaborative projects among the property holders, has been received with skepticism in some quarters, and therefore has attracted the attention of empirically-minded economists. A recent study by Walsh, Arora, and Cohen (2002) set out to look for evidence on the question of whether serious anti-commons effects had materialized in the biotechnology sector, which is where Heller and Eisenberg's article warned it was likely to appear. Their methodology was the survey of participants in business and academic research organizations, and the thrust of their findings was that while there were a few isolated instances of serious difficulties in working out the IPR arrangements among firms, and between firms and universities, their interviews disclosed nothing resembling a "tragedy." Rather more serious reservations were sounded, however, regarding the impediments to research discoveries that might be caused by the patenting of fundamental research tools, an issue that Walsh, Arora, and Cohen treated as distinct from that of the anti-commons effect—even though Eisenberg (2001) evidently regards the two as closely related.

While this study was carefully carried out and reported in a balanced and qualified fashion, it gave little attention to the issue of what sort of indications might be elicited by interviews that would establish the existence of anti-commons tragedies. The interview protocols followed in questioning managers, lawyers, and researchers in biotechnology and pharmaceutical companies, and also a much smaller number of university scientists (numbering 10 among the 70 interviews conducted) are not described in any detail by Walsh, Arora, and Cohen. In particular, the question of exactly what constitutes a "blockage" or "breakdown" to a biomedical research project is never specified. So it is difficult to evaluate the responses that the study reports elicited, but there clearly are potential problems in interpreting some of the reported testimony. (The same must be said with regard to the interview material from the National Institutes of Health Working Group on Research Tools, discussed by Eisenberg [2001].)

To illustrate the methodological problem, let us suppose the procedure involved asking interviewees: "Have you experienced serious obstacles to research you have tried to undertake, due to conflicts arising over intellectual property rights, difficulties that prevented you from pursuing worthwhile projects?" The reported responses, by and large, would be consistent with the informants having replied: "Not really, we are quite able to do our thing." One might wonder what the investigators thought they might be told. Would the heads of research departments say that their research performance was no longer as good as it once had been (no matter what the alleged cause)? The point here is a simple and familiar one: the way that interview questions are phrased is a delicate matter in such an inquiry.

In the case of the university biomedical researchers interviewed by Walsh, Arora, and Cohen (2002), a typical response to the question of whether patenting of research tools was impeding their research seems to have been something along the lines: "Not really. We don't pay attention to it, and the firms seem reluctant to enforce their patents against us." Now it is quite correct that if property rights are granted people do not voluntarily comply with the intent of the law establishing those rights, and the right-holders do not enforce their legal claims, it cannot be said that there will be any effect of the statutory change. But, surely, to conclude that "there is no problem" in such a case is rather misleading. There is no effect because the cause has yet to happen. The proper conclusion is: We can't say, what the effect of the IPR regime will be in this instance, except that when a cease and desist injunction is brought against one of these professors, and her university is charged with patent infringement and sued for damages (perhaps by another university that holds the patent on one of those research tools), it is going to be a big shock.

The problem of interpreting the evidence of the survey responses in other respects is a bit more subtle than was suggested by the previous remarks about possible reporting biases, arising from the framing of the questions. The anti-commons argument can be given a naive interpretation or one that is economically sophisticated, and correspondingly, the evidence can be read either naively or in a sophisticated manner. Let us start by consider the alternative styles of interpretation of the putative "Anti-Commons effect":

- *Naïve*. Parties to a potentially productive coalition will see only the value of cooperating for a common benefit and will ignore the possible costs of contracting. So, if IPR has the effect of raising the parties' valuation of their own contribution to the collective project, and makes it possible for them to deny others access to that contribution, the negotiation of cooperative agreements will be surprisingly difficult, and frequently these will fail. One should be able to find records, or elicit testimony of such failures. Look for them in order to test the anti-commons hypothesis.

- *Sophisticated*. A well-known essay by Ronald Coase, the Nobel Laureate in Economics, pointed out that the institutional arrangements that assigned property rights to some agents would only affect the efficiency of resource allocation among them if there were zero "costs of transacting," of arriving at a contract in which the gains from trade would be secured for the collectivity and distributed among them. Agents understanding this should consider ex ante the likely benefit of a contract for the exchange of assets (entitlement to rent streams), and the costs of negotiating such a contract. If they do this, they will take account of changes in institutions that alter the property rights of the parties, and the likely affect of this on the nature of the contracting process and its costs. So a change in property rights that raises everyone's assessment of what they should get from the same cooperative project would be perceived as raising the costs of contracting sufficiently to render it foolish to pursue some projects, namely those at the lower marginal utility end of the ranking. Note that I use "utility" to cover both projects with lower expected rates of return, and those where there is higher intrinsic technical or commercial risk (in the sense of higher variance relative to the expected rate of return).

Rational agents, therefore, will discard projects as not being worth serious consideration. If they are asked for examples of projects that were "blocked" or "abandoned" because of high transactions costs, they should not report any higher frequency of such events following the institutional change than they reported before. There might be an initial "learning" period in which mistakes were more frequent, but there is no reason why the equilibrium level of failed negotiations should be raised.

What the sophisticated view of the matter suggests is that to find the "effect" of the institutional change, one has to estimate what would happen in the counter-factual world: what projects that were not seriously considered would have been considered. The foregoing line of thought might suggest that the nature of the projects that would be undertaken following the change in the IPR regime would be found to have shifted toward those with the following characteristics:

1. The distribution of initial property rights among the participating was already highly asymmetric and the relative disparity in bargaining power would not be materially changed by the altered property rights regime, so the estimated transaction costs would not be significantly affected;

2. The private expected rate of return was higher than the norm for the previously undertaken projects, and so could justify the higher contracting costs of putting the collaboration together;

3. The risk-return ratios for the projects undertaken are found to have been lower than those among the projects previously undertaken.

In other words, an institutional change that raises the marginal costs of transactions of a particular kind need not actually increase the amount of resources consumed by such transactions; rather it may push resources into other channels, and leave a gap between the marginal rates of return that the realized projects in that area yield (gross of negotiation expenses) and the rates of return on other kinds of projects. That gap is a measure of the social burden of "royalty stacking," "blocking patents," etc.

Unfortunately, ideas are extremely heterogeneous. It is rare that the same options for exploration present themselves at successive points in time; in the world of research the dictum of Heraclitus holds—one cannot dip into precisely the same stream of ideas twice. Because the set of projects to create knowledge will not present themselves after the legal system protecting IPR changed will not be exactly the set that was available before, the gap of discarded opportunities can never be measured exactly. What one could do, in principle, is to examine the characteristics of the entire portfolio of research projects that were being undertaken before, and after the institutional innovation. Of course, it would be necessary in doing so to control for the influence of other temporally correlated changes that would affect those portfolio characteristics.

The hunt for evidence of a "tragedy" in the form of a lost opportunity is inextricably encumbered by the problem of documenting a counterfactual assertion of the form: if we had not done that, the world would now be different. In discussion of such propositions, rhetorical victories tend to go to the side that can shift the burden of proof to the shoulders of their opponents, simply because conclusive proof of a counterfactual assertion will be elusive. Thus, for those who see the system of IPR protection as fundamentally benign, the debating strategy is to demand that the critics show the social efficiency losses that they claim it is causing. But why should that assignment of the burden of proof be accepted? Although in the case of ordinary, physically embodied commodities there is a theoretical presupposition that competitive markets and well-defined private property rights can support a socially-optimal equilibrium in the allocation of resources, that presumption ceases to hold in the realm of information and knowledge.

CONCLUSION

The elision effected by the application of the metaphor of "property" to the domain of ideas has been fruitful in many regards. But there is a danger in permitting those who are enthusiastic for more and stronger IPR to employ this as rhetorical device as a way of avoiding the burden of proof. They should be asked to show that the moves already made in that direction have not been economically damaging, that further encroachments into the public domain of scientific data and information would not be still more harmful, and that society would not benefit by adopting a policy that was just the opposite of the one they support.

REFERENCES

Arrow, Kenneth J. 1962. "Economic Welfare and the Allocation of Resources for Inventions," in *The Rate and Direction of Inventive Activity: Economic and Social Factors*, R.R. Nelson ed., Princeton University Press, Princeton.

Cowan, R., P.A. David and D. Foray. 2000. "The Explicit Economics of Knowledge Codification and Tacitness," *Industrial and Corporate Change*, 9(2), (Summer): pp. 211-253.

Dasgupta, P. and P.A. David. 1987. "Information Disclosure and the Economics of Science and Technology," in *Arrow and the Ascent of Modern Economic Theory*, G. Feiwel, ed., New York University Press, New York, pp. 519-542.

Dasgupta, P., and P.A. David. 1994. "Toward a New Economics of Science," *Research Policy*, vol. 23: pp. 487-521.

David, P.A. 1993. "Intellectual Property Institutions and the Panda's Thumb: Patents, Copyrights, and Trade Secrets in Economic Theory and History," in *Global Dimensions of Intellectual Property Rights in Science and Technology*, (M. Wallerstein, M. Mogee, and R. Schoen, eds.), National Academy Press, Washington, D.C.

David, P.A. and D. Foray. 1995. "Accessing and Expanding the Science and Technology Knowledge Base," *STI Review: OECD—Science, Technology, Industry*, No.16, Fall: pp. 13-68.

David, P.A. 1998a. "Common Agency Contracting and the Emergence of 'Open Science' Institutions," *American Economic Review*, 88(2), May.

David, P.A. 1998b. "Reputation and Agency in the Historical Emergence of the Institutions of 'Open Science'," *Center for Economic Policy Research, Publication No. 261*, Stanford University (March 1994, revised December 1998).

David, P.A. 2000. "The Digital Technology Boomerang: New Intellectual Property Rights Threaten Global 'Open Science'," (Presented at the World Bank ABCDE Conference in Paris, 26-28 June, 2000.) *Department of Economics Working Paper 00-16*, Stanford University, October 2000. (*Proceedings of the World Bank Annual Conference on Development Economics- Europe, 2000*, J. Bas and J.E. Stiglitz, eds., CD-ROM Edition, Fall 2001.)

David, P.A. 2001. "The Political Economy of Public Science," in *The Regulation of Science and Technology*, Helen Lawton Smith, ed., Palgrave, London.

David, P.A., D.C. Mowery, and W.E. Steinmueller. 1992. "Analyzing the Payoffs From Basic Research," (with) *Economics of Innovation and New Technology*, vol. 2(4): pp. 73-90.

Eisenberg, R.S. 2001. "Bargaining over the transfer of proprietary research tools: Is this market failing or emerging," Ch. 9 in *Expanding the Boundaries of Intellectual Property*, R. Dreyfuss, D. L. Zimmerman and H. First, eds., Oxford University Press, New York.

Foray, D. and A. Kazancigil. 1999. "Science, economics and democracy: selected issues." *MOST-UNESCO Discussion Paper No 42*, UNESCO, Paris.

Heller, M.A. and R.S. Eisenberg. 1998. "Can Patents Deter Innovation: the Anticommons in Biomedical Research," *Science* 280 (1 May): pp. 698-701.

Holton, G. 1996. *Einstein, History, and Other Passions: The Rebellion Against Science and the End of the Twentieth Century*, Addison-Wesley Publishing, Reading, MA.

Merton, R.K. 1973. *The Sociology of Science: Theoretical and Empirical Investigations*. N.W. Storer, ed., Chicago University Press, Chicago.

Nelson, R.R. 1959. "The Simple Economics of Basic Scientific Research," *Journal of Political Economy*, 67.

Walsh, J.P., A. Arora, and W.M. Cohen. 2002. "Research Tool Patenting and Licensing and Biomedical Innovation," Forthcoming in *The Operation and Effects of the Patent System,* The National Academies Press, Washington, D.C. in 2003.

Ziman, John. 1994. *Prometheus Bound: Science in a Dynamic Steady State*. Cambridge University Press, Cambridge.

5

Scientific Knowledge as a Global Public Good: Contributions to Innovation and the Economy

Dana Dalrymple[1]

"The preeminent transnational community in our culture is science," Richard Rhodes, 1986

"If the potential of modern science is to be realized, there is no alternative to global public goods and institutions," Lawrence Summers, 2000

INTRODUCTION

Scientific knowledge in its pure form is a classic public good. It is a keystone for innovation, and in its more applied forms is a basic component of our economy. Although recent technical advances have stimulated its generation and greatly accelerated its spread, other forces may limit its public-domain characteristics.

The concept of public goods is not new. Although it is being applied in an increasing number of areas of social importance, this does not yet seem to be true of the natural sciences. Science is seldom mentioned in the public goods literature, or public goods in scientific literature. Yet the combination is a logical and useful one. A few economists and health specialists have recognized this, but the same cannot be said of the scientific community more generally.[2]

The related concept of public domain is also not new. Although it is even broader conceptually (see Drache, 2001), it has found a more specific meaning, particularly among lawyers, in the context of intellectual property rights (IPRs). Lawyers, however, seem to assume the availability of public goods and scientific knowledge and their focus may be limited to national and local legal systems.

Why such limited or partial attention to what should seem a most appropriate and useful common concept? Is it because the three professional groups most likely to be involved—scientists, economists, and lawyers—have not viewed public goods in a broader and more integrated light? This symposium provides a most appropriate opportunity to begin to try to bridge the gap.

In doing so, a few definitions might help set the stage. Data and information are at once both key components in the generation of scientific knowledge and among its major products: they are both inputs and outputs (Arrow, 1962, p. 618). However, knowledge in general is broader, less transitory, and more cumulative. It is derived from perception, learning, and discovery. Scientific knowledge, in particular, is organized in a systematic way and is testable and verifiable. It is used to provide explanations of the occurrence of events (Mayr, 1982, p. 23).

---

[1]Acknowledgments: The author benefited greatly from the advice and assistance of a number of individuals during the preparation of this chapter, particularly John Barton, Paul David, and Vernon Ruttan.

[2]Among economists, the most prominent proponent of scientific goods in the international arena is Jeffrey Sachs, who has been primarily concerned with expanding health and agricultural research in and for developing nations (Sachs, 1999, 2000a). His most recent effort, as part of a World Health Organization (WHO) study, has led to a proposal for a $1.5 billion annual expenditure for a new Global Health Research Fund (WHO, 2001, pp. 81-86; Jha, et al., 2002). A previous WHO report (1996) argued that research and development expenditures in health were an important international public good. Dean Jamison, who was involved in that report, has also written on the subject elsewhere (2001).

The topic itself is, of course, quite ambitious. Moreover, this may be the first attempt to take it on in a fairly comprehensive way. Hence I will only attempt to provide an introduction. Three main topics will be taken up, in varying proportion: (1) principal concepts, (2) provision and use, and (3) implementation. The focus will be on scientific knowledge at the international level, particularly with respect to developing countries. My perspective is that of an agricultural economist and sometime historian. My approach involves a rather wide-ranging review of literature blended with long personal experience in international agricultural research. Others might well follow quite different routes and illustrate different dimensions. I encourage them to do so.

PRINCIPAL CONCEPTS

In examining scientific knowledge as a global public good, I will start by building on several venerable concepts and components. Some of them have been partially woven together before; others have not. Each has its own history and is important to understanding the whole. And they need to be combined with some contemporary economic perspectives.

Historical Perspectives

The starting point is public goods, which were long considered, at most, at the national level and for public institutions and services. Hence there is a need to expand the definition in several directions: to knowledge as a global public good, to global scientific knowledge, and to recognition of the role played by IPRs.

Knowledge as a Public Good

Adam Smith laid the basis for the concept of public goods in The Wealth of Nations in 1776 when he stated:

> The third and last duty of the sovereign is that of erecting and maintaining those public institutions and those public works, which, though they may be in the highest degree advantageous to a great society, are, however, of such a nature, that the profit could never repay the expense to any individual or small number of individuals, and for which it cannot be expected that an individual or small number of individuals should erect or maintain.

The development of more sophisticated theories of public goods began in the last quarter of the 19th century (Machlup, 1984, p. 128). Recent use of the term by economists is usually traced back to two short articles by Paul Samuelson in the mid-1950s (1954, 1955). It became a central concept in public finance, in part due to the writings of Musgrave and Buchanan (Machlup, 1984, pp. 128-129; Olson, 1971; Buchanan, 1968). Public goods, as they have generally come to be known, have two distinct characteristics: (1) they are freely available to all and (2) they are not diminished by use. These properties are often expressed by economists, as we shall see later, in terms of non-excludability and non-rivalry.

In the context of scientific knowledge, a "good" is viewed here, following some dictionary variants, as having or generating two key qualities: (1) it is tangible in the sense that it is capable of being treated as a fact, or understood and realized; and (2) it has intrinsic value in terms of relating to the fundamental nature of a thing. It is neutral with respect to the "good" effect on society, although that also is usually presumed to be good (to be discussed later), and excludes money.

The public goods characteristic of ideas and knowledge has long been noted, first by St. Augustine, sometime between 391 and 426 (Wills, 1999), and then by Thomas Jefferson, in 1813 in a frequently cited letter on patents (1984).[3] Their views were carried further by Powell in 1886 when he stated: "The learning of one man does not subtract from the learning of another, as if there were to be a limited quantity to be divided into exclusive holdings.... That which one man gains by discovery is a gain to other men. And these multiple gains become invested capital...."

---

[3] "He who receives an idea from me, receives instruction himself without lessening mine; as he who lights his taper at mine, receives light without lessening mine."

The Nature and Spread of Scientific Knowledge[4]

The adjective "scientific" can be traced to Aristotle and at some point entered the Romance languages. Its use in English, however, dates only to about 1600 and was synonymous with knowledge. In its earlier incarnations, it referred to demonstrable knowledge as compared with intuitive knowledge. It was at first referred to as natural philosophy in English. Emphasis was on deductive logic, which was useful for confirming what was already known, but not for original discovery.

The situation began to change in the early 1600s with the writings of Francis Bacon, who was a believer in inductive logic and the experimental method. The latter made it possible to discover and understand new facts about the world. From about 1620, with the publication of his Novum Organuum, there was a shifting of the philosophical point of view toward Bacon's interpretation. This process reached its full realization by 1830.

With the new meaning of science—"natural philosophers" would no longer do—there was an increasing need for a new word for its practitioners. In 1834, William Whewell of Cambridge University rather casually proposed the term "scientist." In 1840 he more seriously but very briefly said: "We need very much a name to describe a cultivator of science in general. I should incline to call him a Scientist" (Whewell, 1840/1996, p. cxii).[5]

Bacon, among his other insights, was perhaps the first to record his views on the wider nature of knowledge when he wrote: "For the benefits of discoveries may extend to the whole human race" and "for virtually all time" (Bacon, 1620/2000, p. 99). He clearly saw the benefits of attempting to reach beyond national boundaries, as was evident in his treatment of three levels of ambition, the third of which was put in these terms:[6] "But if a man endeavor to establish and extend the power and dominion of the human race itself over the universe, his ambition . . . is without a doubt both a more wholesome thing and more noble than the other two" (Henry 2002, p. 16).[7]

The age of global exploration that followed Columbus did much to bring about a global exchange of biological material and associated information. In the view of one historian, "Nothing like this global range of knowledge had ever been available before," and it proved to be "a boost to Europe's incipient 'scientific revolution'." "In this way, the exchange "made a major contribution to the long-run shift in the world balance of knowledge and power as it tilted increasingly toward the West" (Fernández-Armesto, 2002, p. 167).

Bonaparte gave a different twist to the process. As part of his "expedition" to Egypt in 1798, which included some of the finest scientific minds in France, he founded the "Institute of Egypt." The essential goals of the institute's researchers, in contrast to the military side, "were the progress and propagation of sciences in Egypt" and "the conquest of knowledge and the application of knowledge to man's lot" (Herold, 1962, pp. 28, 151, 164-176; also see Solé, 1998).

Other approaches were undertaken by other European colonial powers during the next century. Botanical gardens were to play a particular role (Drayton, 2000; Plucknett et al., 1987; Pardey et al., 1991). Brazilian rubber seeds and the Peruvian cinchona tree, a source of quinine, were prime targets (Alvim, 1994, p. 426; Drayton, 2000, pp. 236, 249; Honigsbaum, 2000). In the latter case, collecting expeditions were defended—with some justification—as a "duty to humanity" (Honigsbaum, 2001, p. 81). Science and research played a larger role later, notably in the Spanish Caribbean, in helping improve the production of export crops (McCook, 2002).

---

[4]The first part of this subsection is largely based on a little-known article by Ross (1962)—reprinted in Ross (1991)—and to a lesser degree on Henry (2002) and Bacon (2000, introduction by Jardine). For a further discussion, see "Scientific Knowledge" in Machlup (1980, pp. 62-70). The subject is also briefly noted by Barzun (2000, pp. 191, 544) and Bernal (1965, p. 32).

[5]Merton (1997) indicates that Whewell, evidently stung by the hostile reception of the term by his English colleagues ("man of science" prevailed until about 1910), used the word only once during the remainder of his career.

[6]With respect to the first two, Bacon wrote: "The first is of those who desire to extend their own power in their native country; which kind is vulgar and degenerate. The second is of those who labor to extend the power of their country and its dominion among men. This certainly has more dignity, though not less covetousness."

[7]Thus it is not surprising that when the Royal Society of London was established in the 1660s, "Foreign corresponding members—'The Ingenuous from many considerable parts of the world'—were eagerly recruited" (Barzun, 2000, pp. 210-211).

Development of Intellectual Property Rights[8]

Many forms of knowledge are, of course, presently linked with IPRs. IPRs may seem a product of modern times, but they had their roots in the Middle Ages, when less attention may have been given to the social dimension. Concern about their effect on social welfare appears to have been of more recent and domestic origin.

Patents have existed in various forms since the late 1400s. The first general promise of exclusive rights to inventors was made in a statute enacted in Venice in 1474 (Machlup, 1984, p. 163). They became more widely adopted in Europe in the 1600s. The U.S. Congress first passed a patent statute in 1790 and in 1836 the Patent Office was established. It had a trained and technically qualified staff. Kahn (2002) states that the system was based on the presumption that social welfare coincided with the individual welfare of inventors. Courts subsequently "explicitly attempted to implement decisions that promoted economic growth and social welfare."

Copyrights have also existed in various forms since the late 1400s. In the United States, the earliest federal law to protect authors was passed in 1790. Policymakers, according to Kahn, felt that copyright protection would serve to increase the flow of learning and information, and by encouraging publication would contribute to the principles of free speech. Moreover, the diffusion of knowledge would also ensure broad-based access to the benefits of social and economic development.

In practice, patents were fairly narrowly construed and copyright was interpreted more casually. Kahn views this as appropriate: social experience shows that they warrant quite different treatment if net social benefits are to be realized.

Contemporary Economic Perspectives

The next step in building a conceptual base is to move to some more recent perspectives, largely by economists, about the role played by knowledge in thinking about economic growth and then, in a more applied way, in international development programs.

Knowledge and Economic Growth

The role of knowledge, and particularly of scientific knowledge, in economic growth has received relatively little concerted study.[9] Kenneth Boulding was one of the first to draw attention to the connection.[10] In 1965, he stated at a meeting of the American Economic Association:

> The recognition that development, even economic development, is essentially a knowledge process has slowly been penetrating the minds of economists, but we are still too much obsessed by mechanical models, capital-income ratios, and even input-output tables, to the neglect of the study of the learning process (Boulding, 1966, p. 6).

Only Machlup (1980, 1984) appears to have taken up the subject in a comprehensive manner, but even he did not get very far into the development side.

Why? Part of the problem is that it is difficult to reduce knowledge to numerical form so that it can be used in economic models. As Boulding also said: "One longs, indeed, for a unit of knowledge, which might perhaps be called a 'wit,' analogous to the 'bit' used in information theory; but up to now at any rate no such practical unit has emerged" (Boulding, 1966, pp. 2-3; also see Desai 1992, p. 249 and Romer, 1994, p. 20). Another problem was that there was an uneasy relationship between growth economics that was macroeconomic in orientation and development economics that was more microeconomic and multidisciplinary in its approach (Ruttan, 1998; Altman, 2002).

---

[8]This subsection is largely drawn from Kahn (2002). Additional historical information is provided in Machlup (1984, p. 163) and David (1993, pp. 44-54).

[9]Knowledge plays an implicit role in other studies, especially of those relating to the effects of research and development or information (see, for example, Machlup, 1984, pp. 179-182).

[10]Boulding might seem to have been preceded, on the basis of titles, by Hayek in 1937 and again in 1945, but Hayek was focused on economic or market knowledge and information (similar to information about attributes, to be noted below).

Romer, a macroeconomist, did go on to do some seminal work (first reported in 1986, with further contributions in 1990, 1993, and 1994). It was based on the assumption that long-run growth is driven primarily by the accumulation of knowledge. Knowledge (harking back to Powell, 1886) was considered the basic form of capital. New knowledge is the product of research and will grow without bound. In the industrial sector its production may have an increasing marginal product, in part because the creation of new knowledge by one firm can have a positive effect on the production possibilities of other firms because knowledge cannot be perfectly patented or kept secret. Hence, knowledge, even if generated for private gain, has an important public good characteristic. Romer suggested that an intervention that shifts the allocation of current goods away from consumption and toward research is likely to improve welfare.

Knowledge for Development

The importance of knowledge for development was reflected by the World Bank during the period when Joseph Stiglitz was chief economist. At that time the bank devoted its World Development Report for 1998-1999 to "Knowledge for Development." It distinguished two types of knowledge: technical and attributes.[11] The uneven distribution of the former was referred to as the knowledge gap and the latter as an information problem. Both are more severe in developing than in developed countries.

Closing the knowledge gap was viewed as involving three steps: (1) acquiring knowledge from the rest of the world and creating it locally through research and development, (2) absorbing knowledge, and (3) communicating knowledge. The Green Revolution, "the decades-long worldwide movement dedicated to the creation and dissemination of new agricultural knowledge," was presented as "A paradigm of knowledge for development."

In a separate article, Stiglitz (1999) reiterated some of these points, but went on to note that although "research is a central element of knowledge for development," it is also a "global public good requiring public support at the global level." The latter requires collective action, and "The challenge facing the international community is whether we can make our current system of voluntary, cooperative governance work in the collective interests of all." This indeed is a central question.

PROVISION AND USE

The provision and use of scientific knowledge for the benefit of society are inviting to dream about, but are of course more difficult to realize. In this section an attempt will be made to identify and relate some of the more important steps and considerations. They start with some relevant aspects of the nature of knowledge, then move to the generation and embodiment of knowledge, factors influencing embodied knowledge, and finally consideration of the regulatory structure (as it applies to IPRs) to the provision and use of knowledge-based goods.

Nature of Knowledge

As the World Bank (1998-1999, p. 1) put it in one of its more lyrical moments, "Knowledge is like light, weightless and intangible, it can easily travel the world, enlightening the lives of people everywhere." There is, of course, more to the story as the bank quickly goes on to admit. Although easy to transmit, knowledge can be expensive to produce. And scientific knowledge must normally be transformed into some more tangible form, sometimes referred to as embodied knowledge (processes, products, policies) to be of social benefit. The latter are often grouped under the category of technology.

---

[11] Examples given of technical knowledge, or know-how, are nutrition, birth control, software engineering, and accounting. Knowledge about attributes includes the quality of a product, the credibility of a borrower, or the diligence of a worker and are critical for effective markets.

Relationship of Science and Technology

Science is traditionally viewed as providing the insights needed for the development of technologies. Barzun (2000, p. 205) notes that earlier in history, technology—in the form of the practical arts—"came earlier and was for a long time the foster mother of science." He continues, "Inventors made machines before anybody could explain why they worked . . . practice before theory." Similarly, David (1992, p. 216) notes that "Technological mastery may run far ahead of science and is in many regards both a stimulus to scientific inquiry and the means whereby such inquiries can be conducted."

Over time, the relative role of science has probably increased and in the minds of many, the relationship has reversed. In Barzun's words, "science finds some new principle and applied science . . . embodies it in a device for industry for domestic use." The unqualified version of this concept is sometimes termed the simplest linear model and is probably greatly oversimplified (see David, 1992, p. 216). In reality, many feedback loops are involved.

The role of the public and private sectors in this process has also changed somewhat. Traditionally, the public sector has been associated with the provision of basic knowledge and the private sector with more applied technology and products. The former might be exemplified at the extreme end by federal support for high-energy physics and, more usefully, to basic health research. But there are exceptions to this pattern: one is agricultural research, originally one of the most important areas of federal support for research, which is relatively applied in nature (for background, see Dupree, 1957). And, more recently, certain portions of the private sector have become very involved in some important basic research, such as on the human genome, which has at least partially found its way into the public domain. Whether the private sector will find a satisfactory profit in some of this remains to be seen, but it has altered the basic paradigm.

The conventional model has also been altered by the increasing number and variety of interactions, formal and informal, between the public and private-public sectors. The usual pattern has the public sector financing the private sector to produce what will become public goods (however, some intellectual property issues may complicate this picture). But some arrangements, and these are more unusual, have the private sector providing support to universities, in part to gain access to their scientists and scientific knowledge.

Scientific Goods and Bads

The term public "good," despite a more neutral definition adopted at the outset of this chapter, usually implies a positive social benefit. In the case of science, however, there is often considerable question about the degree to which some "goods" are good.

This was not always the case. During the first flush of science in the 1800s, it would appear that it was viewed quite positively. In 1884, John Wesley Powell, possibly reflecting the more ebullient spirit of the times stated "The harvest that comes from well-directed and thorough scientific research has no fleeting value, but abides through the years, as the greatest agency for the welfare of mankind." The belief that science could be harnessed for the benefit of all continued and perhaps peaked in the 1950s (Watson, 2002, p. 375).

This viewpoint, to the extent that it was shared, began to change with the advent of atomic energy and its unfortunate direct and indirect long-term effects. For other products of science, the outcome was more mixed. *Silent Spring* revealed the dark side of DDT, but it is still one of the best and lowest-cost methods of mosquito control in developing countries (Honigsbaum, 2001, p. 286; UNDP, 2001, p. 69). As Freeman Dyson (1979, p. 7), a physicist, has written: "Science and technology, like all original creations of the human spirit, are unpredictable. If we had a reliable way to label our toys as good and bad, it would be very easy to use technology wisely."

This, of course, is not the case, and so the presumed and actual "goods" and "bads" of science continue to be reported and debated daily. The agricultural area has shared fully (and perhaps more than proportionally lately), particularly with respect to biotechnology and genetically modified organisms, which can be, or become, all things—good, bad, and indifferent—depending on timing and circumstances.

This uncertainty, along with the possibility that the research may not accomplish even what it was originally intended to do, also has economic implications. It means that not all scientific research activities are in retrospect necessarily a good investment. This is true of both public and private research.[12]

Generation and Embodiment of Scientific Knowledge

Scientific knowledge is usually the product of a process. Sometimes it is informal and individual, but most often it is more formal and communal. Here we will look very briefly at the linkage of invention and scientific theory, the introduction of the research laboratory, and of the generalized process of moving from ideas and scientific knowledge to public goods.

Linkage of Invention and Scientific Theory

Invention is as old as mankind. But it often consisted of chance, trial and error, and individual intuition. Such activities seldom led to further innovative activity or to sustained economic growth. This was, according to Mokyr (2002), because invention without a scientific base was difficult to duplicate and quickly ran into diminishing returns. And such science as existed was not very concerned with practical applications.

The situation began to change around 1750 with the growth of what Mokyr calls "propositional" knowledge, which includes both scientific and artisanal or practical knowledge. During a subsequent period, which he calls the "Industrial Enlightenment," a set of social changes occurred that resulted in growth of scientific knowledge, an increase in the flow of information, and an attempt to connect technique with theory. The process was facilitated because it was a period of relatively open science: knowledge was a public good. But up until about 1850, the contribution of formal science to technology remained modest.

Establishment of Research Laboratories

It has been stated that in the 1800s, a century that was differentiated from its predecessors by technology, "The greatest invention . . . was the invention of the method of invention" (Whitehead, 1925, pp. 140, 141). Research laboratories "institutionalized the process of transforming intellectual and physical capital into new knowledge and new technology" (Ruttan, 2001, p. 82). The paths were somewhat different in agriculture and industry.

The early stages of innovation were fairly simple in agricultural societies. In the late 1700s and early 1800s, more progressive (and often more affluent) farmers experimented and expanded the envelope of knowledge. However, by the mid-1800s, the advantages of a more structured approach became increasingly evident in the United States. The House Committee on Agriculture, in its report establishing the U.S. Department of Agriculture in 1862, stated that accurate knowledge of nature "can be obtained only by experiment, and by such and so long continued experiments as to place it beyond the power of individuals or voluntary associations to make them" (Congressional Globe, 1862). Thereafter, public-sponsored research at the federal and state level gradually began to accelerate in the United States.[13]

Industrial research also dates from the latter half of the 1800s. The first corporate research laboratories were established in Europe by the chemical industry, particularly the dyestuffs sector, in the late 1860s and 1870s (Homburg, 1992; Mokyr, 2002, p. 85). Edison was the first individual in the United States to establish a significant

---

[12]The evaluation of the benefits of any public enterprise is complicated by the need to take into account what is vividly known as the excess burden or deadweight loss of taxation. I have discussed this concept elsewhere in terms of public agricultural research (Dalrymple, 1990).

[13]Some European nations, especially England and Germany, initially moved more quickly than the United States. For further details on the development of public agricultural research in the United States, see Dupree (1957, pp. 149-183); OTA (1981); and Ruttan (2001, pp. 207-211). The U.S. Department of Agriculture established its first field trials on what is now the Mall between 12th and 24th streets and Independence and Constitution avenues in 1865.

research organization when he did so at Menlo Park, New Jersey, in 1876. The first U.S. industrial laboratory was established by General Electric in Schenectady, New York, in 1900 (Reich, 1985).

Since the turn of the 20th century, public and private research laboratories have played a key role in innovation and creation of new knowledge. It has not been, in historical terms, a long period.

From Ideas to Public and Private Goods

Given some thoughts about the nature of knowledge and research structures, it may now be useful to attempt to show how they and other elements interact in the process of moving from ideas to public and private goods and to social benefit. My view of the highly interactive process is outlined in Figure 5-1.

There are roughly four stages. The first is the generation of ideas and concepts by researchers and others. The second stage is the process of moving, through an interactive research process, from (a) ideas and concepts to (b) disembodied or pure knowledge, and from there to (c) embodied or applied knowledge (technology for short). The third stage involves the intellectual property dimension, which I will return to in greater degree below. The fourth stage is the movement of the "goods"—public, public-private, and private—into use in society and the resulting, one hopes, social benefits. In the first three stages, actions taken will likely be influenced, although to a differing degree, by perceptions of social needs and opportunities.

Throughout, there may be considerable crossover between the public and private sectors. Research initiated in the public sector can end up being utilized by the private sector and vice versa. Numerous other forms of interaction may occur. The result can be a hybrid good (van der Meer, 2002). This pattern, which is growing, can be very productive and useful for society, but it may also complicate the intellectual property dimension.

Factors Influencing Use and Adoption

It is one thing to provide knowledge-based public goods. Their adoption and productive use is another. Many factors may be involved. Three important categories follow from the previous section.

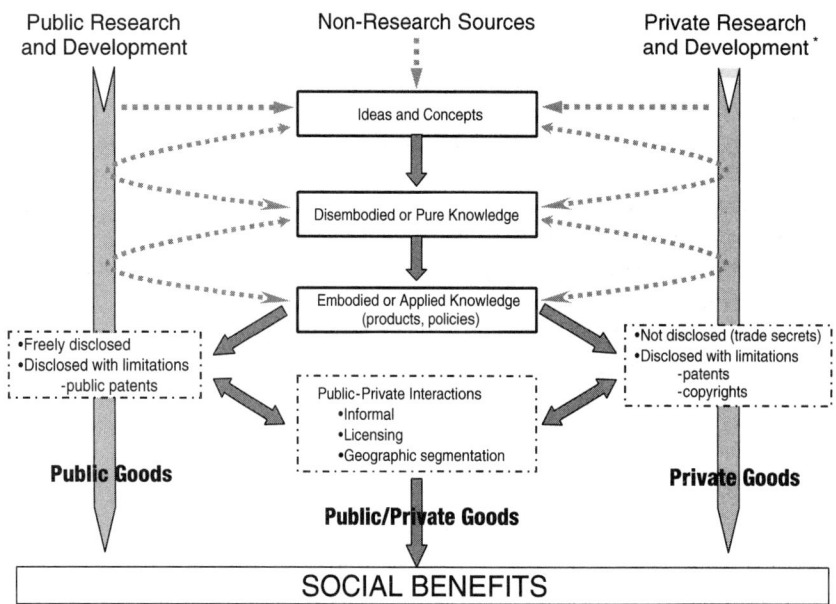

* Knowledge includes both data and information derived from perception, discovery, and learning. Scientific knowledge is normally cumulative and verifiable.

FIGURE 5-1 The evolution of scientific knowledge into public and private goods.

User Characteristics

Although the availability of knowledge-based public goods is a necessary condition, it is not a sufficient one. Potential users must themselves have a sufficient degree of knowledge to identify, understand, and possibly adapt the public goods available to them. They must also have an appropriate policy and legal environment, adequate infrastructure, and adequate financial resources. The more one gets into technology, the greater the degree of need for adaptation to local conditions. This is particularly true in the agricultural area where environmental and natural resource conditions are important and may vary sharply. Adaptation may well involve a further research and development capacity at the regional, national, and local level.

"Good" Characteristics

The main identifying characteristics of public goods, in economic terms, are their nonrival nature and their non-excludability. Nonrival means that use by one firm or individual does not limit use by another. Nonexcludable means use is not denied to anyone. These conditions are seldom found in totally pure form and may be limited for most purposes to (a) disembodied or pure knowledge; (b) noncopyrighted publications; and (c) some products of public research programs, such as those produced by federal laboratories.[14]

Virtually everything else, strictly speaking, is an impure public good or a private good. But the degree of impurity varies widely and in most cases contains a significant public goods dimension. Moreover, the impurity, insofar as it involves a useful private-sector contribution, may play an important role in improving the quality and usefulness of the product and maximizing social welfare. And purely private products may contribute significantly to social welfare.

The Path to Market

I have attempted to integrate these economic characteristics with intellectual property considerations in a market power context in Figure 5-2. Clearly, the left side represents the relatively pure public goods situation and the right side the relatively pure private good situation. The extreme forms of each variant are not very common. This leaves the middle as the most prevalent category composed of partially rival and excludable goods. Partial excludability is maintained with governmental participation—principally through copyrights, patents, or trade secrets.[15]

The differing paths result in differing degrees of market power or control, ranging from none (or pure competition) to complete (or a monopoly). Again the extremes are uncommon, but may be personified on the pure competition end by, as I have mentioned, some products of public research in agriculture and on the monopolistic end by actual trade secrets by the private sector. Most everything else results in partial market power. But in any case, as a recent study concluded, "... knowledge has become to an even greater degree than before the principal source of competitive advantage for companies and countries" (Commission on Intellectual Property Rights, 2002, p. 13).

---

[14] Many state universities are beginning to patent and license their products as a source of revenue.

[15] Both Figures 5-1 and 5-2 contain reference to public patents. These are patents taken out by a public entity to ensure that the patented item stays in the public domain. It is made available under license at no cost or at a very nominal charge. This was, for instance, standard practice at Tennessee Valley Authority during the 1950s (personal communication from Vernon Ruttan). Such patents are now uncommon. The principal reason is that the Bayh-Dole Act of 1980 allowed federal government contractors and grantees to take title to any subsequent inventions. Moreover, as a result of the Technology Transfer Act of 1986, the federal government may keep royalties from licensing its inventions. Defensive patents under 35 U.S.C.157, which provides for invention registration, are still possible, but probably would also be limited by the more general profit focus. (The previous four sentences are based on a personal communication from Richard Lambert, the National Institutes of Health, August 9, 2002.)

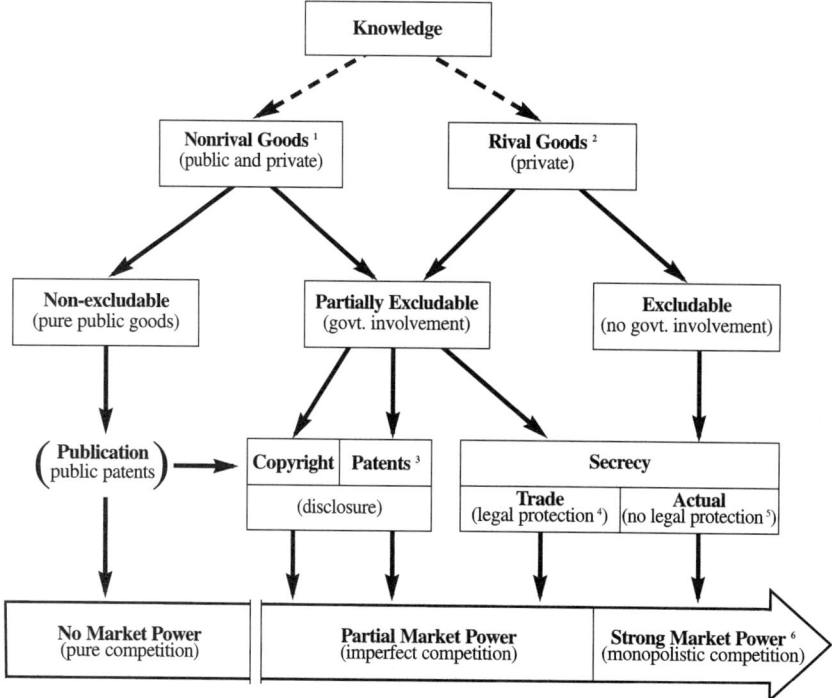

FIGURE 5-2 Knowledge-based public and private goods and the path to market: an economic and intellectual property perspective. **Sources:** Romer (1990, 1994) and Eisenberg (1987).

Intellectual Property Systems and Scientific Information

Clearly the various forms of intellectual property play a key role in the innovation process for knowledge-based public goods, especially those with some degree of private-sector involvement. What is their likely effect on the generation, flow, and use of scientific information? Fortunately, Eisenberg (1987) and a more recent Commission on Intellectual Property Rights (CIPR) (2002) have examined several of these issues.[16]

Secrecy

Secrecy may seem like an odd item to include under intellectual property systems, but one of the principal types—trade secrets—relies on a legal system for enforcement. Legal trade secrecy affords a remedy in court when individuals disclose information to others who subsequently breach this confidence or who otherwise misappropriate the information acquired. Actual trade secrecy is a strategy for protection in circumstances where not all the requirements for legal trade secrecy have been met. Eisenberg (1987) thinks that, although both types involve substantial nondisclosure, legal trade secrecy may be more disruptive of scientific communication than actual secrecy.

---

[16]The commission was established by Claire Short, the Secretary of State for International Development in the United Kingdom, in May 2001. John Barton of Stanford served as chair. Its emphasis was on developing countries.

Patent Law

Patent laws may be more congruent with scientific norms because they are based on disclosure.[17] The fit, however, may be less than perfect because (a) patents may operate to delay the dissemination of knowledge to other researchers, and (b) the granting of rights to exclude others from using patented inventions for a period of years threatens the free use and extension of new discoveries. Eisenberg (1987, p. 180) concludes that, although there are substantial parallels between patent laws and scientific norms, "the conjunction may nevertheless cause delay in the dissemination of new knowledge and aggravate inherent conflict between the norms and reward structure of science." Resolution, she thinks, will involve adjustments on both sides.

Although the purpose of the patent system was to stimulate invention and provide an incentive to technical progress, the CIPR (2002, p. 123) states that over time, "The emphasis has shifted toward viewing the patent system as a means of generating resources required to finance R&D and to protect investments." The system, in their view, fits best a model of progress where the patented product is the result of a discrete outcome of a linear research process. By contrast, they note that for "many industries, and in particular those that are knowledge-based, the process of invention may be cumulative, and iterative, drawing on a range of prior inventions invented independently . . . ." (p. 124). They suggest that the cumulative model fits more current research than the discrete model. Hence a patent system that was developed with a discrete model in mind may not be optimal for a more knowledge-based cumulative model.

Copyright

The degree to which copyrights have played a significant role in stimulating or limiting the dissemination of scientific information to date seems to be uncertain, but this may be in the process of changing, and not for the better. As the CIPR (2002, p. 18) has noted, copyright protects the form in which ideas are expressed, not the ideas themselves. The form, however, may have a significant impact on their use, and this is becoming more of an issue in view of changes in information technology.

In any case, copyrights probably have been of much greater importance to developed than to developing nations. This position may also change: the CIPR (2002, p. 106) stated, "We believe that copyright-related issues have become increasingly relevant and important for developing countries as they enter the information age and struggle to participate in the knowledge-based global economy." In this case, "The critical issue . . . is getting the right balance between protecting copyright and ensuring access to knowledge and knowledge-based products" (p. 106).

In addition to economic questions, there can also be more philosophical concern about basic rights. As Kahn (2002, p. 53), in a paper also prepared for the commission, put it, "Even in cases where a strong copyright may be necessary to provide the incentives to create, it might be advisable to place limits of the power of exclusion in order to promote social and democratic ends such as the diffusion and the progress of learning." We will doubtless hear much more about these issues in the future.

IMPLEMENTATION: AN EXAMPLE

To this point, I have largely dealt with concepts relating to scientific knowledge as an international public good. Although many of the individual components have an extended history, heretofore they do not seem to have been linked together to the degree that one might expect. If so, this might suggest that the overall concept is fairly theoretical and untested—possibly not implementable. Fortunately, that is not the case. There is a substantial instance of, as Barzun was earlier quoted as saying, "practice before theory" (Barzun, 2000, p. 205).

In the early 1970s, the Consultative Group on International Agricultural Research (CGIAR) was established and has operated until recently without any knowledge of many of the more conceptual issues discussed here. And

---

[17]There is a complication in the case of plant germplasm. As Ronald Cantrell, the Director General of the International Rice Research Institute, puts it, "Germplasm is unlike other types of intellectual property in that the average practitioner cannot completely benefit from the disclosure unless a cross can be made with the patent protected germplasm" (personal communication, August 19, 2002). This underlines the need for research exemptions (see Figure 5-2, footnote 3).

yet, in retrospect, it conforms very closely to what one might envisage a group oriented to science-based global public goods to be like and to do. It provides both confirmation of the concept and insights about what is involved in implementing it.[18]

Origin and Nature

The CGIAR is an informal group of donors—numbering about 60—who together sponsor 16 international agricultural research centers largely located in, and working on, problems of importance in agriculture and natural resources in developing nations. The centers are established as independent organizations with international status, boards, and staffs. They are both regional and global in their orientation and as public entities produce public goods. Their focus is on applied research and the development of improved technologies and policies that can be widely adopted, although they generally require local adaptation.

The CGIAR system includes a chair who is a vice president of the World Bank, a small group of international organizations as cosponsors (the Food and Agriculture Organization of the United Nations throughout and some others that have come and gone[19]), a general secretariat housed at the World Bank, and the Technical Advisory Committee (TAC) housed at the Food and Agriculture Organization. TAC has been a particularly useful component: it was recognized at the outset that the donor representatives to the CGIAR would not necessarily be scientists and that some outside source of continuing scientific and technical advice would be needed.[20] The TAC members, totaling about 14, were half drawn from developed countries and half from developing counties. The group is currently being transformed into a Science Council with somewhat modified duties.

A group such as the CGIAR was vitally needed by developing nations for several reasons. The principal one was that, as of the 1960s, little research attention had been given to food crops for domestic consumption. The principal emphasis of colonial powers was on tropical export crops, often grown under plantation conditions. Thus national research systems, with a few exceptions, had relatively little capacity in commodities targeted to domestic food use. Moreover, the private sector was almost completely inactive in this area. Thus the developing countries found it difficult to increase agricultural productivity to meet the needs of an expanding population. Early efforts to simply bring in foreign technology virtually all failed because of the previously mentioned need for adaptation. Training programs proved to have much more value.

The CGIAR and its centers provided the opportunity to generate the global public scientific goods that could be adapted to regional and national needs (and in some instances used directly). The research efforts were and are carried out in a collaborative manner with national scientists so there is a built-in feedback loop. Concurrently, during the 1970s and 1980s, donors such as the U.S. Agency for International Development, invested heavily in the improvement of national research programs, both with respect to facilities and training. By the early 1980s, a vast improvement had taken place in the national research programs in many developing countries. Since then, however, the funding situation in many developing countries has stagnated or declined and the CGIAR itself began to experience similar difficulties.

---

[18]This section draws on my involvement with the CGIAR system since 1972 and material that has been reported in greater detail in Dalrymple (2002). Historical background on the CGIAR is provided in Baum (1986). Current information on the CGIAR and its centers can be obtained from the group's Web site (www.cgiar.org). An introduction to other international and bilateral agricultural research programs is contained in Gryseels and Anderson (1991, pp. 329-335). Further information on public-private relationships in agricultural research is provided in Byerlee and Echeverria (2002).

[19]The relationship of the World Bank with the CGIAR is reviewed by Andersen and Dalrymple (1999). The United Nations Development Program was a charter member of this group but in recent years has sharply diminished its support for the provision of many public goods (including the CGIAR); instead it has, somewhat paradoxically, initiated a series of studies and reports that elucidate the virtues of global public goods (see, for example, Kaul, Grunberg and Stern, 1999). The United Nations Environment Program was a member for a while but withdrew for financial reasons. Recently the International Fund for Agricultural Development has become a member.

[20]The TAC model has reportedly been emulated by the Global Environmental Facility and the Global Water Partnership.

Promise and Perils

The basic original premise of the CGIAR has been realized to a remarkable, but variable, degree. The system has worked quite well, although it has had its problems, and the center organizational pattern has proved sound. Synergies and spillovers of collaborative research in the crop area have been substantial. In addition to their own research, the centers accumulate, utilize and organize, and pass on data and information in every direction. They have become focal points for knowledge in their particular areas of work and national programs have been stimulated. Increases in productivity and in turn the food supply have resulted in a lowering of prices to consumers below what they would have been otherwise. The overall result is a significant contribution to innovation and to the economies of the developing nations (Evenson and Gollin, 2002).

A group such as the CGIAR, however, continually faces program and management challenges and financial perils. These are not unique to public research, but are multiplied at the international level. The array of pressing problems is immense; views on what should be tackled vary (local views may vary from those at the national, regional, and global level); and the operation and management of research facilities in a developing country, with a mixture of international and local staff, can be challenging. The task can be further complicated by a variety of external or exogenous issues such as nationalistic inhibitions about sharing biodiversity, concerns about IPRs, and the global debate about genetically modified organisms.[21]

The budget for all of this is rather modest—only $337.3 million in 2001, and has stagnated for the past decade. It represents only a small portion (less that 5 percent) of all public funding for agricultural research in developing nations. And it is totally dependent on voluntary contributions from year to year. Unlike many other multilateral programs, the CGIAR has not been established by treaty. This is an advantage in that it can operate more informally and has considerable flexibility. It is, however, a disadvantage in that when some donor nations face a budget crunch, recently the case for Japan, treaty commitments take first place. Also, long-term support for multilateral programs in science and technology is probably not near the top of the list for many donors. There have been recent efforts to develop an endowment to provide stability of funding for the center genebanks at least, but this has a way to go.

This situation is compounded by a shift in the nature of the funding. There are two major categories: (a) unrestricted (core institutional support) and (b) restricted (special projects). The unrestricted provides most of the support for the longer-term and more globally oriented research, which is where the system has its strongest comparative advantage. Much of the restricted funding is for more localized and shorter-term research or development activities that are generally specified by the donor. Because the CGIAR centers have no other source of institutional support, as they are not governmental agencies, unrestricted funding is very important. Yet an increasing proportion of the funding is restricted—reaching about 57 percent in 2001. Two forces appear to be at play. The first, largely external, occurs when a large donor of unrestricted funding faces severe budget cuts, or as new donors come in with constraints on the use of their funds. The second is more internal and may be a combination of donor fatigue with long-term institutional support, a desire to adapt to emerging or changing priorities (or simply something new or flashier), and a wish to more clearly identify with specific projects.[22]

Clearly the CGIAR needs to broaden its funding base beyond developmental groups where science has, at best, a somewhat tenuous hold. Most of the national programs that support science, such as the National Science

---

[21]The CGIAR system established a small Central Advisory Service on Intellectual Property in 1999 to facilitate the exchange of experience and knowledge among its centers and to provide advice on a wide range of intellectual property issues. Some recent issues are noted in Commission for Intellectual Property Rights (2002, p. 144). More general issues are discussed in Byerlee and Fisher (2002) and Longhorn et al. (2002).

[22]There was an earlier parallel in the Spanish Caribbean. "During the early nineteenth century, governments and agricultural interest had funded specific research projects but had been reluctant to provide sustained funding for research institutions" (McCook, 2002, p. 3).

Foundation in the United States, focus on domestic project grants.[23] If such grants could be expanded to provide more of an international dimension, it could certainly provide help in terms of project support, but it would do nothing for the fundamental problem of institutional support.[24] That may well prove to be the Achilles heel of international research efforts.

Prospects

Thus it seems that the 30 years of the CGIAR experience confirms the promise of the concept of knowledge-based global public goods. It demonstrates that, although it is not an easy task, it is possible to establish a structure for carrying out a global program in science and technology. It illustrates that such programs can be operated, again not without difficulty, over a wide area and an extended period of time. It shows that these programs can be productive, stimulate innovation, and make many contributions to society. But it also illustrates the perils of maintaining such a program, and particularly retaining a global focus, even in times that call out for it. The problems of funding long-term public scientific programs at the global level, in the absence of an international funding mechanism (which undoubtedly would have its own problems), are indeed the heart of the matter.

CONCLUDING REMARKS

Scientific knowledge in its relatively pure form is, as stated at the outset, the epitome of a global public good. It is normally freely available to all and is not diminished by use—indeed it may grow with use. Moreover, due to the miracles of modern communication, it can be transmitted around the world almost instantly. It can provide the basis for major contributions to the innovation process and to economic growth.

But to play these roles, a number of conditions must be met. First, there must be a process for generating knowledge somewhere, and this may not be an inexpensive or simple process. Second, knowledge must be embodied in some sort of socially useful technology, which also requires effort and resources. Third, both knowledge and technology must retain some sort of public goods dimension in terms of being freely available to be of maximum social benefit. Fourth, there must be some ability on the part of recipients or users to adapt the technology to their conditions and needs.

These steps—and others may also be involved—involve an interplay of research of various types and IPRs in the form of patents and copyrights. The research process takes many forms but where it is publicly financed, the products traditionally have been public goods. Where they were sponsored by the private sector, proprietary rights are involved. And when this happens, as is increasingly the case with IPR in both sectors, the public domain dimension is certainly complicated and even diminished in quantitative terms. There may also be a qualitative effect in that research investment may be directed into areas where the social rates of returns are below the private rates of return (see David, 1992, p. 230; van der Meer, 2002, p. 126).

While IPRs are, so far, less a problem in developing countries, those nations generally suffer a more basic restraint—weak research programs in both the public and private sectors. Over 30 years of experience with the CGIAR has demonstrated that it is possible to help fill this gap, but not without some effort and resources. The

---

[23]International center scientists may participate in grant proposals submitted to, say, the National Science Foundation (NSF), but there are few such examples. One basic problem is that CGIAR centers are focused on applied research, whereas NSF grants are usually for more advanced research, normally headed by a domestic institution. This orientation, however, would be significantly broadened by a provision in the authorization bill for the NSF for fiscal year 2002-2007 (H.R. 4664), which cleared Congress on November 15, 2002, and was sent to the White House. Section 8, Specific Program Authorizations, item 3C on Plant Genome Research, provides for "Research partnerships to focus on — (i) basic genomic research on crops grown in the *developing world* . . . (iv) research on the impact of plant biotechnology on the social, political, economic, health, and environmental conditions in countries in the *developing world* . . . Competitive, merit-based awards for partnerships under this subparagraph . . . shall include one or more research institutions in one or more *developing nations* . . . " (italics added). If approved by the President, the next, and probably more difficult, step is the appropriation process.

[24]The International Foundation for Science, headquartered in Stockholm, has rather limited resources and is oriented to relatively small grants to developing country scientists. (For more information see www.ifs.se/). A Global Research Alliance, with a secretariat in South Africa and an evident interest in industrial research, was formed in mid-2002 (www.research-alliance.net).

principal problem is, not surprisingly, the maintenance of long-term public funding in the face of a seemingly endless array of other urgent calls on their use. These forces are particularly prevalent in foreign assistance programs (the key source of funding for the CGIAR), and may be complicated by political issues at both ends and natural disasters and civil disturbances in developing nations. Some promising international health research activities currently are getting under way with the support of a major foundation, but the narrow base of support may provide a problem in the future.

Thus, there is a substantial gap between potential and reality. On one hand, both scientific promise and communication opportunities were never greater. On the other hand, funding for the provision of public goods needed by much of humankind, particularly those in developing nations, is seriously constrained. IPRs, meant to facilitate innovation and economic growth, may, in some cases, be coming to have a less benign effect due to their quantitative and qualitative effects on the public domain.

The situation calls for a wider and deeper understanding of the international public goods dimension of scientific and technical knowledge. As Stiglitz (1999, p. 320) has commented: "The concept of public goods is a powerful one. It helps us think through the social responsibilities of the international community." Similarly, Sachs (2000b) has stated: ". . . international public goods are not just a nice thing that we need to add on. They are the fundamental thing that has been missing from our template for the past 30 years."

But to bring this wider understanding about, there will have to be—for a start—a closer and more interactive relationship between scientists, economists, and lawyers. Much could be gained if bridges could be built between them and with policy makers. This paper has been an initial attempt to begin to do so and to provide a framework for further thought. I would be delighted if it prompted further consideration of this most vital subject.

REFERENCES

Altman, D. 2002. "Small-Picture Approach to a Big Problem: Poverty," *New York Times*, August 20, C2.

Alvim, P. 1994. "Non-Chemical Approaches to Tropical Tree Crop Disease Management: the Case of Rubber and Cacao in Brazil," in Anderson, J. R., *Agricultural Technology: Policy Issues for the International Community*. CAB International, Wallingford (U. K.), 426.

Anderson, J., and D. Dalrymple. 1999. *The World Bank, The Grant Program, and the CGIAR; A Retrospective Review*. The World Bank (Washington), Operations Evaluation Department. OED Working Paper Series No. 1.

Arrow, K. 1962. "Economic Welfare and the Allocation of Resources for Invention," in *The Rate and Direction of Inventive Activity: Economic and Social Factors* (A Report of the National Bureau of Economic Research). Princeton University Press, Princeton, 609-626.

Bacon, F. 2000 (1620). *The New Organon*. Ed. by L. Jardine and M. Silverthorne. Cambridge University Press, Cambridge, ix, xiv, 66.

Barzun, J. 2000. *From Dawn to Decadence: 1500 to the Present, 500 Years of Western Cultural Life*, Harper/Collins, New York.

Baum, W. C. 1986. *Partners Against Hunger: The Consultative Group on International Agricultural Research*. The World Bank, Washington, D.C.

Bernal, J. D. 1965. *Science in History: The Emergence of Science* (Vol. 1). M.I.T. Press, Cambridge, 32.

Boulding, K. E. 1966. "The Economics of Knowledge and the Knowledge of Economics" in *American Economic Review* 56 (2), 1-13.

Buchanan, J. 1968. *The Demand and Supply of Public Goods*. Rand McNally & Company, Chicago.

Byerlee, D. and K. Fisher. 2002. "Assessing Modern Science: Policy and Institutional Options for Agricultural Biotechnology in Developing Countries" in *World Development* 30, 943-944.

Byerlee, D. and R. Echeverria. (Eds.), 2002. *Agricultural Research Policy in an Era of Privatization*. CABI Publishing, Wallingford and New York.

CIPR/Commission on Intellectual Property Rights. 2002. *Integrating Intellectual Property Rights and Development Policy: Report of the Commission on Intellectual Property Rights*. Commission on Intellectual Property Rights, London. See [www.iprcommission.org]. Reviewed in "Intellectual Property: Patently Problematic," *The Economist*, September 14, 2002, 75-76.

Congressional Globe. 1862. "U.S. Department of Agriculture," Vol. 33, February 17, 855-856. See Office of Technology Assessment (1981, 31).

Dalrymple, D. G. 1990. "The Excess Burden of Taxation and Public Agricultural Research" in Echeverria, R. G. (Ed.), *Methods for Diagnosing Research System Constraints and Assessing the Impact of Agricultural Research*, Vol. II. International Service for National Agricultural Research, The Hague, 117-137.

Dalrymple, D. G. 2002. "International Agricultural Research as a Global Public Good: A Review of Concepts, Experience, and Policy Issues." U.S. Agency for International Development, Office of Agriculture and Food Security, Washington, unpublished paper, September 2002.

David, P. 1992. "Knowledge, Property, and the System Dynamics on Technological Change." *Proceedings of the World Bank Annual Conference of Development Economics*. World Bank, Washington, D.C., 215-255.

David, P. 1993. "Intellectual Property Institutions are the Panda's Thumb: Patents, Copyrights, and Trade Secrets in Economic theory and History," in *Global Dimensions of Intellectual Property Rights in Science and Technology*, National Academies Press, Washington, D.C.

Desai, A., 1992. "Comment on 'Knowledge, Property, and the System Dynamics of Technological Change' by David," *Proceedings of the World Bank Annual Conference on Development Economics*. World Bank, Washington, D.C.

Drache, D. (Ed.). 2001. *The Market or the Public Domain? Global Governance and the Asymmetry of Power*. Routledge, London and New York.

Drayton, R. 2000. *Nature's Government: Science, Imperial Britain, and the 'Improvement' of the World*. Yale University Press, New Haven and London.

Dupree, A. H. 1957. *Science in the Federal Government: A History of Policies and Activities to 1940*. The Belknap Press of Harvard University Press, Cambridge.

Dyson, F. 1979. *Disturbing the Universe*. Harper & Row, New York.

Eisenberg, R. S. 1987. "Proprietary Rights and the Norms of Science in Biotechnology Research," *Yale Law Journal* 97 (December), 177-231.

Evenson, R. and D. Gollin, D. (Eds.). 2002. *Crop Variety Improvement and its Effect on Productivity: The Impact of International Agricultural Research*. CAB International, Wallingford and New York (forthcoming).

Fernández-Armesto, F. 2002. *Near a Thousand Tables: A History of Food*. The Free Press, New York.

Gryseels, G. and J. Anderson, J. 1991. "International Agricultural Research" in Pardey, P., Roseboom, J., Anderson, J. (Eds.), *Agricultural Research Policy: International Quantitative Perspectives*. Cambridge University Press, Cambridge, 309-339.

Hayek, F. A. 1937. "Economics and Knowledge," *Economica* 4, 33-54.

Hayek, F. A. 1945. "The Use of Knowledge in Society," *American Economic Review* 35, 519-530.

Henry, J. 2002. *Knowledge is Power; Francis Bacon and the Method of Science*. Icon Books, Cambridge, 6, 8, 16 (quote).

Herold, J. C. 1962. *Bonaparte in Egypt*. Harper & Row, New York.

Homburg, E. 1992. "The Emergence of Research Laboratories in the Dyestuffs Industry, 1870-1900," *British Journal for the History of Science* 25, 91-111.

Honigsbaum, M. 2001. *The Fever Trail: In Search of the Cure for Malaria*. Farrar, Straus and Giroux, New York.

Jefferson, T. 1984. Letter to Isaac McPherson, August 13, 1813. In: *Thomas Jefferson, Writings*. The Library of America, New York, 1291.

Jamison, D. 2001. "WHO, Global Public Goods, and Health Research and Development" in *Global Public Policies and Programs: Implications for Financing and Evaluation*, ed. by C. Gerrard, M. Ferroni, A. Mody. World Bank, Operations Evaluation Dept., Washington, D.C., 107-111.

Jha, P., et al. 2002. "Improving the Health of the World's Poor," *Science* 295, 2036-2039.

Kahn, B. Z. 2002. "Intellectual Property and Economic Development: Lessons form American and European History." Commission on Intellectual Property Rights, London., 21-24 (patents), 37-38 (copyright), 48. Source: www.ipcommission.org.

Kaul, I., Grunberg, I. and M. Stern (Eds.). 1999. *Global Public Goods: International Cooperation in the 21st Century*. Published for the United Nations Development Program by Oxford University Press, Oxford.

Longhorn, R., Henson-Apollonio, V., and J. White, J. 2002. *Legal Issues in the Use of Geospatial Data and Tools for Agriculture and Natural Resources Management: A Primer*. International Maize and Wheat Improvement Center (CIMMYT), Mexico, D. F.

Machlup, F. 1980. *Knowledge: Its Creation, Distribution, and Economic Significance*, vol. I, *Knowledge and Knowledge Production*. Princeton University Press, Princeton.

Machlup, F. 1984. *Knowledge: Its Creation, Distribution, and Economic Significance*. vol. II, *The Economics of Information and Human Capital*. Princeton University Press, Princeton.

Mayr, E. 1982. *The Growth of Biological Thought: Diversity, Evolution, and Inheritance*. The Belknap Press of Harvard University Press, Cambridge, MA.

McCook, S. 2002. *States of Nature: Science, Agriculture, and Environment in the Spanish Caribbean, 1790-1940*. University of Texas Press, Austin, 27-49.

Merton, R. K. 1997. "De-gendering 'Man of Science': The Genesis and Epicine Character of the Word Scientist" in *Sociological Visions*. Ed. by Kai Erikson. Rowman & Littlefield, Lanham (Md.), 225-253.

Mokyr, J. 2002. *The Gift of Athena: Historical Origins of the Knowledge Economy*. Princeton University Press, Princeton.

Office of Technology Assessment. 1981. "The Role and Development of Public Agricultural Research," *An Assessment of the United States Food and Agricultural Research System*. Office of Technology Assessment, Congress of the United States, Washington, D.C., 27-49. (Chp. III, prepared by D. Dalrymple.)

Olson, M. 1971. *The Logic of Collective Action: Public Goods and the Theory of Groups*. Harvard University Press, Cambridge, 14.

Pardey, P., Roseboom J. and J. Anderson. 1991. "Regional Perspectives of National Agricultural Research" in Pardey, Roseboom, Anderson. Eds., *Agricultural Research Policy: International Quantitative Perspectives*. Cambridge University Press, Cambridge, 215, 236, 247.

Plucknett, D., Smith, N., Williams, J. and N. Anishetty. 1987. *Gene Banks and the World's Poor*. Princeton University Press, Princeton, 41-58.

Powell, J. W. 1884. In: *Testimony Before the Joint Commission*. U.S. Congress (49th Congress, 1st Session). Senate Misc. Doc. 82, 1886 (testimony on December 18, 1884), 180.

Powell, J. W. 1986. In: *Testimony Before the Joint Commission* (letter to W. B. Allison, February 26, 1886), 1082.

Reich, L. S. 1985. *The Making of American Industrial Research: Science and Business at GE and Bell, 1876-1926*. Cambridge University Press, New York, 43, 62-71.

Rhodes, R. 1986. *The Making of the Atomic Bomb*. Simon & Schuster, New York, 749-788 (opening quote from 788).

Romer, P. 1986. "Increasing Returns and Long-Run Growth," *Journal of Political Economy* 94, 1002-1037.

Romer, P. 1990. "Endogenous Technological Change," *Journal of Political Economy* 98, 71-102.

Romer, P. 1993. "Idea Gaps and Object Gaps in Economic Development," *Journal of Monetary Economics* 32, 543-573.

Romer, P. 1994. "The Origins of Endogenous Growth." *Journal of Economic Perspectives* 8, 3-22.

Ross, S. 1962. "*Scientist*: The Story of a Word," *Annals of Science* 18, 65-85.

Ross, S. 1991. *Nineteenth-Century Attitudes: Men of Science*. Kluwer Academic Publishers, Boston, 1-39.

Ruttan, V. W. 1998. "The New Growth Theory and Development Economics: A Survey," *The Journal of Development Studies* 35, 1-26.

Ruttan, V. W. 2001. *Technology, Growth, and Development: An Induced Innovation Perspective*. Oxford University Press, New York, Oxford, 82-85, 436-437.

Sachs, J. 1999. "Helping the World's Poorest," *The Economist*, August 14, 17-19.

Sachs, J. 2000a. "New Map of the World," *The Economist*, June 24, 81-83.

Sachs, J. 2000b. "Globalization and the Poor." Unpublished talk presented to the CGIAR on October 23 and recorded in the "Transcript of Proceedings," CGIAR Secretariat Library, World Bank, Washington, D. C.,194-216.

Samuelson, P. 1954. "The Pure Theory of Public Expenditure," *Review of Economics and Statistics* 36, 387-389.

Samuelson, P. 1955. "Diagrammatic Exposition of a Theory of Public Expenditure." *Review of Economics and Statistics* 37, 350-356.

Smith, A. 2000 (1776). *The Wealth of Nations*. The Modern Library, New York, 778.

Solé, R. 1998. *Les Savants de Bonaparte*. Editions du Seuil, Paris.

Stiglitz, J. 1999. "Knowledge as a Global Public Good," in Kaul, Grunberg and Stern (1999), Eds., 308-325 (listed above).

Summers, L. 2000. Speech to the United Nations Economic and Social Council (ECOSOC), New York, July 5. (Text available from Federal News Service, Washington; via Lexis-Nexis database service.)

UNDP (United Nations Development Program). 2001. *Human Development Report 2001: Making New Technologies Work for Human Development*. Oxford University Press, New York.

Van der Meer, K. 2002. "Public-Private Cooperation in Agricultural Research: Examples from the Netherlands" in Byerlee and Echeverria (Eds.), 125-126.

Watson, P. 2002. *The Modern Mind: An Intellectual History of the 20th Century*. Harper/Collins (Perennial Edition), New York

Whewell, W. 1840 (1996). *The Philosophy of the Inductive Sciences*. Rotledge/Thoemmes Press, London, Vol. 1 (Vol. 2 in 1846 edition, 560).

Whitehead, A. N. 1925 (1957). *Science and the Modern World* (Lowell Lectures 1925). Macmillan, New York.

Wills, G. 1999. *Saint Augustine*. Viking/Penguin, New York, 145.

World Bank. 1998-1999. *World Development Report: Knowledge for Development*. Oxford University Press, New York, iii-v, 1-7, 131-133.

World Health Organization (WHO). 1996. *Investing in Health R&D*. Geneva, WHO/TDR 96-1.

WHO. 2001. *Macroeconomics and Health: Investing in Health for Economic Development*. Geneva, 23-14, 81-85. See www.cmhealth.org. Summary in Jha, et al. (2002), listed above.

6

Opportunities for Commercial Exploitation of Networked Science and Technology Public-Domain Information Resources

Rudolph Potenzone

I would like to address, the importance of having access to data, both data that are proprietary and data that are in the public domain. I certainly agree with Paul David's comments that it is a mixture of relationships that is vital to the good of increasing our scientific knowledge.[1]

Why are data so important to us, and why do we worry so much about them? Clearly, when you start a research program, no matter what the topic or field, you have to know what information is available on the subject. You have to know all the data that have been accumulated on a particular subject to do good research. You have to be able to get your hands on them, access them, and actually use them. Ultimately, the process of research is really about generating new data, and so as you study the existing information, you build your own concepts and ideas and generate new data. The research process is one of data generation.

Hopefully, if the research is successful, and you are able actually to accomplish something, the data turn into some new knowledge. Depending on your particular area or affiliation, you may want to invent the greatest new thing since sliced bread, maybe a new pharmaceutical or you have discovered some whole new unifying theory of whatever. Yet it is all based on the ability to get access to the right information at the right time.

Information resources are vitally important to every project, and it does not matter as you work on your project where relevant data come from. In fact, it is hard to predict where the best data will be found for any particular project. Even at the time the new data are generated, it is not certain what those data will be used for. How often do you find a bit of information or data that are in the most obscure place, not related to the topic you are studying, but in fact are relevant to the new project that you are working on? The availability of data is ultimately important to the ability to generate good science.

What happens to these good data once they are generated? The reality is that most data never enter a database that is available publicly, whether it is a fee-based database or a free database. Data are often locked in papers, reports, or lab notebooks. Maybe they get published and are available in a peer-reviewed journal or are archived electronically. However, by and large, those data are not accessible and are not able to be found because they actually have never been brought into a database. In particular, if the data have a particular commercial interest, and the pharmaceutical industry has been quite aggressive about this, they may find their way into a commercial

---

[1] See Chapter 4 of these *Proceedings,* "The Economic Logic of 'Open Science' and the Balance between Private Property Rights and the Public Domain in Scientific Data and Information: A Primer," by Paul David.

database and become quite accessible. Otherwise, it really depends on someone's personal initiative either to get funding to create a database through one of the government sources of funding or finding some support so that the information can be put into a form that is actually accessible.

I will describe two different scenarios for the pharmaceutical industry. One is chemical information (which is where I have spent much of my career), compared with what I call the bioinformatics movement; they differ in how the data have been handled and collected. Historically, chemical informatics has been a very commercialized activity. Three of the larger data repositories—Beilstein, Derwent, and Chemical Abstracts Service (CAS)—are organizations that all started out building printed repositories and then ultimately turned into electronic sources. They are highly commercialized, profitable activities that have served a truly critical role in the preservation and availability of chemical information.

Beilstein Institute, founded in 1881, first published its handbook with 1,500 compounds. The final version was printed in 1998, with the oldest references going back to 1771. Interestingly, this was converted in electronic form with some assistance by the German government and today holds 9 million compounds, a lot of information and data, and is distributed on a commercial basis by a Reed Elsevier company.

Derwent, which is today a Thompson company, was founded from ideas initiating from Monty Hyams in 1948. He was trying to make some sense out of the patent literature and began writing some simple abstracts about what was being published at the time. This information became useful. People started getting more interested, and it turned into a commercial operation. Today, Derwent's world patent index is global in scope, covering 40 different patent-issuing authorities and details in over 8 million separate inventions. It is a very large repository. If you work in the area of pharmaceutical research and development, you have to go to the Derwent database to understand what the patents are about.

CAS has a similar history. They were founded in 1907, with the goal of monitoring and abstracting the world's chemical-related literature. Today, with the Internet and all of the information that we have to deal with, it is mind-boggling to think that in 1907 this operation was formed because there was too much information to handle. CAS is a subsidiary of the American Chemical Society. There are 20 million organic and inorganic compounds registered at CAS; 21 million biological sequences; and almost 42 million separate and unique chemical entities registered in their system, all of them complete with names and references to the published literature, allowing scientists to find more information about these items. It is an incredible amount of information.

My observations are that chemical informatics has been quite commercialized and brings in quite a bit of money. In the area of pharmaceuticals, this has all been organized and put out there to be used because of its high commercial interest. These three companies and others look at scientific journals, books, patents, conferences, and dissertations; they do the work, and they extract a significant fee for it. The data are organized and then made available to the community, but at a price. The reason these data are not free is not because the underlying information is not free, because in many cases it is. But the fact that these companies have organized it and brought this together in a searchable (i.e., more useful) fashion gives these databases a very high value. This makes life much easier in terms of the chemical information community.

There is clearly value in that much of the original research was publicly funded research. There has also been a significant cost of creating these data sources. Although the pricing is fairly significant for these groups, they certainly have provided a service to the community. Frankly, whether these operations could have continued to exist in an electronic form without the current funding support is doubtful.

There are certainly hundreds, maybe thousands, of other databases and data repositories in chemistry that do not get picked up by these services. Are these less important to us as a scientific society? There are certainly cost barriers that prevent all the good data from being collected and organized in a reasonable fashion. The public funding has not been available to make these data generally available. I think that the funding tends to go toward collecting more data, and yet the funding for making the repositories of these data and making them available has not been there. My personal opinion is that funding authorities should consider the utility of the resulting data they pay to have generated from the start of these projects.

An interesting contrast to this illustration is in the bioinformatics arena, which in some ways is the antithesis of the chemical data franchise. Here, largely publicly funded projects have been formed by different highly motivated groups to put together what have become literally hundreds of sequence databases. A few of these are being commercialized, and so there are some annotation systems in place, and there are a number of companies who provide commercial databases. Yet the vast majority are still collected and made available for free, and with little support.

Figure 6-1 provides an illustration. If you look at the increase in information and amount of the sequencing speed, coupled to how much information is out there in terms of known sequences, and the growing complexity of information as we get into systems biology and expression information, there is absolutely an astounding amount of information that is suddenly becoming available in this arena.

From LION Biosciences' perspective, we have been the beneficiary of some of the public work. The Structure Retrieval System has been in the community for over 10 years, largely driven by scientists in terms of its capability, with over 500 parsers to search different kinds of databases. It was developed at the European Molecular Biology Laboratory (EMBL). This has now been turned over to LION Bioscience to commercialize and to keep the product growing and maturing. In terms of differential pricing, we continue to make this freely available to academic institutions, but are charging a fair price to the industrial community. We have merged this with our new technology called Discovery Center.

There are over 800 data sources that are relevant in bioinformatics research, which interrelate to each other. Examples include GenBank, SWISS-PROT, EMBL, and all the various pieces. Even if we believe that the cost of disseminating information on the Internet is free, which it really is not, the working scientist has to navigate through this network of data. He or she must sort through these many sources to find out the particular kinds of information they are looking for. This complexity suggests that they are most likely missing something in terms of the kinds of capabilities that we are making available through the proliferation of databases.

These databases are the fuel for the research projects in genomics and proteomics. These data sources came from the broader community, but they lack the financial incentives because they are all free to motivate the commercial services in terms of bringing these together. The users are individual scientists, who have to look at the local unpublished data that they are generating in their laboratory or calculations that they are doing. They have to

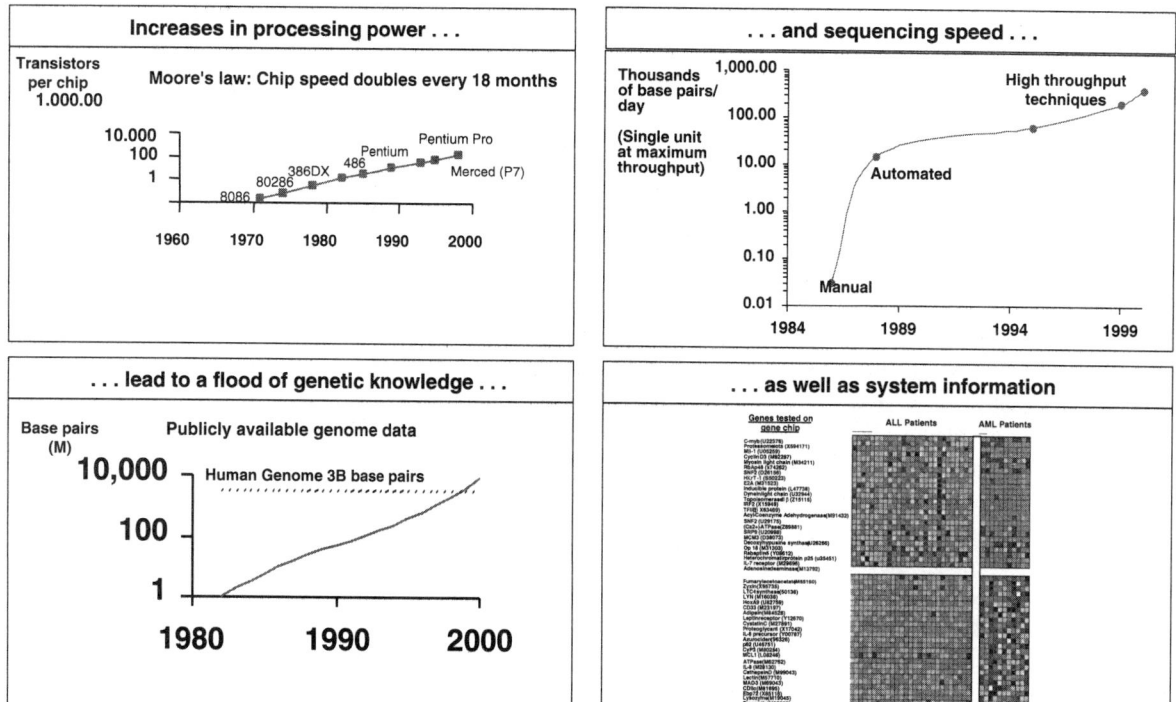

FIGURE 6-1 Genomics generates a flood of information. **Sources:** Genbank; Sequenom; 3700.com; Applied Biosystems; Human Genome Project; Food and Agriculture Organization of the United Nations; Database on Genome Size; and Intel Corp.

look at the other data around the company or research institution. These data could be at other sites around the world. They have to look at the Internet and all the various databases that are out on the Internet. They have to look at commercial databases, where they are loaded internally or externally. These researchers need to be able to have all of this synthesized for them in some fashion so that it ultimately facilitates their work, which is the research and creation of the new knowledge that we all like to talk about.

Having access to all these data is essential, whether they come from one's own lab or from some small college. If those particular data are relevant to the project you are working on, it is absolutely critical that all these data continue to be collected and made available in some reasonable form. Commercial databases provide an essential part of the information chain that we have to consider. However, noncommercial sources of data are also vital for the scientific community. These often fill huge gaps that, for whatever the reason in the commercial sector, have not been funded and have not been supported and actually fill out what I like to call our "data portfolio."

It is the integration of all this information that ultimately will enable us to continue to assist the working scientific community, to push back the frontiers of science, and to expand human knowledge into the future.

7

The Role, Value, and Limits of S&T Data and Information in the Public Domain for Education

Bertram Bruce

I want to focus on education, which I define broadly to include learning in K-12 schools and universities, informal learning, as well as learning in the workplace. I want to make an argument that attention to the role of the public domain in education is not only important for education, but also can give us a better understanding of what the public domain means and how to think about it in a larger society. I am going to group my comments into four areas. The first is to lay out a context for thinking about education and its relation to information today. The second is to describe Biology Workbench, a tool that has been used increasingly in K-12 and college education. Third is a classroom example to show some of the things that students are doing with this tool. Finally, I will address some of the implications.

RESPONDING TO A CHANGING WORLD

We hear a lot of talk about societal change and change in the workplace. One of the constants of education has been how we cope with change. We see today new technologies for communication and transportation, business, industry, manufacturing, medicine, and so on; globalization; immigration; evolving languages and incorporating words from around the world; a shift to knowledge work; and changing social values in organizations. Many people call this a paradigm shift. They say it calls for a new kind of 21st century education, which responds to a new kind of world that we are living in. We might call this the turn of the century problem in education. But it is important in thinking about these changes to realize there have been changes in the past. Many people argue that the change from the 19th to the 20th century led to greater changes on all of these dimensions, at least in the United States. For example, the technology changes during the late 19th and early 20th centuries included the telegraph and telephone, the phonograph, radio, motion pictures, mass printing, and other inventions with profound consequences for science and education.

All of these changes in the larger society led to many kinds of changes in education. First of all, there was a huge expansion of schooling as the waves of immigrants in the late 19th century were incorporated into the school system. The progressive education movement developed. New subjects were created. In 1880, one study complained that there were over 30 high school subjects taught and that was too many. In one recent count, there were 4,000 different subjects offered in high school today.

The beginning of research universities and what people call the American Library Movement, all of these things happened early in the last century in response to changes similar to those we see today. During this time, one of the key notions that developed was inquiry-based learning. It is the idea that in a world that is rapidly changing, where there are vast amounts of information—conflicting and redundant information—it is often hard to find what is crucial. In this context, we need to have learning tools that are open ended, inquiry based, group and teamwork oriented, and relevant to new careers. This is something the National Science Foundation has been pushing, as have the Boyer Commission on Undergraduate Education and a wide variety of other groups. It contrasts with the textbook-oriented learning that many of us experienced.

One model for this starts with the idea that it is not just important for students to be able to solve problems. They need to learn how to ask good questions; to find problems as well as solve them. Second, they need to learn how to investigate complex domains of knowledge, not just to read the chapter and answer questions at the end, but to integrate multiple sources of information. Third, they need to learn to be active creators of meaning, to construct knowledge, not just to follow directions. Fourth, they need to learn how to work with others, to discuss and to understand different perspectives. Finally, they need to reflect on what they have learned and articulate those meanings for themselves and others.

BIOLOGY WORKBENCH: OPEN-WORLD LEARNING

We could spend a long time talking about inquiry-based education, but one way to convey that and to bring it back to the public-domain data and information is to take one concrete example in the area of bioinformatics. Dr. Potenzone talked about the vast amounts of information that are available now for doing molecular biology, for investigating gene sequences, diseases, and so on.[1] Bioinformatics is developing as a distinct science and, in fact, many people are arguing that biology itself is being transformed into an information-driven science. Biology Workbench is one of the tools that has been built to address this. It is a web-based interface to a set of tools and databases, which researchers can use to access information stored throughout the world. Investigations that might previously have taken two years in the lab can now be done in a day sitting at the computer. There are tools for sequence alignment of proteins and genes, visualization tools, a digital library of articles, and so on. New knowledge has come out of using the Workbench. People in pharmaceutical companies, universities, and other places have made discoveries that would certainly have taken much longer without a tool like this.

In addition to looking at sequences and sequence alignment, a user can use the Workbench to visualize the structure of molecules, for example, that of hemoglobin in both its normal and the sickled form that causes sickle cell anemia. This visualization shows a mutated region of the molecule, in which it is easy to understand how one sickled molecule can hook into other molecules and create the sickling phenomenon.

Researchers also use this tool to investigate relationships among species. So, for example, users can compare horses, chickens, cows, vultures, dogfish, tuna, and moles to examine their degree of relatedness. By looking at the similarity, researchers can build phylogenetic trees or cladistic diagrams. These show that the tuna and the dogfish are more closely related to each other than to the other organisms, such as the horse, cow, and mole. The mammals are all more closely related, and the horse and cow are more closely related than either is to the mole, and so on.

Using Biology Workbench, a user can become an active investigator of the kinds of studies reported regularly in *The New York Times* science section. For example, when some new discovery about relatedness of organisms comes out, a reader could verify or challenge those results using a home computer connected to the Web.

A tool like this creates great possibilities for education. It also poses challenges. Many educators feel uncomfortable with tools like this, or what my group has called open-world learning, in which there are open, dynamically changing data, computational tools, and community interactions.

Imagine an instructor who prepares a lesson, checks it out the night before, and goes in the next day to teach about it. By the time the class begins, the data have changed. When students look at the computer, they find a different answer, a different set of information, because these databases are being constantly changed. This scenario reflects the first characteristic of open-world learning, that the set of data is open and changing. The

---

[1]See Chapter 6 of these *Proceedings*, "Opportunities for Commercial Exploitation of Networked Science and Technology Public Domain Information Resources," by Rudolph Potenzone.

Biology Workbench also implies an open computational environment: Many of the tools are open source and new tools are being created all the time. Third, it is an open community, which encourages direct communication between industry and university and between researchers and schools.

Classroom Examples

As an example of this approach, a recent article in *Nature* argues that Neanderthals and humans could not have interbred because of genetic differences. Students investigated this claim and found that Neanderthals and humans were actually more similar than other organisms, which are known to interbreed, such as horses and llamas. Another example focused on fungi being more closely related to animals than to plants. In another case, students looked at cetaceans, whales, porpoises, and so on and how they are related to hippos.

Students often take on these investigations as part of their class experiences. As one teacher said, this enables them to do projects in which they have to learn things that are not covered in the textbook. In addition, they get access to technologies that professional scientists are using. This means that students are not only using the tools to learn things, but also learning about the tools and the practices of science. They are also learning how to collaborate and how to articulate their knowledge.

This kind of investigation is not possible without access to the Workbench. Students could get books, but the number of books needed would be too expensive in most schools. Not only that, the information is rapidly changing. Through the Biology Workbench, students were able to find articles online that talked about research that they were investigating. In effect, they entered the scientific community, became participating, practicing scientists, and potentially could make their own contributions to the larger scientific literature. Instead of simply being recipients of knowledge created elsewhere, the students become creators of knowledge and participants in the knowledge-making community.

CHALLENGES AND OPPORTUNITIES

I want to make a few comments here about challenges and opportunities and then some closing comments about the public domain, education, and democracy. Inquiry-based learning is not a universal approach among educators. Moreover, many educators do not view public-domain data and information as an unvarnished good. Where most of us here today say we need more access to information, different models of education more or less accommodate and welcome use of public domain or information.

The fact that information is becoming more abundant, more complex, and rapidly changing is exciting to some people and scary to others. There is a challenge in any case. Even if you think, as I do, that it is absolutely crucial to education today, it is a challenge to think about how to make this kind of information not only available but truly accessible to students, particularly students in marginalized groups and with less than the latest equipment. The reason it is important is it creates so many kinds of opportunities. One is access to resources for inquiries. Students can now investigate questions that they could only pose before. Now they can engage seriously in carrying through an investigation to seek answers, which then generate new questions for further inquiries.

The very fact that this information is in multiple forms is exactly a reason students should be given the opportunity to engage with it. They need to learn how to cope with this abundance of information, media, and genres of representation. By using tools like Biology Workbench, which is but one example in one domain, students can become part of a larger community of inquiry: they learn not only the concepts or the skills of biology, but also learn what it means to be a biologist. This kind of activity elides many of the distinctions between practice and research, students and teachers, learners and researchers, and learning and research.

James Boyle spoke about the generational difference in how people think about public domain information.[2] He also spoke about how environmental studies in the 1960s changed the way we thought about the environment.

---

[2] See Chapter 2 of these *Proceedings*, "The Role, Value, and Limits of S&T Data and Information in the Public Domain in Society," by James Boyle and Jennifer Jenkins.

Those are educational issues and they exemplify why education should be at the heart of the debate about public domain knowledge. How do each of us acquire the beliefs and values that tell us what is just, what is feasible, what is desirable, independent of any particular law or policy of the moment?

It is common to talk about schooling and society as two separate realms. We think of school as the place where ideas from society go, once they are well formulated, well worked out. We think of society as the place where students go once they are fully prepared. But we treat school and society as two different worlds, which just touch on graduation day.

John Dewey, who did much of his writing during that revolution in education of a century ago, challenged people to rethink dichotomies, such as that of school and society. As he did with similar analyses of public and private, individual and social, or child and curriculum, Dewey pointed out that treating those terms as oppositional leads to an impoverished understanding of both. He went on to argue that education was fundamentally about democracy and that a democratic society cannot exist without an educational system, which encourages and fosters the development of individuals who are capable of self-government. At the same time students cannot learn about democracy and about a democratic society if they do not have the chance to participate in it, both in the classroom and in the larger society.

Because data and information are inherent to meaningful communication, the public domain is absolutely crucial, not only for the development of knowledge in general and not only for learning, but ultimately for the development of a just and equitable society.

8

The Role, Value, and Limits of S&T Data and Information in the Public Domain for Research: Earth and Environmental Sciences

Francis Bretherton

This presentation is based in part on a National Research Council report called *Resolving Conflicts Arising from the Privatization of Environmental Data*, which is available on the National Academies Web site.[1] First, I want to emphasize that I am a scientist, an environmental scientist insofar as that ever exists. I am a meteorologist with some experience in oceanography, but I have made it my business over the past 20 years to learn about all my other colleagues, what they do as geologists, chemists, ecosystem people, people interested in the cryosphere, and so on. They are all environmental scientists, whether they recognize it or not. The environment sciences are not a homogeneous domain. There are many different sorts of environmental scientists. Twenty years ago, they never used to talk to each other at all. One of the great changes that has happened in the past 15 to 20 years is that there is a group of people trying to look at how the system functions as a whole and how the various pieces are interrelated.

My priority in this presentation is to explain to nonscientists the special data needs of environmental science. There are some differences from the bioinformatics area. In particular, there are few startups in the environmental sciences. There are some, but not very many. The other important difference is that our topic is fundamentally international. Many environmentalists' views are global and other governments and countries are partners in that enterprise. To come to sensible public policies about the environment, we have to work collaboratively with those other nations. It simply is not feasible to devise a strategy for the United States and expect the rest of the world to follow that strategy.

There are a number of issues surrounding my presentation, which I am not going to deal with directly. In particular, we have heard already about some sui generis intellectual property rights in databases, which have been introduced in the European Union. That is of great concern to environmental scientists in the United States because the Europeans are our collaborators, and the Database Directive has led to restrictions on the availability of environmental data of various sorts. We are very concerned about what would happen if the United States also went the same way. If it did, I have no doubt that the world would follow.

Another issue here is that some foreign governments are trying to sell their data, typically in Europe, but not universally. Those are government agencies acting as quasi-commercial enterprises. That, likewise, gives us great

---

[1] National Research Council. 2001. *Resolving Conflicts Arising from the Privatization of Environmental Data*, National Academy Press, Washington, D.C. Available on the National Academies Press Web site at http://www.nap.edu/catalog/10237.html?se_side.

concern. I do not believe they are actually going to succeed. It turns out that the market is very thin and there are no indications that any of them are even beginning to cover their costs for reasons that I will come to later.

Additional issues are that the United States has a policy of encouraging public–private partnerships in the information provision area. These partnerships have to be thought through very carefully as to what the respective roles are of the partners and, in particular, what are the relative data rights. That has to be done on a case-by-case basis; it is not something that can be written into legislation.

Finally, the U.S. policy on commercialization of space is also introducing tensions into this area. Many applications, such as satellite observations, provide useful data for environmental science and potentially have commercial applications. That interface is, in fact, troublesome.

Let me now focus on the imperatives for environmental research and education, which is the primary purpose of this talk. As I have already indicated, there has been a movement over the past 20 years among the scientific community to face up to the fundamental problem, which is understanding human interactions with the natural environment. That is a huge canvas and I am certainly not going to touch on more than small pieces of it today. What I will assert, however, is that long-term global data, by that I mean many decades, are essential to document what is going on and to unravel a lot of the interconnections that exist between, for example, the ecosystems and climate. These data are also important in the distillation of interconnections to enhance the understanding of what is occurring, which can be conveyed not only among the specialists, but also to our children and grandchildren. If we do not understand what we are doing to their futures, they are not going to be in a position to do very much about it. A central requirement within all of this is a dependable, coherent observing and information system through which researchers can synthesize core information products. I am going to come back to this again and again, but let me just introduce an analogy.

The key analogy here is to a tree, as you can see in Figure 8-1.[2] The roots are where the data are actually collected in many different countries with many different types of instruments. As you move into the trunk and go up the trunk, data are being collated, cross-checked, and put into higher-order information products. That conversion of data to information is really seamless. There is a key point in this process, which I have labeled "core products," that get distributed by a whole variety of mechanisms to the end uses and that are represented by the

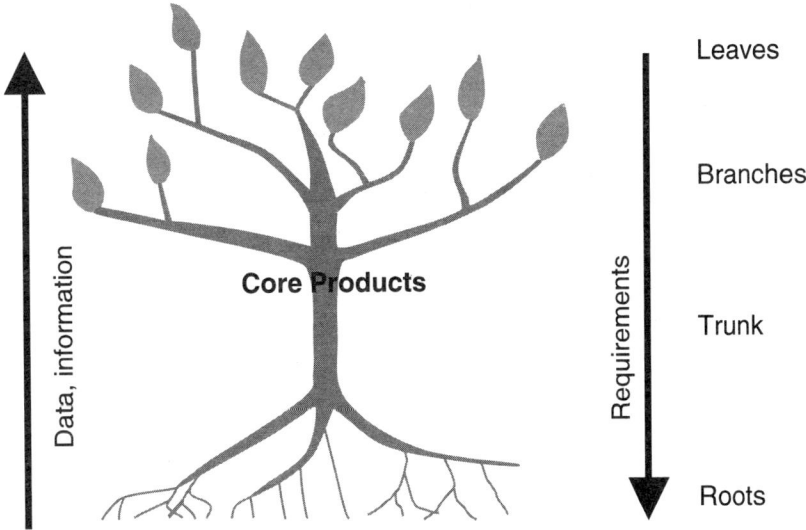

FIGURE 8-1 An environmental information system. Source: National Research Council. 2001. *Resolving Conflicts Arising from the Privatization of Environmental Data,* National Academy Press, Washington, D.C.

---

[2]The tree analogy was originally proposed in the NRC *Resolving Conflict* report previously cited. For further explanation of the various components, please see Chapter 3 of that report.

leaves in the tree. Core products, for example, can include calibrated and verified data derived from a single rain gauge.

When considering such an enterprise, one has to take the systems point of view and start with what are the end uses, what are we trying to cater to, what are the priorities for that, and then come back to what are the core products that might be produced and what are the implications.

However, it is also clear that you do not pay for a system like that, which is expensive, simply on the basis of research. There is a lot of research money going into this, but it is not nearly enough to pay for the complete systems that we have and need. Indeed, you have to serve multiple users and applications, which will generate a broader social return for the taxpayers as a whole to justify the large costs. We also need to foster consensus on scientific understanding and policy action. This implies that other countries have to be involved in what we are talking about. They have to participate actively in the system. That includes building research capacity, particularly in developing nations that may not yet have it. That is the benefit that they get out of a system of this sort in exchange for participating in the data collection.

The public requires reliable information that is properly interpreted. If the public does not believe what is coming out of a system of this sort, it is a waste of money. I already have mentioned that many environmental issues are international and global in scope. I would like to emphasize that the contributions of foreign governments come in kind, rather than through direct payments. They are based on what those governments do within their own borders or with their own systems because money is not easily transferred internationally, as I think we all understand.

Finally, the natural environment is very complex and uncontrollable, and describing its behavior requires many observations from different places. No single scientist or group conceivably can accomplish this alone. These are the absolute imperatives for pooling resources and for sharing the data effectively.

I am now going to provide some brief examples of information systems, starting with a most familiar one of weather and climate. Think of a data buoy out in the tropical Pacific. It is measuring the winds and the atmospheric temperature, and there is a 300-meter cable below that is measuring the temperatures in the ocean. All of these data are being telemetered back through a satellite and are available on the Web. Researchers, students, and people in many other sectors who have a need for such information can look up these data on the Web site and get the complete picture of the ocean and atmospheric temperatures for the past five days.

Another example is a processed satellite image to give the type of vegetation that is present. This provides information about land use, which is a fundamental part of the environment. It is frequently socially determined.

Another type of information system includes one used to predict and assess fish stocks in fisheries around the world. This is a major concern because, of course, a lot of the world's people depend on fish for their protein. The take is increasing, but the stocks are rapidly decreasing and more species are being fished out. As is the case with data about other natural resources, the same information can be used to both deplete and protect them.

Earthquake hazards provide yet another example. There is a worldwide seismological network measuring earthquakes. Of particular interest to this group of researchers is that proprietary data from the big oil and gas exploration companies are now being donated into the public domain. These are data that had significant commercial value when they were collected, but are now outdated and are being donated in the public domain. There are costs of assimilating and storing those data, but they can be very valuable for research purposes.[3]

Let us return to the tree analogy and start to fill out some of the details (see Figure 8-2). In the roots, there is a mix of systems. One is the international networks, the contributions that are being made by different countries and telemetered around the world as needed. There are also national networks doing the same things and there are also different types of measurements being made, some of which are satellite and others in situ measurements. To get a successful system, we need all of them, and they have to work together seamlessly. That is a major enterprise.

The main point is that the total cost is mostly in the roots. Collecting the data and pulling them together is where the cost lies. As you move up the roots to the trunk, it is the preparation of core data products, which is the primary function, and they have to be made available in the public domain at marginal cost. Otherwise we are

---

[3] See Chapter 27 of these *Proceedings*, "Corporate Donations of Geophysical Data," by Shirley Dutton.

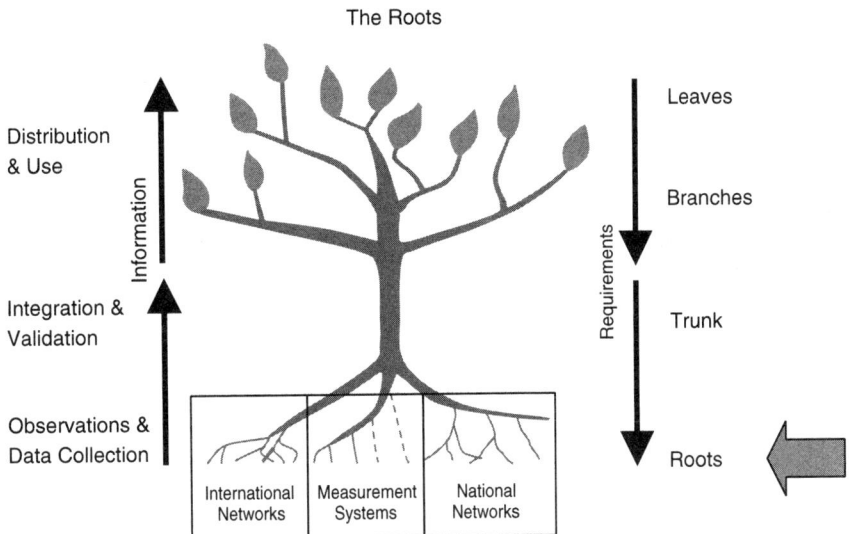

FIGURE 8-2 Analogy of how the networks come together seamlessly. Source: National Research Council. 2001. *Resolving Conflicts Arising from the Privatization of Environmental Data,* National Academy Press, Washington, D.C.

cheating ourselves by investing all that government money in the roots and not taking full advantage of it. The crucial thing about the trunk is that both the input data and the algorithms must be open to scientific scrutiny, because without that, the outputs are not credible. If they are not credible, then we are wasting our money.

Finally, moving up to the branches and leaves, we have the distribution and use of the data and information. It is as complex as the roots. Each U.S. federal science and technology agency sponsors a system like this, such as the National Oceanic and Atmospheric Administration for weather, but it is different for different agencies. They will have their own internal agency requirements and also have their own distribution system.

The leaves represent the end uses. For example the energy, forestry, and insurance industries are all big users of this information, as is education through the integration of data in textbooks and things like that. The general public is very interested in a lot of these data for recreational purposes. There also is a whole set of issues about setting environmental policy on regulations.

Finally, the branch represents a distribution system tailored to identifiable user groups by reformatting core products, adding additional information, or otherwise increasing value to that group. These branches are always developing and changing. Diversity of the branches is another major feature, which is not always fully appreciated.

The fundamental premise is that those products at the top of the trunk have to be in the public domain at marginal cost. Having said that, the branches do not have to be public domain. In fact, many of them currently are not. They are in the form, for example, of value-added weather data. That is perfectly in order, provided that all those products are starting from the same base of public-domain information coming from the core.

There are also opportunities for the private sector down in the roots. There are now commercial satellites that provide 1-meter resolution of what is going on in your own backyard. The satellite companies are selling these data. There are various legitimate purposes for these data, some of which are needed for environmental studies. The point is, if the purchase of such data from the private sector is the cheapest way to get what the government needs, it is entirely fair that they should buy it from the commercial concerns. However, I want to emphasize that what the government buys has to include the rights to the data they actually purchase. If they cannot afford to do that, then they have to reduce the amount of data they purchase. You cannot mix restricted data with public-domain data in the trunk because it ruins the transparency and essentially compromises the whole research enterprise.

To conclude, publicly funded, shared-use, long-term observational information is essential for sound public policy concerning human interactions with the natural environment. Core products of the trunk of such systems must be in the public domain and available at marginal cost of reproduction. Value-added, private-sector distribution systems may enhance the enterprise if they show benefits. They do not have to be, however, in the private sector. For example, scientists have their own climate and weather data distribution system, which is paid for out of research funds and justified on that basis. Finally, purchase from private vendors of all rights to a limited amount of data may under certain circumstances be cost-effective.

There is one concluding issue that I would like to note. I have presented a model of various environmental systems. It turns out that not a single one includes a recognizable mechanism by which the different stakeholders—the government agencies, the policymakers, the scientists, and the private-sector participants—can actually get together and work out mutually satisfactory win-win situations of some of the conflicts that are arising. A key conflict that tends to arise is just what is the right definition of the core products at the top of the trunk. That is the place where conflicts most come into focus, but the point is we have no forum for working those out. That needs to be established.

9

The Role, Value, and Limits of S&T Data and Information in the Public Domain for Biomedical Research

Sherry Brandt-Rauf[1]

I will explore the boundaries and tensions between public and private data in the domain that has been referred to earlier in this symposium as "small science;" that is, the arena of biomedical scientists doing research individually or in small groups. For the theoretical underpinnings of the talk, I am relying on a paper that Stephen Hilgartner and I published some years ago in the journal *Knowledge*.[2]

I recently read the book *Brunelleschi's Dome* by Ross King, the interesting story of the radical design for the Santa Maria del Fiore cathedral in Florence.[3] It describes the architect, Filippo Brunelleschi, doing research for his designs in the vast ruins of ancient Rome. To this day, what he sought in the ruins is unknown because, fearful of losing priority in his architectural work, he recorded his notes on strips of parchment in a series of symbols and Arabic numerals. In effect, he withheld his data from his compatriots as well as from those in later generations who would have liked to understand the classical principles and discoveries on which he relied. And that was hardly the first episode of a researcher withholding data. Hundreds of years earlier, Roger Bacon advised all scientists to use what he called "concealed writing" in recording their discoveries. Even in olden times, the dark side of withholding data was evident: use of such cryptic methods for recording data sometimes interfered with scientists' voluntary exchanges. For example, as King also describes,[4] when Galileo wanted to communicate to Kepler that he had discovered the rings around Saturn, he did so in an anagram that, unscrambled, read, "Observo Altissimum Planetam Tergeminim" (I have observed the most distant of planets to have a triple form). Unfortunately, Kepler read the anagram as saying "Salve Umbistineum Geminatum Martia Proles" (Hail twin companionship, children of Mars), which puts quite a different spin on this scientific communication.

It is clear, then, that the question of under what circumstances and with whom to share data has long been one of some interest among scientists. At one time, the issue tended to revolve around the question of priority in discovery. A classic social science perspective on science, first laid out by the sociologist Robert K. Merton and

---

[1] The author gratefully acknowledges the National Science Foundation, Grant No. BIR-9112290, for its support for the research on x-ray crystallographers and data sharing described in this talk.

[2] Hilgartner S. and S. I. Brandt-Rauf. 1994. "Data Access, Ownership, and Control: Toward Empirical Studies of Access Practices," in *Knowledge: Creation, Diffusion, Utilization* 15(4):355-72. See also Hilgartner, S. 1997. "Access to Data and Intellectual Property: Scientific Exchange in Genome Research," pp 28-39, in National Research Council, *Intellectual Property Rights and Research Tools in Molecular Biology*, National Academy Press, Washington, D.C.

[3] King, R. 2000. *Brunelleschi's Dome*. New York: Penguin Books.

[4] Id. at 25.

elaborated by quite a number of subsequent social scientists, points to the dominant norms of science, which he indicated were organized skepticism, universalism, disinterestedness, and what he called communism, which is of most relevance to this symposium: the idea that findings belong not to the individual but to the entire scientific community, that is, they become part of the public domain.[5] This notion of collective ownership was always, in some sense, prescriptive rather than descriptive of the behavior of scientists. One has to look no further than James Watson's book *The Double Helix*[6] to know that, but nowadays, with commercial interests so permeating the scientific process, even a pretense of normativeness is often gone. At one time, scientists who were unwilling to share data were often responding to concerns that other scientists would steal their findings to get credit for discoveries rightly their own. It might be said that they wanted credit more than ownership. These types of concerns have certainly survived. But a change in recent years has been the extent to which commercial interests have affected the desire to establish ownership over biomedical research data. Under these circumstances, researchers and their commercial entities want not only credit but also ownership. Because of this desire for both ownership and credit, scientists often restrict access to their data by keeping them privatized, either by their own choice or at the insistence of their commercial collaborators. These restrictions often revolve around publication of the data, which may be delayed or suppressed entirely. However, they sometimes affect informal exchanges of data as well. The magnitude of these concerns will be discussed in later sessions, as will the concerns that arise with disputes over data access. Here, it is sufficient to note that supporters of a free flow of scientific data believe that resistance to data sharing and disputes over data sharing can:

- waste resources by leading to duplication of efforts,
- slow the progress of science because scientists cannot easily build on the efforts of others or discover errors in completed work, and
- lead to a generalized level of mistrust and hostility among scientists in place of what should be a community of scientists.

Let me give one example from a study I conducted several years ago together with Stephen Hilgartner, who will be speaking with you later in this meeting. We studied data-sharing practices among x-ray crystallographers. One of these scientists reported to us that an industry group had published a paper with an incomplete structure, containing just what he referred to as "the juicy parts of the analysis." He wrote to ask them for their coordinates, and they responded, "Well, maybe in a couple of years after we look at it a little bit more." Three years later, he finally gave up waiting and went ahead and did the structure for a homologous substance, for which he intended to deposit coordinates and to publish. Not only was he looking forward to a significant publication, but he was especially gleeful about the possibility of harming the first group by putting into the public domain the very data they sought to keep private. This is surely not the most productive way for science to proceed. One could not even regard this as productive from the standpoint of replication because the original data were not made accessible to be replicated. It does, however, provide an example of the "disappearing property rights" referred to by Paul Uhlir and Jerry Reichman.

Exploring these issues requires a detailed analysis of what the basic terms mean. At the least, we need to understand what we mean by data and what we mean by sharing. In addition, we need to consider how, with whom, and under what circumstances and conditions scientists share and withhold data, recognizing that sharing and withholding constitute a spectrum of entities. Few scientists can afford the time and resources involved in sharing everything with everyone and few, if any, refuse to share anything. The hypothetical scientist who shares everything could never be productive, since she is spending all her time e-mailing and talking on the phone. The chimera who shares nothing would have a career that is nasty, brutish, and short. He would never publish or speak at meetings and probably would never even talk to colleagues. Indeed, he could not really be said to have colleagues. Nobody makes everything public and nobody keeps everything private. Data sharing constitutes a flexible concept

---

[5]Merton, R.K. 1973 [1942]. "The Normative Structure of Science," in *The Sociology of Science*. Edited by N.W. Storer. University of Chicago Press, Chicago.

[6]Watson, J.D. 1968. *The Double Helix: a Personal Account of the Discovery of the Structure of DNA*. Atheneum, New York.

that incorporates a variety of actions at different points in the scientific process. I will come back to this flexible concept of sharing a bit later.

The concept of data also is flexible. For purposes of our data-sharing study, Hilgartner and I found that it was necessary to define data broadly and fluidly. Unfortunately, these issues are often explored using an atomistic approach that imposes artificial distinctions between the input and output of scientific work. In this approach, which might be called the "produce and publish model," scientists first produce data or findings—the output of the process. Second, these findings are disseminated through publication or more informal channels. And third, the data, then in the public domain, become the input for other scientists in their own research projects. In this way, the original findings become evaluated, certified, and incorporated within, or perhaps rejected from, the public corpus of scientific knowledge.

According to this model, which I have oversimplified, restrictions on access constitute departures from the normal and normative course of science described by the Mertonian norms and by many who are interested in data sharing. In our research, on the other hand, Hilgartner and I came to believe in the need for a more process-oriented model that was directed more toward continuity and flow and less toward the notion of data as a clearly defined and fixed entity. This more flexible approach, which receives support from the ethnographic literature of science,[7] we came to call the "data stream model." Within the framework of this model, data are not classified as discrete and atomistic "input" or "output," but are rather seen as part of an evolving stream of scientific production.

These data streams have several important properties for purposes of our analysis. First, they are composed of a heterogeneous collection of entities. Hilgartner and I include within the rubric of data any of the many different things that scientists use or produce during the process of scientific research. Scientists use a variety of terms for these entities that represent the contributions to and by-products of their work, including findings, preliminary results, samples, materials, laboratory techniques and know-how, protocols, algorithms, software, instrumentation, and the contents of public databases: any and all information and resources that are used in or generated by scientific work. The meanings of these terms vary across fields and subfields and therefore can become confusing. The elements of the data stream are, by their very nature, situational in character. The fact that they are heterogeneous means that access to them comes in different forms and brings along different practical considerations. Providing access to a reagent differs from providing access to a lab technique. In addition, as a further dimension of their heterogeneity, these elements vary as to a variety of characteristics, among them perceived factual status, scarcity, novelty, and value. Some entities may be well established and others more novel. Some may be easily accessible and others quite rare. Some may be accepted by most of the scientists in a field and others may be regarded as less reliable. Over time, of course, these attributes, each of which is related to access, shift and change flexibly. For example, as data become better established and enter the core of accepted science, decisions about access—to whom, how, what, and when—change as well.

A second critical characteristic of data streams after heterogeneity is the fact that they are composed of chains of products. Elements of the data stream are modified over the course of research and laboratory practice and assume different forms as samples are modified and converted to statistics, which, in turn, find their positions in tables and charts and eventually in scientific papers. These changes alter not only form but also utility, affecting in turn decisions about access. This notion that elements of the data stream are connected as chains underlines the fact of the data stream's continuousness. And, as a continuous stream, it can be diverted in whole or part in any of an infinite number of different directions. The consequence of this for data sharing is that there cannot be a correct single way to provide access to data. Therefore, it is unlikely that a single definitive statement can apply, as a policy matter, to all elements of the data stream at all points in time.

Using this fluid concept of the data stream, some exchanges of data will be formal ones such as by publication in peer-reviewed journals. However, many of the most critical exchanges of data will be informal. Indeed, these

---

[7]See, e.g., Knorr-Cetina, K. 1992. "The Couch, the Cathedral, and the Laboratory: on the Relationship Between Laboratory and Experiment in Science," in *Science as Practice and Culture*, A. Pickering, ed., University of Chicago Press, Chicago; Knorr-Cetina, K. 1981. *The Manufacture of Knowledge,*Pergamon, New York; Latour, B. 1987. *Science in Action,* Harvard University Press, Cambridge, MA; Latour, B. and S. Woolgar. 1979. *Laboratory Life*, Sage, Beverly Hills, CA; and Lynch, M. 1985. *Art and Artifact in Laboratory Science*. Routledge & Kegan Paul, Boston.

informal exchanges may be far more significant for the progress of science than is publication, as important as it may be. To lay out the nature of some of these informal exchanges, I interviewed a biomedical researcher working in the area of genetics, asking him to articulate some of these informal exchanges and to lay out as comprehensively as possible the ways in which he gave access to his own data and got access to other people's data. The exchanges he spoke about could be divided into three broad categories: sending people, sending things other than results, and sending results. These are, of course, arbitrary and overlapping categories, but within the broadly inclusive meaning of the word "data" that Hilgartner and I have used, they all involve data sharing.

In the category of sending people, he spoke of having members of his group visit other labs to learn a new technique or having people come to his lab to learn such a technique. Sometimes, if the technique is extremely complex or critical, he might himself visit another lab. This is perhaps more common with junior researchers. Nevertheless, a fairly senior crystallographer spoke of spending his sabbatical after he was already tenured in someone else's lab learning how to produce an enzyme he wanted to crystallize. Another crystallographer indicated that, although he had been working in a related field, when he decided to move into crystallography, he went to one of the most active programs and worked there for several years to master the techniques. Sometimes the researcher might accept or send a graduate student who is to learn not a single technique but the entire research process. This might be an explicit exchange or it might occasionally be a bit covert, such as when someone hires a graduate from one lab to do a postdoc in another lab, motivated by the fact that the postdoc is familiar with the techniques used in the first lab. The same process might occur with someone who had simply worked in the first lab and was now looking for another job. As an example of this, one of the crystallographers told of trying to grow different substances and of being overwhelmed by all of the details in the new techniques he was trying to master. One of his colleagues in a related field suggested, "Why don't you hire one of our grad students to sterilize the media for you and show you how to inoculate media and that sort of thing?" This strategy worked famously for him and he was able eventually to master the techniques on his own. In these ways, a portion of the data stream is diverted by virtue of the movement of people, a very important informal mode of access to data.

Within the category of sending people, we might also include the giving of talks. Our informant scientist categorized the different types of talks he had been asked to give within the previous six months. In one instance, he spent a day with another research group, describing his work, his lab techniques, and where he intended to go next in his research. He viewed that the quid pro quo for this experience was the expectation that someone from the recipient group likely would be asked to visit his group in the future. In addition, he had been asked to give a variety of talks, some of which he delivered and some of which he did not, depending on "what was in it for" him. If, for example, he would be speaking in a place to which he had wanted to get access for some reason (including the fact that their work seemed interesting to him), he would give the talk. He had also been invited to address small groups of different sizes, including seminars of graduate students.

The extent to which these processes proceed smoothly or at all depends in part on the attributes of the elements of the data stream that I discussed earlier. Where fields are competitive and samples and techniques are rare, there may be less inclination to undertake some of these informal modes of sharing. The potential for commercialization may affect the process as well. Some of the crystallographers expressed the feeling that these processes had combined to result in a loss of openness in the field. One scientist who had been in the field for many years bemoaned the loss of less competitive times. At one time, this scientist said, "If someone had a problem, they'd call up and say, look, I'm interested, can we collaborate, or do you mind if I work [on this problem] or something like that." Now, with the increasing competition in that field, this crystallographer believed that these overtures were less likely to be made, and, if made, were less likely to be successful. Another believed that the most important aspect of his program, accounting for the high level of productivity of the participants, was the fact that there was what he called a "whole catalytic mass" of individuals who collaborated rather than competed. Compare this with one crystallographer who told me that he was moving to a new institution in which, supported by a pharmaceutical company, he would not even be permitted to speak to the members of his own department about his work. Each of these instances reflects differences in the processes of access to data, sharing of data, and diversion of the data stream.

In the category of sending things other than results, the geneticist indicated that some lab groups did not wish to participate in the exchange of people. Instead, they used their know-how, reagents, or instrumentation as a kind of currency in the exchange of data. Groups unwilling to export their protocol might be willing to import another scientist's samples and perform their protocol on them, providing that scientist with their findings. Often, this occurs as part of a contract process. Other motivations for a group to treat another lab's samples include the desire to

- maintain quality control over their procedure by being the only ones to employ it,
- have their names on additional papers,
- work with a particularly interesting data set, or
- corroborate the accuracy of their method using a new data set.

Like the movement of people, these processes also represent modification of the data stream.

Other kinds of "things other than results" that are shared include unique samples, clones, or reagents. This kind of sharing can occur informally or formally. The informant scientist described two instances in which this occurred. In one case, he was given access to a rare reagent, but was required to sign a Material Transfer Agreement in which he promised to use the materials for research only, to provide the donor lab with access to his results, and not to pass the reagent on without express permission. Such agreements may require the addition of the donors' names to future scientific papers, although this particular agreement did not. In a second instance, the requests for samples became so onerous that when a commercial company became involved and handled the distribution, it was a great relief not to have to handle it any more. One of the crystallographers, on the other hand, complained of being unable to get a critical reagent from a pharmaceutical company that refused him on the grounds that they were already collaborating with another group. This refusal stopped this thread of the scientist's research entirely, and he had been, at the time of the interview, unable to get the reagent from any other source. Another crystallographer told me that he was unable to get a certain enzyme. "Unless you are well known, a Nobel Prize winner," one has to make the substance oneself. He indicated that if a scientist is a notable person in the field, other scientists would be more inclined to give him or her materials, hoping something dramatic would be done with them that would bring reflected glory on their producer. Still another crystallographer ended up paying an academic colleague in another lab thousands of dollars to produce the material he needed.

Computer programs are often treated in similar ways, sometimes with the stricture attached that the developer's name be on further papers or that further sharing be with the approval of the developer of the program, sometimes without restrictions, either with or without a financial cost attached. Other sharing within this category relates to instrumentation. Some instruments are small enough and inexpensive enough so that every lab will have their own. One example would be glassware. Other instruments are large, not portable, and very expensive. They must be shared in situ, with the samples—with or without people attached—coming to the instruments. One of the crystallographers described this process with respect to a magnetic device at another institution. Complaining about having to queue up for access to the magnetic device, this scientist regarded wealthy labs as fortunate in being able to send personnel to do the experiments themselves. Smaller labs that could not spare the personnel to do so were perceived as achieving lower positions in the queue as a result. In our parlance, the characteristics of this element of the data stream, the fact that it was heavily in demand, rare, and expensive, colored the access process. However, it is important to note that this crystallographer had been successful in completing dozens of experiments over the period of the relationship and recognized the process as a sharing of data on both of their parts, with one lab providing the samples and the other providing the instrumentation. In our terms, it would be considered a merging of the data stream.

Which pattern is followed in these situations is shaped by the characteristics of the data entities in question—rare, proliferate, easy and cheap to make, expensive or difficult to make—as well as on those of the scientists—junior or senior, part of a large lab or small lab, part of a common network created by past experience such as a common postdoc or academic institution or strangers. Do they each have something that the other wants such as the materials and the reflected glory discussed above or is the exchange more unequal? Each of these factors will color the access pathways and the results will vary in each instance.

What is perhaps most often referred to as data sharing is the sharing of findings. This process is also shaped by the attributes of the data stream, the nature of the findings and of the actors. One of the crystallographers spoke of releasing data to a scientist—not a crystallographer—in another country who was working on the same problem from a different angle. If this scientist could corroborate the crystallographer's data using his own theoretical model and methods, it would strengthen the crystallographer's findings and suggest new directions for research for both of them. This possibility, compounded perhaps by the fact that the two scientists were in different fields and therefore were not in direct competition, led to a comfortable and extensive collaboration between the two.

Our informant geneticist has been the recipient of requests for findings that can be pooled with other data to increase the significance of findings and to test the reliability and accuracy of findings. Such requests are fairly common in certain areas of research such as epidemiology. I asked him what happens. His response was a short version of the whole long story: "Some people share and some don't." Some of this willingness or aversion to sharing is chargeable to personal idiosyncrasy. After all, even back in nursery school, some people shared the Legos and some people did not. But the more interesting issues involve the social structural and economic considerations that shape the choices people make. Without a full understanding of such considerations, it is difficult to alter these choices by fiat. For example, the geneticist indicated that at times findings are released because of mandates by either journals or funders. He made the point, however, that even where disclosure was mandatory, findings were often released without important details whose lack rendered the data significantly less helpful for downstream users. Sometimes, data were incomplete. One of the crystallographers reported an occasion in which the coordinates of a structure were released for publication purposes omitting a water, without which the coordinates were not terribly helpful. Another told of the release of only the central chain. In some instances, results may be coded in such a way that the critical information cannot be accessed.

The informant geneticist reported his belief that some scientists intentionally modify their data so as to make them less useful to subsequent users, but other times the data are simply not in usable form because of the format in which they were originally collected. If the data producers must do any kind of real work in terms of modifying these findings to make them more usable to the downstream user, that person may well expect to be rewarded. For example, when this scientist requested that a data set to which he had been given access be updated, he was asked to include the names of the original data producers on subsequent scientific papers.

Sometimes, scientists are not averse in principle to releasing their data but believe that because of the nature of the data set, they must delay release. One crystallographer reported that colleagues had said, "you can't have it, it's a mess, so please, now, it's not good enough, so they didn't give it to [him] for six, eight months, but they gave [him] enough of the overall orientation, [so that he] could do work with it even at that . . . point." He further stated that, "Never has anyone said, no, you can't have it, but if it isn't finished and if it's not there, you can't blame them." Although this delay was presumably temporary, another crystallographer said that he often refrains from sharing the source code from his self-developed software because it takes too long to explain how he deals with each of the many glitches in it. In these cases again, the characteristics of the elements of the data stream contribute to shaping the timing and circumstances of access.

My point in laying out these details is that much of the significant sharing of data occurs not through publication but in these less formal contexts that I described. As one of my informants put it, "Being in touch replaces abstracts and publications. The most interesting stuff I hear is either presented at meetings or heard on an e-mail. Even the fastest publication is slow compared to that and if you have to wait until you see stuff in print, you're out of the loop." One of the crystallographers referred to presenting an abstract as a "little trick" in the interests of "one-upmanship," but it also reflects the way structural incentives in science can result in sharing. The way to influence the smartest scientists, one of the crystallographers said, is through "talks at national meetings that they happen to be at, discussions, interactions with high-profile people who they happen to run into at a meeting." Too much of a focus on data sharing through formal publication and the incentives and disincentives that exist to publish at time A as opposed to time B will miss much of this critical sharing of data.[8]

The publication process is, indeed, one important mechanism for the sharing of data and for entering scientific information into the public domain. What is equally if not more crucial for the progress of science, however, is the effect of social and economic pressures on the informal sharing of data by scientists and on the flow of data through the different scientific fields. As we consider the solutions to the problems of access and of the privatization of scientific data, we need to keep a close eye on these informal mechanisms of data sharing and on the ways in which they are shaped by the climate of science and society.

---

[8]Ironically, the publication process itself may lead to a certain amount of unintended informal data sharing as when a colleague reviewing an article for a major journal called one crystallographer to report on the progress of another group working on the same structure as he was.

SESSION 2: PRESSURES ON THE PUBLIC DOMAIN

10

Discussion Framework

Jerome Reichman[1]

The digital revolution has made investors acutely aware of the heightened value that collections of data and information may acquire in the new information economy. Attention has logically focused on the incentive and protective structures for generating and disseminating digital information products, especially online. Although most of the legal and economic initiatives have been focused on—and driven by—the entertainment sector, software, and large publishing concerns, significant focus has been devoted to the possibility that commoditization of even public-sector and public-domain data would stimulate substantial investments by providing new means of recovering the costs of production. Moreover, investors have increasingly understood the economic potential that awaits those who capture and market data and information as raw materials or inputs into the upstream stages of the innovation process.

What follows focuses first on pressures to commoditize data in the public sector and then on legal and technological measures that endow database producers with new proprietary rights and with novel means of exploiting the facts and data that copyright law had traditionally left in the public domain. These pressures arise both from within the research community itself and from forces extraneous to it. How that community responds to these pressures will over time determine the future metes and bounds of the information commons that support scientific endeavors.

If, as we have reason to fear, current trends will greatly diminish the amount of data available in the public domain, this decrease could initially compromise the scientific community's ability to fully exploit the promise of the digital revolution. Moreover, if these pressures continue unabated and become institutionalized at the international level, they could disrupt the flow of upstream data to both basic and applied science and undermine the ability of academia and the private sector to convert cumulative data streams into innovative products and services.

The pressures discussed here also pose serious conflicts between the norms of public science and the norms of private industry. We contend that failure to resolve these conflicts and to properly balance the interests at stake in preserving an effective information commons could eventually undermine the national system of innovation.

---

[1]This presentation is based on an article by J. H. Reichman and Paul F. Uhlir. *"A Contractually Reconstructed Research Commons for Scientific Data in a Highly Protectionistic Intellectual Property Environment,"* 66 *Law and Contemporary Problems* (Winter-Spring 2003), and is reprinted with the permission of the authors.

COMMODITIZATION OF DATA IN PUBLIC SCIENCE

During the past 10 years, there has been a marked tendency to shift the production of science-relevant databases from the public to the private sector. This development occurs against the background of a broader trend in which the government's share of overall funding for research and development vis-à-vis that of the private sector has decreased from a high of 67 percent in the 1960s to 26 percent in 2000. Furthermore, since the passage of the Bayh-Dole Act in 1980, the results of federally funded research at universities have increasingly been commercialized either by public–private partnerships with industry or directly by the universities themselves.

Reducing the Scope of Government-generated Data

The budgetary pressures on the government are both structural and political in nature. On the whole, mandated entitlements in the federal budget, such as Medicare and Medicaid, are politically impossible to reduce; as their costs mount, the money available for other discretionary programs, including federally sponsored research, has shrunk as a percentage of total expenditures.

This structural limitation is compounded by the rapidly rising costs of state-of-the-art research, including some researcher salaries, scientific equipment, and major facilities. With specific regard to the information infrastructure, researchers earmark the lion's share of expenses to computing and communications equipment, with the remainder devoted to managing, preserving, and disseminating the public-domain data and information that result from basic research and other federal data collection activities. The government's scientific and technical data and information services are thus the last to be funded and are almost always the first to suffer cutbacks.

For example, the National Oceanic and Atmospheric Administration's (NOAA) budget for its National Data Centers remained flat and actually decreased in real dollars between 1980 and 1994, whereas its data holdings increased exponentially and the overall agency budget doubled (mostly to pay for new environmental satellites and a ground-based weather radar system that are producing the exponential data increases). Information managers at most other science agencies have complained about reductions in funding, for both their data management and scientific and technical information budgets.

These chronic budgetary shortfalls for managing and disseminating public-domain scientific data and information have been accompanied by recurring political pressures on the scientific agencies to privatize their outputs. Until recently, for example, the common practice of the environmental and space science agencies was to procure data collection systems, such as observational satellites or ground-based sensor systems, from private companies. Such procurements were made under contract and pursuant to government specifications based on consensus scientific requirements recommended by the research community. Private contractors would thus build and deliver the data collection systems, which the agencies would then operate pursuant to their mission. All data from the system would then belong to the government and would enter the public domain.

Today, however, industry has successfully pursued a strategy of providing an independent supply of the government's needs for data and information products rather than building and delivering data collection systems for government agencies to operate. This solution leaves the control and ownership of the resulting data in the hands of the company, and allows it to license them to the government and to anyone else willing to pay. Because of this new-found role of the government agency as a cash cow, there has recently been a great deal of pressure on the science agencies, particularly from Congress, to stop collecting or disseminating data in-house and to obtain them from the private sector instead.

This approach previously resulted in at least one well-documented fiasco, namely, the privatization of the NASA–NOAA Landsat program in 1985, which seriously undermined basic and applied research in environmental remote sensing in the United States for the better part of a decade. More recently, the Commercial Space Act of 1998 directed NASA to purchase space and earth science data collection and dissemination services from the private sector and to treat data as commercial commodities under federal procurement regulations. The meteorological data value-adding industry has directed similar lobbying pressures at NOAA. The photogrammetric industry has likewise indicated a desire to expand the licensing of data products to the U.S. Geological Survey and to other federal agencies.

Efforts have also been made by various industry groups to limit the online information dissemination services of several federal science and technology agencies. In the cases of the patent database of the U.S. Patent and Trademark Office, the PubMed Central database of peer-reviewed life science journal literature provided on a free and unrestricted basis by the National Institutes of Health National Library of Medicine, and certain types of weather information disseminated by the National Weather Service, such efforts have proved unsuccessful to date. However, publisher groups did succeed in terminating the Department of Energy's PubScience Web portal for physical science information.

Commercial Exploitation of Academic Research

Turning to government-funded research activities, the trend of greatest concern for purposes of this chapter is the progressive incorporation of data and data products into the commercialization process already under way in academia. The original purpose of the Bayh–Dole Act and related legislation was primarily to enable universities to obtain patents on applications of research results. More recently, this activity has expanded to securing both patents and copyrights in computer programs. Now, databases used in molecular biology have themselves become sources of patentable inventions, and the potential commercial value of these databases as research tools has attracted considerable attention and controversy.

These and other databases have increasingly been subject to licensing agreements prepared by university technology transfer offices, which may be prone to treat databases like other objects of Material Transfer Agreements. The default rules that such licensing agreements tend to favor are exclusive arrangements under onerous terms and conditions that include restrictions on use, and even grant-back and reach-through clauses claiming interests in future applications.

Moreover, there is a growing awareness in academic circles generally that data and data products may be of considerable commercial value, and individual researchers have become correspondingly more wary of making them as available as before. This trend, together with the pressures on government agencies described previously, would pose serious problems for the research community's abilities to access and use needed data resources under any circumstances. In reality, these problems could become much greater as new legal and technological fencing measures become more broadly implemented.

INTELLECTUAL PROPERTY, E-CONTRACTS, AND TECHNOLOGICAL FENCES

Traditional copyright law was friendly to science, education, and innovation by dint of its refusal to protect either facts or ideas as eligible subject matter; by limiting the scope of protection for compilations and other factual works to the stylistic expression of facts and ideas; by carving out express exceptions and immunities for teaching, research, and libraries; and by recognizing a catchall, fall-back "fair-use" exception for nonprofit research and other endeavors that advanced the public interest in the diffusion of facts and ideas at relatively little expense to authors. Reinforcing these policies were various judicial decisions and partially codified exceptions for functionally dictated components of literary works, which take the form of nonprotectible methods, principles, processes, and discoveries. On the whole, these principles tended to render facts and data as such ineligible for protection and to allow researchers to access and use facts and data otherwise embodied in protectible works of authorship without undue legal impediments.

In contrast, recent legal developments in intellectual property law and contracts law have radically changed the preexisting regime. These and other related developments now make it possible to assert and enforce proprietarial claims to virtually all the factual matter that previously entered the public domain the moment it was disclosed.

Some of the earliest changes were intended to bring U.S. copyright law into line with longstanding norms of protection recognized in the Berne Convention. For example, the principle of automatic copyright protection, the abolition of technical forfeiture due to lack of formal prerequisites, such as notice, and the provision of a basic term of protection lasting for the life of the creator plus 50 years were all measures adopted in the predigital era for this reason.

Beginning in the 1980s, however, the United States took the lead in reshaping the Berne Convention itself to accommodate computer programs, which many commentators and governments had preferred to view as "electronic information tools" subject to more procompetitive industrial property laws, including patents, unfair competition, and hybrid (or *sui generis*) forms of protection. By the 1990s, a coalition of "content providers" concerned about online copying of movies, music, and software in the new digital environment had persuaded the U.S. government to press for still more far-reaching changes of international copyright and related laws. These efforts led to the codification of universal copyright norms in the Trade-Related Intellectual Property Rights (TRIPS) Agreement of 1994 and to two 1996 World Intellectual Property Organization (WIPO) treaties on copyrights and related rights in cyberspace, which endowed authors with a bevy of new exclusive rights tailor-made for online transmissions and imposed unprecedented obligations on participating governments to prohibit electronic equipment capable of circumventing these rights. All of these new norms and obligations, ostensibly adopted to discourage market-destructive copying of literary and artistic works, then became domestic law, often with no regard for their impact on science and sometimes with deliberate disregard of measures adopted to safeguard science and education at the international level.

At the same time, and as part of the same overall movement, the coalition of content providers that had captured Congress' attention took aim at two closely related areas in which much more than market-destructive copying was actually at stake. The first of these was to validate the uncertain status of standard-form electronic contracts used to regulate online dissemination of works in digital form. Because traditional contracts and sales laws can be interpreted in ways that limit the kinds of terms that can be imposed through "shrinkwrap" or "click-on" licenses, and the one-sidedness of the resulting "adhesion contracts," the coalition pushing the high-protectionist digital agenda has also sponsored a new uniform law, the Uniform Computer Information Transactions Act, to validate such contracts in the form they desire, and it has lobbied state legislatures to adopt them.

The last major component of the high-protectionists' digital agenda was an attempt by some of the largest database companies to obtain a *sui generis* exclusive property right in noncopyrightable collections of information, even though facts and data had hitherto been off-limits even to international copyright law as reformed under the TRIPS Agreement of 1994. These efforts culminated in the European Community's Directive on the Legal Protection of Databases adopted in 1996; in a proposed WIPO treaty on the international protection of databases built on the same model, which was barely defeated at the WIPO Diplomatic Conference in December 1996; and in a series of database protection bills that have been introduced in the U.S. Congress and that attempt to enact similar measures into United States law.

Most of the developments outlined above resulted from efforts that were not undertaken with science in mind, although publishers who profit from distributing commercialized scientific products promoted some of the changes that appear most threatening for scientific research, especially database protection laws. The following subsections show that all these measures—whatever their ostensible purpose—have the cumulative effect of shrinking the research commons.

I will first briefly note the impact of selected developments in both federal statutory copyright law and in contract laws at the state level. I then discuss current proposals to confer strong exclusive property rights on noncopyrightable collections of data, which constitute the clearest and most overt assault on the public domain that has fueled both scientific endeavors and technological innovation in the past.

Expanding Copyright Protection of Factual Compilations: The Revolt Against *Feist*

The quest for a new legal regime to protect databases was triggered in part by the U.S. Supreme Court's 1991 decision in *Feist Publications, Inc. v. Rural Telephone Service Co.*, which denied copyright protection to the white pages of a telephone directory. As discussed above, that decision was notable for reaffirming the principle that facts and data as such were ineligible for copyright protection as "original and creative works of authorship." It also limited the scope of copyright protection to any original elements of selection and arrangement that otherwise met the test of eligibility. In effect, this meant that second-comers who developed their own criteria of selection and arrangement could in principle use prior data to make follow-on products without falling afoul of the copyright owner's strong exclusive right to prepare derivative works. Taken together, these propositions supported the customary and traditional practices of the scientific community and facilitated both access to and use of research data.

In recent years, however, judicial concerns about the compilers' inability to appropriate the returns from their investments have induced leading federal appellate courts to broaden copyright protection of low authorship compilations in ways that significantly deform both the spirit and the letter of *Feist*. At the eligibility stage, so little in the way of original selection and arrangement is now required that the only print media still certain to be excluded from protection are the white pages of telephone directories.

More tellingly, the courts have increasingly perceived the eligibility criteria of selection and arrangement as pervading the data themselves to restrain second-comers from using preexisting datasets to perform operations that are functionally equivalent to those of an initial compiler. In the Second Circuit, for example, a competitor could not assess used car values by the same technical means as those embodied in a first-comer's copyrightable compilation, even if those means turned out to be particularly efficient.[2] Similarly, the Ninth Circuit prevented even the use of a small amount of data from a copyrighted compilation that was essential to achieving a functional result.[3]

Copyright law provides a very long term of protection, and it endows authors, including eligible database proprietors, with strong rights to control follow-on applications of the protectible contents of their works. Stretching copyright law to cover algorithms and aggregates of facts (and even so-called "soft ideas") as these recent decisions have done conflates the idea-expression dichotomy and indirectly extends protection to facts as such.

Opponents of *sui generis* database protection in the United States cite these and other cases as evidence that no *sui generis* database protection law is needed. In reality, these cases suggest that, in the absence of a suitable minimalist regime of database protection to alleviate the risk of market failure without impoverishing the public domain, courts tend to convert copyright law into a roving unfair competition law that can protect algorithms and other functional matter for very long periods of time and that could create formidable barriers to entry. This tendency, however, ignores the historical limits of copyright protection and ultimately jeopardizes access to the research commons.

The Digital Millennium Copyright Act of 1998: An Exclusive Right to Access Copyrightable Compilations of Data?

With regard to copyrightable compilations of data distributed online, amendments to the Copyright Act of 1976, known as the Digital Millennium Copyright Act of 1998 (DMCA), may have greatly reduced the traditional safeguards surrounding research uses of factual works. Technically, Section 1201(a) establishes a right to prevent the direct circumvention of any electronic fencing devices that a content provider may have employed to control access to a copyrighted work delivered online. Section 1201(b) then perfects the scheme by exposing manufacturers and suppliers of equipment capable of circumventing electronic fencing devices to liability for copyright infringement when such equipment can be used to violate the exclusive rights traditionally held by copyright owners.

In enacting these provisions, Congress seems to have detached the prohibition against gaining unauthorized direct access to electronically fenced works under Section 1201(a) from the balance of public and private interests otherwise established in the Copyright Act of 1976. As Professor Jane Ginsburg interprets this provision, a violation of Section 1201(a) is not an "infringement of copyright" because it attracts a separate set of distinct remedies set out in Section 1203 and because it constitutes "a new violation" for which those remedies are provided.[4] On this reading, unlawful access is not subject to the traditional defenses and immunities of the copyright law, and one is "not. . .permitted to circumvent the access controls, even to perform acts that are lawful under the Copyright Act," including presumably the user's rights to extract unprotectible facts and ideas or to invoke the "fair use" defense.[5] On the contrary, "Congress may in effect have extended copyright to cover 'use' of

---

[2]*CCC Info. Services, Inc. v. Maclean Hunter Market Reports, Inc.*, (44 F.3d 61 (2d Cir. 1994)

[3]*CDN Inc. v. Kapes* (197 F.3d 1256 (9th Cir. 1999)

[4]Jane C. Ginsburg, "U.S. Initiatives to Protect Works of Low Authorship," in *Expanding the Boundaries of Intellectual Property: Innovation Policy for the Knowledge Society*, by Rebecca S. Eisenberg, at 63-64.

[5]Ibid., at 62-64.

works of authorship, including minimally original databases...because 'access' is a prerequisite to 'use,' [and] by controlling the former, the copyright owner may well end up preventing or conditioning the latter."[6]

While the precise contours of these provisions remain to be worked out in future judicial decisions, they could potentiate the ability of both publishers and scientists to protect online collections of data that were heretofore unprotectible in print media. If, for example, a database provider combined the noncopyrightable collection of data with a nominally copyrightable component, such as an analytical explanation of how the data were compiled, the "fig leaf" copyrightable component might suffice to trigger the "no direct access" provisions of Section 1201(a).[7] In that event, later scientific researchers could not circumvent the electronic fence in order to extract or use the noncopyrightable data, even for nonprofit scientific research, because Section 1201(a) does not recognize the normal exceptions to copyright protection that would allow such use and scientific research is not one of the few very limited exceptions that were codified in Section 1201(d)-(j).

Later researchers would thus have to acquire lawful access to the electronically fenced database under Section 1201(a) and then attempt to extract the noncopyrightable data for nonprofit research purposes under Section 1201(b), which does in principle recognize the traditional users defenses as well as the privileges and immunities codified in Sections 107-122 of the Copyright Act of 1976. Even here, however, later scientists could discover that the technical devices they had used to extract nonprotectible data from minimally copyrightable databases independently violated Section 1201(b) of the DMCA because those devices were otherwise capable of substantial infringing uses.[8] In practice, moreover, the posterior scientists' theoretical opportunity to extract noncopyrightable data by technical devices that did not violate Section 1201(b) could already have been compromised by the electronic contracts these scientists will have accepted in order to gain lawful access to the online database in the first place and thus to avoid the crushing power of Section 1201(a). In that event, the scientists would almost certainly have waived any user rights they had retained under Section 1201(b), unless the electronic contracts themselves became unenforceable on one ground or another, as discussed below.

In effect, the DMCA allows copyright owners to surround their collections of data with technological fences and electronic identity marks buttressed by encryption and other digital controls that force would-be users to enter the system through an electronic gateway. To pass through the gateway, users must accede to non-negotiable electronic contracts, which impose the copyright owner's terms and conditions without regard to the traditional defenses and statutory immunities of the copyright law.

The DMCA indirectly recognized the potential conflict between proprietors and users of ineligible material, such as facts and data, that Section 1201(a) of the statute could thus trigger, and it empowered the Copyright Office, which reports to the Librarian of Congress, to exempt categories of users whose activities could be adversely affected.[9] While representatives of the educational and library communities have petitioned for relief on various grounds, including the need of researchers to access and use noncopyrightable facts and ideas transmitted online, the authorities have so far declined to act. It is too soon to know how far owners of copyrightable compilations can push this so-called "right of access" at the expense of research, competition, and free speech without incurring resistance based on the misuse doctrine of copyright law, on the public policy and unconscionability doctrines of state contract laws, and on First Amendment concerns that have in the past limited copyright protection of factual works. For the foreseeable future, nonetheless, the DMCA empowers owners of copyrightable collections of facts to contractually limit online access to the pre-existing public domain in ways that contrast drastically with the traditional availability of factual contents in printed works.

ONE-SIDED ELECTRONIC LICENSING CONTRACTS

Data published in print media traditionally entered the public domain under the classical intellectual property regime described above. Further ensuring that result is an ancillary copyright doctrine, known as "exhaustion" or

---

[6] Ibid., at 63.
[7] Ibid., at 63.
[8] Ibid., at 65-67.
[9] 17 U.S.C. § 1201(a)(1)(C)(D).

"first-sale doctrine," which limits the authors' powers to control the uses that third parties can make of copyrighted literary works distributed to the public in hard copies.

Under this doctrine, the copyright owner may extract a profit from the first sale of the copy embodying an original and protectible compilation of data, but cannot prevent a purchaser from reselling that physical copy or from using it in any way the latter deems fit, say, for research purposes, unless such uses amount to infringing reproductions, adaptations, or performances of the expressive components of the copyrighted compilation. In effect, copyright law not only made it difficult to protect compilations of data as such, it denied authors any exclusive right to control the use of a protected work once it had been distributed to the public.

The first-sale doctrine thus complements and perfects the other science-friendly provisions described above, unless individual scientists, libraries, or scientific entities were to contractually waive their rights to use copies of purchased works in the manner described above. Such contractual waivers always remain theoretically possible, and publishers have increasingly pressed them upon the scientific and educational communities in the online environment for reasons discussed below. Nevertheless, it was not generally feasible to impose such waivers against scientists who bought scientific works distributed to the public in hard copies, and even when attempts to do so were made, such contracts could not bind subsequent purchasers of the copies in question. The upshot was that, precisely because authors and publishers could not rely on contractual agreements, they depended on the default rules of copyright law, which are binding against the world. These default rules, in turn, impose legislatively enacted "contracts," which balance public and private interests by, for, example, defining the uses that libraries can make of their copies and by further allowing a set of "fair uses" that scientists and other researchers can invoke.

Against this background, online delivery of both copyrightable and noncopyrightable productions possesses the inherent capabilities of changing the preexisting relationship between authors and readers or between "content providers" and "users." By putting a collection of data online and surrounding it with technological fencing devices, publishers can condition access to the database on the would-be user's acquiescing to the terms and conditions of the former's "click-on," standard-form, nonnegotiable contract (known as a "contract of adhesion"). In effect, online delivery solves the problems that the printing press created for contractually restricting the use of published works and it thus restores the "power of the two-party deal" that publishers lost in the sixteenth century.

The power of the two-party deal that online delivery makes possible is conceptually and empirically independent of statutory intellectual property rights, which makes it of capital importance for the theses discussed here. It means that anyone who makes data available to the world at large can control access to them and control their use by contract in ways that were inconceivable only a few years ago. Nevertheless, statutory intellectual property rights can reinforce the contractual powers of online vendors to prohibit would-be users from disarming encryption devices to gain entry or to limit the ability of would-be users to extract uncopyrightable facts and ideas from copyrightable works delivered online, or even to limit their ability to invoke the statutory defense of fair use. The DMCA lends itself to these ends, ostensibly with a view to impeding market-destructive copying, but with the result of strengthening the copyright monopoly at the expense of the public domain.

Online delivery, coupled with technological fencing devices, potentially confers these same contractual powers on content providers in the absence of supporting intellectual property regimes, such as the DMCA discussed above, and the new database protection rights to be discussed below. The highly restrictive digital rights management technologies that are being developed include hardware- and software-based "trusted systems," online database access controls, and increasingly effective forms of encryption. These emerging technological controls on content, when combined with the statutory intellectual property and contractual rights, can supersede long-established user rights and exceptions under copyright law for print media and thereby eliminate large categories of data and information from public-domain access.

Moreover, because electronic contracts are enforceable in state courts, they provide private rights of action that tend to either substitute for or override statutory intellectual property rights. Electronic contracts become substitutes for intellectual property rights to the extent that they make it infeasible for third parties to obtain publicly disclosed but electronically fenced data without incurring contractual liability for damages. They may override statutory intellectual property rights, for example, by forbidding the uses that libraries could otherwise make of a scientific work under federal copyright law, or by prohibiting follow-on applications or the reverse engineering of a computer program that both federal copyright law and state trade secret law would otherwise permit.

To the extent that these contracts are allowed to impose terms and conditions that ignore the goals and policies of the federal intellectual property system, they would establish privately legislated intellectual property rights unencumbered by concessions to the public interest. By the same token, a privately generated database protected by technical devices and electronic adhesion contracts is subject to no federally imposed duration clause and, accordingly, will never lapse into the public domain.

Whether electronic contracts—especially the nonnegotiable, standard-form "click on" and "shrinkwrap" contracts—are in fact enforceable remains an open and controversial question. In addition to technical obstacles to formation based on general contracts law principles, courts may deem such contracts unenforceable under the "public policy" defense of state contracts law, under the preemption doctrine that supports the integrity of the federal intellectual property system, or under some combination of the two. In practice, however, courts appear reluctant to exercise such powers even when their right to do so is clear. The most recent line of cases, led by the Seventh Circuit's opinion in *ProCD* v. *Zeidenberg*,[10] has tended to validate such contracts in the name of "freedom of contract."

In this same vein, the National Council of Commissioners for Uniform State Law has proposed a Uniform Computer Information Transactions Act (UCITA), which, if state legislatures enacted it, would broadly validate electronic contracts of adhesion and largely immunize them from legal challenge. For example, UCITA permits vendors of information products to define virtually every transaction as a "license" rather than a "sale," and it tolerates perpetual licenses. It could thus override the first-sale doctrine of copyright law and any analogous doctrine that might be embodied in the proposed database protection laws discussed below. The proposed uniform law would then proceed to broadly validate mass-market "click-on" and "shrink-wrap" licenses that imposed all the provisions vendors could hope for, with little regard for the interests of scientific and educational users, or the public in general.

A detailed analysis of UCITA's provisions is beyond the scope of this discussion. Suffice it to say, however, that its less than transparent drafting process so favored the interests of sellers of software and other information products at the expense of consumers and users generally that a coalition of 16 state attorneys general vigorously opposed its adoption, and the American Law Institute withdrew its cosponsorship of the original project. Nonetheless, two states—Maryland and Virginia—have adopted nonuniform versions of UCITA, and major software and information industry firms continue to lobby assiduously for its enactment by other state legislatures.

If present trends continue unabated, privately generated information products delivered online—including databases and computer software—may be kept under a kind of perpetual, mass-market trade secret protection, subject to no reverse engineering efforts or public-interest uses that are not expressly sanctioned by licensing agreements. Contractual rights of this kind, backed by a one-sided regulatory framework, such as UCITA, could conceivably produce an even higher level of protection than that available from some future federal database right subject to statutory public-interest exceptions. The most powerful proprietary cocktail of all, however, would probably emerge from a combination of a strong federal database right with UCITA-backed contracts of adhesion.

New Exclusive Property Rights in Noncopyrightable Collections of Data

The challenge of protecting collections of information that fail to meet the technical eligibility requirements of copyright law poses a hard problem that has existed for a half-century or longer, and at least three different approaches have emerged over time. One solution was to allow a domestic copyright law to accommodate "low authorship" literary productions, with some adjustments to the bundle of rights at the margins. A second approach, adopted in the Nordic countries, was to enact a short-term *sui generis* regime, built on a distinctly copyrightlike model, that would protect catalogs, directories, and tables of data against wholesale duplication, without conferring on proprietors any exclusive adaptation right like that afforded to authors of true literary and artistic works. A third approach, experimented with at different times and to varying degrees in different countries, including the

---

[10] 86 F.3d 1447 (7th Cir. 1996)

United States, was to protect compilers of information against wholesale duplication of their products under different theories rooted in the "misappropriation" branch of unfair competition law.

What changed in the 1990s was the convergence of digital and telecommunications networks, which potentiated the role of electronic databases in the information economy generally and which made scientific databases in particular into agents of technological innovation whose economic potential may eventually outstrip that accruing from the patent system. Notwithstanding the robust appearance of the present-day database industry under freemarket conditions, analysts asked whether inadequate investment in complex digital databases would not inevitably hinder that industry's long-term growth prospects if free-riding second-comers could appropriate the contents of successful new products without contributing to their costs of development and maintenance over time. In other words, if copyright, contract law, digital rights management technologies, residual unfair competition laws, and various protective business practices inadequately filled a gap in the law, then regulatory action to enhance investment might be justified. This utilitarian rationale, however, raised new and still largely unaddressed questions about the unintended social costs likely to ensue if intellectual property rights were injudiciously bestowed upon the raw materials of the information economy in general and on the building blocks of scientific research in particular.

Any serious effort to find an appropriate *sui generis* solution to the question of database protection accordingly should have engendered an investigation of the comparative economic advantages and disadvantages of regimes based on exclusive property rights as distinct from regimes based on unfair competition laws and other forms of liability rules. This investigation also should have taken account of larger questions about the varying impacts of different legal regimes on freedom of speech and on the conditions of democratic discourse, which, in the United States at least, are of primary constitutional importance. Instead, the Commission of the European Communities cut the inquiry short by adopting the Directive on the Legal Protection of Databases in 1996.[11] This directive required all E.U. member countries (and affiliated states) to pass laws that confer a hybrid exclusive property right on publishers who make substantial investments in noncopyrightable compilations of facts and information.

The European Union Database Directive in Brief

The hybrid exclusive right that the European Commission ultimately crafted in its Directive on the Legal Protection of Databases does not resemble any preexisting intellectual property regime. It protects any collection of data, information, or other materials that are arranged in a systematic or methodological way, provided that they are individually accessible by electronic or other means.[12] To become eligible for protection, the database producer must demonstrate a "substantial investment," as measured in either qualitative or quantitative terms,[13] which leaves the courts to develop this criterion with little guidance from the legislative history. The drafters explicitly recognized that the qualifying investment may consist of no more than simply verifying or maintaining the database.

In return for this investment, the compiler obtains exclusive rights to extract or to utilize all or a substantial part of the contents of the protected database. The exclusive extraction right pertains to any transfer in any form of all or a substantial part of the contents of a protected database;[14] the exclusive reutilization right, by contrast, covers only the making available to the public of all or a substantial part of the same database.[15] In every case, the first-comer obtains an exclusive right to control uses of collected data as such, as well as a powerful adaptation (or derivative work) right along the lines that U.S. copyright law bestows on "original works of authorship,"[16] even though such a right is alien to the protection of investment under existing unfair competition laws. In a recent interpretation of this provision, a U.K. court vigorously enforced this right to control follow-on applications of an original database against a value-adding second-comer.[17] It took this position even though the proprietor was the sole source of the data in question and there was no feasible way to generate them by independent means.

---

[11]Directive 96/9 of the European Parliament and the Council of 11 March 1996 on the legal protection of databases, 1996 O.J. (L77)2.
[12]Ibid., art. 1(2).
[13]Ibid., art. 7(1).
[14]Art. 7(2)(a).
[15]Art. 7(2)(b).
[16]17 U.S.C. §§101 ("derivative works"), 103, 106(2) (U.S.).
[17]*British Horseracing Bd. Ltd. v. William Hill Org. Ltd.*, 201 E.W.C.A. Civ. 1268 (Eng. C.A.).

The directive contains no provision expressly regulating the collections of information that member governments themselves produce. This lacuna leaves European governments that generate data free to exercise either copyrights or *sui generis* rights in their own productions in keeping with their respective domestic policies. This result contrasts sharply with the situation in the United States, where the government cannot claim intellectual property rights in the data it generates and must normally make such data available to the public for no more than a cost-of-delivery fee.

The directive provides no mandatory public-interest exceptions comparable to those recognized under domestic and international copyright laws. An optional, but ambiguous, exception concerning "illustrations for teaching or scientific research" applies to extractions but not reutilization.[18] It may be open to flexible interpretation, and some member countries, notably the Nordic countries, have implemented this broader version. Other countries, notably France, Italy, and Greece, have simply ignored this exception altogether, which defeats the commission's supposed concerns to promote uniform law.

The directive's *sui generis* regime does exempt from liability anyone who extracts or uses an insubstantial part of a protected database.[19] However, such a user bears the risk of accurately drawing the line between a substantial and an insubstantial part, and any repeated or systematic uses of even an insubstantial part will forfeit this exemption.[20] Judicial interpretation has so far taken a very restrictive view of this exemption, and one cannot effectively make unauthorized extractions or uses of an insubstantial part of any protected database without serious risk of triggering an action for infringement.

Qualifying databases are nominally protected for a 15-year period.[21] In reality, each new investment in a protected database, such as the provision of updates, will requalify that database *as a whole* for a new term of protection.[22] In this and other respects, the scope of the *sui generis* adaptation right exceeds that of U.S. copyright law, which attaches only to the new matter added to an underlying, preexisting work and expires at a certain time.[23]

Finally, the directive carries no national treatment requirement into its *sui generis* component. Foreign database producers become eligible only if their countries of origin provide a similar form of protection or if they set up operations within the European Union.[24] Nonqualifying foreign producers, however, may nonetheless seek protection for their databases under residual domestic copyright and unfair competition laws, where available.[25]

The E.C.'s Directive on the Legal Protection of Databases thus broke radically with the historical limits of intellectual property protection in at least three ways. First, it overtly and expressly conferred an exclusive property right on the fruits of investment as such, without predicating the grant of protection on any predetermined level of creative contribution to the public domain. Next, it conferred this new exclusive property right on aggregates of information as such, which had heretofore been considered as unprotectible raw material or as basic inputs available to creators operating under all other preexisting intellectual property rights. Finally, it potentially conferred the new exclusive property right in perpetuity, with no concomitant requirement that the public ultimately require ownership of the object of protection at the end of a specified period. The directive thus effectively abolished the very concept of a public domain that had historically justified the grant of temporary exclusive rights in intangible creations.

The Database Protection Controversy in the United States

The situation in the United States differs markedly from that which preceded the adoption of the E.C.'s Directive on the Legal Protection of Databases. In general, the legislative process in the United States has become relatively transparent. Since the first legislative proposal, H.R. 3531, which was modeled on the E.C. Directive and

---

[18]Op. cit., note 11, arts. 9, 9(b).
[19]Ibid., art. 9(b).
[20]Ibid., arts. 7(2), 7(5), 8(1).
[21]Ibid., art. 10.
[22]Ibid., art. 10(3).
[23]See 17 U.S.C. §§103, 302.
[24]Op. cit., note 11, , art. 11.
[25]Ibid., art. 13.

introduced by the House Committee on the Judiciary in May 1996, this transparency has generated a spirited and often high-level public debate. Very little progress toward a compromise solution had been reached as of the time of writing, however, which is hardly surprising given the intensity of the opposing views, the methodological distance that divides them, and the political clout of the opposing camps.

We are, accordingly, left with the two basic proposals that were still on the table at the end of the legislative session that ended in 2000 at an impasse. These proposals, as refined during that session, represent the baseline positions that each coalition carried into the current round of negotiations. One bill, H.R. 354, as revised in January 2000, embodied the proponents' last set of proposals for a *sui generis* regime built on an exclusive property rights model (although some effort was made to conceal that solution behind a facade that evoked unfair competition law). The other bill, H.R. 1858, set out the opponents' views of a so-called minimalist misappropriation regime as it stood on the eve of the current round of negotiations.

The Exclusive Rights Model. The proposals embodied in H.R. 354 attempted to achieve levels of protection comparable to those of the E.C. Directive by means that are more congenial to the legal traditions of the United States. The changes introduced at the end of the 2000 legislative session softened some of the most controversial provisions at the margins, while maintaining the overall integrity of a strongly protectionist regime.

The bill in this form continued to define "collections of information" very broadly as "information . . . collected and . . . organized for the purpose of bringing discrete items of information together in one place or through one source so that persons may access them."[26] Like the E.C. Directive, this bill then cast eligibility in terms of an "investment of substantial monetary or other resources" in the gathering, organizing or maintaining of a "collection of information."[27] It conferred two exclusive rights on the investor: first, a right to make all or a substantial part of a protected collection "available to others," and second, a right "to extract all or a substantial part to make available to others." Here the term "others" was manifestly broader than "public" in ways that remained to be clarified.

H.R. 354 then superimposed an additional criterion of liability on both exclusive rights that is not present in the E.C. model. This is the requirement that, to trigger liability for infringement, any unauthorized act of "making available to others" or of "extraction" for that purpose must cause "material harm to the market" of the qualifying investor "for a product or service that incorporates that collection of information and is offered or intended to be offered in commerce." The crux of liability under the bill thus derived from a "material harm to markets" test that is meant to cloud the copyrightlike nature of the bill and to shroud it in different terminology.

Here a number of concessions were made to the opponents' concerns in the last public iteration of the bill (January 11, 2000), some of them real, others nominal in effect. The addition of "material" to the market harm test may, for example, have addressed complaints that proponents viewed "one lost sale" as constituting actionable harm to the market.

At the same time, the revised bill contained convoluted and tortuous definitions of "market" that the Clinton administration hoped would reduce the scope of protection in the case of follow-on applications.[28] On closer inspection, however, these definitions provided a static picture of a moving target that amounted to a mostly illusory limitation on the investor's broad adaptation right. Notwithstanding these so-called concessions, the bill effectively assigned most follow-on applications to any initial investor whose dynamic operations expand the range of potentially protectible matter with every update, ad infinitum.

The bill then introduced a "reasonable use" exception that was intended to benefit the nonprofit user communities, especially researchers and educators,[29] and that conveyed a sense of similarity to the "fair-use" exception in copyright law.[30] Once again, this became largely illusory on closer analysis, because under the proposed bill, the

---

[26]H.R. 354, §1401(1), the "Collections of Information Antipiracy Act," 106th Congress (2000). Here the overlap with copyright law is so palpable that it is hard to conceive of any assemblage of words, numbers, facts, or information that would not also qualify as a potentially protectible collection of information.

[27]Ibid., §1402(a).

[28] Ibid.., §§1401(3)(A), (B).

[29]Ibid., §1403(2).

[30] 17 U.S.C. §§102(b), 107-122.

very facts, data, and information that copyright law exclude themselves became the objects of protection, and there were no other significant exceptions. Hence, virtually every customary or traditional use of facts or data compiled by others that copyright law would presumably have allowed scientists, researchers, or other nonprofit entities to make in the past now becomes a prima facie instance of infringement under H.R. 354. These users would in effect have either to license such uses or be prepared to seek judicial relief for "reasonableness" on a continuing basis. Because university administrators dislike litigation and are risk averse by nature, and this provision put the burden of showing reasonableness on them, there is reason to expect a chilling effect on customary uses by these institutions of data heretofore in the public domain.

The bill recognized an "independent creation" norm, which presumably exempts any database, however similar to an existing database that was not the fruit of "copying."[31] This provision codified a fundamental norm of copyright law, and the European Commission made much of a similar norm in justifying its own regulatory scheme. In reality, this "independent creation" principle would produce unintended and socially deleterious consequences when transposed to the database milieu precisely because many of the most complex and important databases are inherently not able to be independently regenerated. Sometimes the database cannot be reconstituted because the underlying phenomena are one-time events, as often occurs in the observational sciences. In other instances, key components of a complex database can no longer be reconstituted with certainty at a later date. Any independently regenerated database suffering from these defects would necessarily contain gaps that made it inherently less reliable than its predecessor.

These problems point to a more general phenomenon that affects competition in large or complex databases. Even when, in principle, such databases could be reconstituted from scratch, the high costs of doing so—as compared with the add-on costs of existing producers—will tend to make the second-comer's costs so high as to constitute a barrier to entry. Meanwhile, the first-comer's comparative advantage from already owning a large collection that is too costly to reconstitute will only grow more formidable over time, an economic reality that progressively strengthens the barriers to entry and tends to reinforce (and, indeed, to explain) the predominance of sole-source data suppliers in the marketplace.

Government-generated data would have remained excluded, in principle, from protection, in keeping with current U.S. practice,[32] which differs from E.U. practice in this important respect. However, there is considerable controversy surrounding the degree of protection to be afforded government-generated data that subsequently become embodied in value-adding, privately funded databases. All parties agree that a private, value-adding compiler should obtain whatever degree of protection is elsewhere provided, notwithstanding the incorporation of government-generated data. The issue concerns the rights and abilities of third parties to continue to access the original, government-generated data sets. The proponents of H.R. 354 have been little inclined to accept measures seeking to preserve access to the original data sets, despite pressures in this direction.

H.R. 354 imposed no restrictions whatsoever on licensing agreements, including agreements that might overrule the few exceptions otherwise allowed by the bill.[33] Despite constant remonstrations from opponents about the need to regulate licensing in a variety of circumstances—and especially with respect to sole-source providers—the bill itself did not budge in this direction. On the contrary, new provisions added to H.R. 354 in 2000 would have set up measures that would prohibit tampering with encryption devices ("anti-circumvention measures") and with electronically embedded "watermarks" in a manner that paralleled the provisions adopted for online transmissions of copyrighted works under the DMCA. Because these provisions would have effectively secured a database against unauthorized access (and tended to create an additional "exclusive right of access" without expressly so declaring), they would only have added to the database owner's market power to dictate contractual terms and conditions without regard to the public interest. These powers were further magnified by the imposition of criminal sanctions in addition to strong civil remedies for infringement.[34]

---

[31]Op. cit., note 20, §1403(c).
[32]Ibid., §1404.
[33]Ibid., §1404(e).
[34]Ibid., §§1406-1407.

The one major concession that was made to the opponents' constitutional arguments concerned the question of duration. As previously noted, the E.C. Directive allows for perpetual protection of the whole database so long as any substantial part of it is updated or maintained by virtue of a new and substantial investment, and the proponents' early proposals in the United States echoed this provision. However, the U.S. Constitution clearly prescribes a limited term of duration for intellectual property rights,[35] and the proponents finally bowed to pressures from many directions by limiting the term of duration to 15 years.[36]

Any update to an existing database would have qualified for a new term of 15 years, but this protection would apply, at least in principle, only to the material added in the update. In practice, however, the inability to clearly separate old from new matter in complex databases, coupled with ambiguous language concerning the scope of protection against harm to "likely, expected, or planned" market segments, could still have left a loophole for an indefinite term of duration.

The Unfair Competition Model. The opponents' bill, the Consumer and Investor Access to Information Act of 1999, H.R. 1858, was introduced by the House Commerce Committee in 1999, as a sign of good faith, in response to critics' claims that the opponents' coalition sought only to block the adoption of any database protection law. H.R. 1858 began with a definition of databases that is not appreciably narrower than that of H.R. 354, except for an express exclusion of traditional literary works that "tell a story, communicate a message," and the like.[37] In other words, it attempted to draw a clearer line of demarcation between the proposed database regime and copyright law, to reduce overlap or cumulative protection as might occur under H.R. 354.

The operative protective language in H.R. 1858 was short and direct, but it relied on a series of contingent definitions that muddy the true scope of protection. Thus, the bill would prohibit anyone from selling or distributing to the public a database that is (1) "a duplicate of another database . . . collected and organized by another person or entity," *and* (2) "is sold or distributed in commerce in competition with that other database."[38] The bill then defined a prohibited duplicate as a database that is "substantially the same as such other database, as a result of the extraction of information from such other database."[39]

Here, in other words, liability would attach only for a wholesale duplication of a preexisting database that results in a substantially identical end product. However, this basic misappropriation approach became further subject to both expansionist and limiting thrusts. Expanding the potential for liability was a proviso added to the definition of a protectible database that treats "any discrete sections [of a protected database] containing a large number of discrete items of information" as a separably identifiable database entitled to protection in its own right.[40] The bill would thus have codified a surprisingly broad prohibition of follow-on applications that make use of discrete segments of preexisting databases, subject to the limitations set out below.

A second protectionist thrust resulted from the lack of any duration clause whatsoever, with the prohibition against wholesale duplication—subject to limitations set out below—*conceivably lasting forever*. This perpetual threat of liability would have attached to wholesale duplication of even a discrete segment of a preexisting database, if the other criteria for liability were met.

These powerfully protective provisions, put into H.R. 1858 at an early stage to weaken support for H.R. 354, were offset to some degree by other express limitations on liability and by a codified set of misuse standards to help regulate licensing. To understand these further limitations, one should recall that liability even for wholesale duplication of all, or a discrete segment, of a protected database would not attach unless the unauthorized copy were sold or distributed in commerce and "in competition with" the protected database.[41] The term "in competition with," when used in connection with a sale or distribution to the public, was then defined to mean that the unauthorized duplication "displaces substantial sales or licenses likely to accrue from the original database" *and*

---

[35] U.S. Constitution, Art. I, Sec. 8, cl. 8.
[36] Op. cit., note 16, at §1409(i).
[37] H.R. 1858, §101(1), the "Consumer and Investor Access to Informatin Act of 1999," 106th Congress (1999).
[38] Ibid., §102.
[39] Ibid., §101(2).
[40] Ibid., §101(1)(B).
[41] Ibid., §102.

"significantly threatens...[the first-comer's] opportunity to recover a reasonable return on the investment" in the duplicated database.[42] Both prongs had to be met before liability would attach.

It follows that even a wholesale duplication that was not commercially exploited or that did not substantially decrease expected revenues (as might occur from, for example, nonprofit scientific research activities) could presumably have escaped liability in appropriate circumstances. Similarly, a follow-on commercial product that made use of data from a protected database might have escaped liability if it were sold in a distant market segment or required substantial independent investment.

H.R. 1858 then further reduced the potential scope of liability by imposing a set of well-defined exceptions and by limiting enforcement to actions brought by the Federal Trade Commission.[43] There were express exceptions comparable to those under H.R. 354 for news reporting, law enforcement activities, intelligence agencies, online stockbrokers, and online service providers.[44] There was also an express exception for nonprofit scientific, educational, or research activities,[45] in case any such uses were thought to escape other definitions that limit liability to unauthorized uses in competition with the first-comer. Still other provisions clarified that the protection of government-generated data or of legal materials in value-adding embodiments would remain contingent upon arrangements that facilitate continued public access to the original data sets or materials.[46] A blanket exclusion of protection for "any individual idea, fact, procedure, system, method of operation, concept, principle or discovery" wisely attempted to provide a line of demarcation with patent law and to ward off unintended protectionist consequences in this direction.[47]

Another important set of safeguards emerged from the drafters' real concerns about potential misuses of even this so-called "minimalist" form of protection. These concerns were expressed in a provision that expressly denied liability in any case where the protected party "misuses the protection" that H.R. 1858 would afford. A related provision then elaborated a detailed list of standards that courts could use as guidelines to determine whether an instance of misuse had occurred.[48] These guidelines or standards would have greatly clarified the line between acceptable and unacceptable licensing conditions, and if enacted, they could have made a major contribution to the doctrine of misuse as applied to the licensing of other intellectual property rights as well.

In summary, the underlying purpose of H.R. 1858 was to prohibit wholesale duplication of a database as a form of unfair competition. It thus set out to create a minimalist liability rule that would prohibit market-destructive conduct rather than an exclusive property right as such, and in this sense, it initially posed a strong contrast to H.R. 354. Over time, however, different iterations of the bill, designed to win supporters away from H.R. 354, made H.R. 1858 surprisingly protectionist—especially in view of its de facto derivative work right.

---

[42]Ibid., §101(5).
[43]Ibid., §107.
[44]Ibid., §§103(b), (c), 104(b), (e), 106(a).
[45]Ibid., §103(d).
[46]Ibid., §§101(b), 104(f).
[47]Ibid., §104(d).
[48]Ibid., §§106(b), 106(b)(1-6).

11

The Urge to Commercialize: Interactions Between Public and Private Research Development

Robert Cook-Deegan[1]

Almost everything that I am going to present will be blindingly obvious. I see my job as synthesizing some of what has been talked about previously and to look at some overall trends. The reason we are here is because we like what science and technology produce. There has been a lot more spending by governments in research and development (R&D) and even faster growth in R&D spending by private entities in the postwar era.

Why are we doing that? Because we buy the products and services that come at the end of that process and we have been buying a lot of them. R&D has been a source of economic growth. Governments like it because it creates wealth, and people who have more money are happier voters and it feeds back on itself. It is a virtuous cycle. We have a robust system of innovation. We are coming to the end of a decade when it seemed like that growth was never going to end.

I am going to go through some of these overall trends, and then I will look at biomedical research as a particular sector of interest. I will spend most of my time talking about genomics because genomics is a poster child, representing an area where intellectual property and the public domain are intersecting, colliding, and causing conflict constantly, and quite conspicuously. The amount of funding going into biomedical research has increased by three to four orders of magnitude in real dollars since World War II. That is an unbelievable amount of growth in five decades. The scale of effort, the number of people doing it, the amount of money, and the commitment of social resources have all mushroomed in a relatively short period. It has happened in both the public and the private sectors, led by government funding, at least in the case of life sciences, and it has been followed with a time lag of several years by investment in the private sector.

The R&D growth in both public and private sectors has led to conflict. The Human Genome Project is especially good at generating conflict. When it does so, it usually is on the front pages of *Time, Newsweek, The New York Times,* and *The Washington Post.* You all have heard the stories I am going to recount. I am just going to try to tease out some of the structures underneath the surface.

What drives the growth of R&D—in science, in academia, in government, and in industry? The practice of generating new knowledge has become more capital intensive. Research costs a lot of money, and it takes a lot of people. The scale and the complexity have increased. We need machines to generate the data. We need computers to keep track of data. This has been increasing on a large scale since World War II in almost every discipline.

---

[1] The author would like to acknowledge the assistance of Stephen McCormack and LeRoy Walters and the DNA Patent Database at Georgetown University.

In the political arena, drawing lessons from World War II, governments woke up to the fact that universities and academic centers seem to be the focal front end of a big process that generates a lot of dollars at the back end. They realized that universities are valuable resources and began to pay attention to the policies that foster their development. In particular, they thought about their role in economic growth, in addition to the traditional university roles of creating knowledge and disseminating knowledge. At the same time, we have seen intellectual property grow in strength and quantity; this has been happening historically during this postwar period, most prominently in the United States.

I am going to shift for a moment to the life sciences, specifically on drug development and biotechnology, the products and therapeutics where most of the money is made. It is about a $200 billion a year enterprise, much larger than at the end of World War II. The overall policy framework has been a simple one that was crafted in the immediate postwar era: The government funds basic research, which spills over and people make useful things out of the new knowledge and techniques, which are turned into products and services, the so-called pipeline model. That scenario actually is not too different from how many drugs have been discovered.

The government has funded a lot of science. The past five decades have seen spectacular budget growth at the National Institutes of Health (NIH). I do not know what is going to happen after this year, but we are at the end of another doubling. There have been many doublings of the NIH budget since World War II.

We have also encouraged patenting. The Bayh-Dole Act solidified policies that were already falling into place in the 1980s. The federal court system was reorganized, creating a single appeals court for patent cases. The presumption in favor of patents increased. The structure of the judicial system presiding over intellectual property decisions reinforced the technology transfer statutes addressed earlier today. Moreover, there have been many other policies that try to foster—at the state level, at the national level, and in private industry—the private development of these public resources to bolster the public domain in science. What have the results of these policies been?

There are many start-up firms. In biotechnology, I do not think anybody knows how many firms there are, but there are more than 2,000. In 1992 you could not have called any firm in the world a genomics firm. There are now approximately 400 plus genomics firms, at least 300 of which are still operating—in less than a decade, all of these companies were created.

There have been many patents issued to academic institutions, 3,000 of them last year. Private R&D investment in universities has grown, as has licensing income from academic intellectual property. Last year, the most recent survey from the Association of University Technology Managers found that universities received about $1.1 billion. That is about 3 percent of the total that they spend on R&D. The patent logjams and intellectual property strictures on research that we are talking about, however, are some of the unintended consequences of the policies that have been largely successful (see Figure 11-1).

- Transaction costs
    - Money costs
    - Material Transfer and database agreements
    - Technical licensing offices and firm lawyers as intermediaries
- Secrecy in academe
    - Nondisclosure agreements
    - Trade secrecy until patents are filed
- Sludge in information pipelines
- Anticommons as an unrealized but worrisome threat: patent thickets, royalty stacks, innovations foregone

FIGURE 11-1 Unintended consequences.

BIOMEDICAL RESEARCH MODELS

I now want to focus on genomics. My background is in human genetics. I did my first research on the genetics of Alzheimer's disease. A long lineage of human clinical research goes all the way back to Pasteur and beyond. Many areas of human clinical research, including human genetics, had only glancing acquaintance with the public domain. Data hoarding, keeping things secret, using your data so that you could get the next publication, and making sure that nobody else got access to data were common behaviors. Within science, particularly within human genetics, there were not strong open disclosure norms. There were other communities doing molecular biology at the same time on yeast and Drosophila that had "open-science" norms. Those norms were the ones adopted as the models for the Human Genome Project.

Within biomedical research, different norms for disclosure pervade different fields, so that open science and secretive science are often working in parallel. I merely want to point out that going all the way back to Pasteur, there have been norms of secrecy. Geison has shown that Pasteur was highly secretive, particularly regarding human experimentation (one provision of his will was to keep his laboratory records secret). Given this history, we cannot think of some golden era that we are trying to return to, at least not in human genetics. One of the goals of the Human Genome Project, as articulated by the 1988 National Research Council Report *Mapping and Sequencing the Human Genome*,[2] was to tilt in favor of open science over the more territorial norms of human genetics and clinical research.

Pharmaceutical development is one of the most patent-dependent sectors in the whole economy. There was a tremendous amount of government and nonprofit funding flowing into genetics and genomics for the better part of a decade, before the private genomics effort began around 1992-1993. There is also an intricate mutualism between the public domain and the private domain in genomics that creates databases, products, and sequence information. Genomics attracted many players, including national and state governments, as well as private firms, because genomics was "hot." It was in the news and everybody wanted a piece of the action. But action toward what end? The policies that we are talking about focus on information flow.

Three different models drive commercial genomics, which were created in that first wave of genomic startups from 1992 to 1994. One business plan is represented by Human Genome Sciences (HGS). It started from owning the intellectual property from a nonprofit organization created in 1992 called the Institute for Genomic Research. The Institute and HGS were initially looking for human genes by picking out the parts of the genome that were known to code for proteins and then looking for those proteins that were most likely to be thrown outside the cell, to span cell surface membranes, to bind DNA, or to serve other known functions. HGS would look for DNA sequences corresponding to proteins that might be valuable to pharmaceutical companies or to themselves to develop into pharmaceuticals, and they focused on characterizing the whole gene, sequencing it, and then doing some biology to figure out what it did. The HGS strategy was very focused on intellectual property. There was a patent lawyer, or somebody with legal training in patents, associated with every project team.

HGS was very careful about how they set up walls of nondisclosure around their arrangements. There was some outlicensing of their technologies to major firms so that they could get money, but there was not much in the way of publications. The main output and the main contribution to the public domain happens when HGS gets a patent because then it is published. Then the sequence information that was part of their patent application is published and it becomes part of the public domain, except that it is constrained by the patent rights that are associated with that work.

Incyte turned seriously to DNA sequencing of genes about a year before HGS. It had a somewhat similar scientific strategy, but a different business plan. Incyte was also looking for the juicy bits of the genome to sequence and characterize genes. Incyte had a somewhat different business model in that it was working with multiple, large pharmaceutical players, and it did not appear to have focused on exactly the same kinds of sequences as HGS. Incyte was creating a database that high-paying customers could access. Through Incyte, large pharmaceutical companies could avoid having to do all the sequencing themselves. They could license access to DNA sequence information on genes from Incyte. Incyte's business plan has changed several times since the early

---

[2]National Research Council. 1988. *Mapping and Sequencing the Human Genome*, National Academy Press, Washington, D.C.

years (as has the company's full name), but gene sequencing was the original core idea. Incyte pursued a model that allowed multiple players to get access to its data. It also had looser boundaries and there was a little bit more leakage into the public domain, and a few more publications.

After these two companies had been doing genomics for 5 years, Celera Genomics Corporation came along in 1998 and proposed to sequence the whole human genome. Celera was going to do it in a slightly different way from the way the public-domain project was sequencing the human genome. Celera's initial goal was to create a database of genomic sequence (that is, of the genome in its native state on the chromosomes before being edited into shorter and more compact known genes). Celera needed powerful informatics, and indeed spent more on computers and programming than on DNA sequencing and laboratory biology. Because of a natural asymmetry, Celera would always have more data than was in the public domain—that is, Celera could draw on the public domain and at the same time create its own proprietary sequencing data. Celera would always have one leg up on the public genome project and they could sell a service, which was access to their data. They were doing that nonexclusively, but even more nonexclusively than Incyte, and they had more academic collaborators. Celera was selling licenses to get access to their data using discriminate pricing—a lower pricing level to academic institutions, to the Howard Hughes Medical Institute, and other academic research and higher prices (and more intensive services) for pharmaceutical companies.

Celera's intellectual property strategy was somewhat different from Incyte and HGS. Celera was filing provisional patent applications that could be converted to patent applications, and a few dozen had ripened into patents (52 as of December 2002 compared with 711 for Incyte and 277 for HGS). The interesting thing about the Celera model is that publication in scientific journals was an inherent part of the business plan. Celera planned to publish in scientific journals, and the information they publish or that they post on their Web site is available for other people to use. I do not think we could use the word "public domain" for these data and information because there are restrictions on their use.

These three companies have three different models. If you think of a continuum with the public domain on one end and the private domain on the other, these companies are at intermediate points along a continuum of contributing data into sources where scientists can use them.

There was a strong ideology that grew out of the nematode and the yeast research communities that access to the data should be free and rapid. The sequencing centers in the Human Genome Project are getting paid by the government or nonprofit organizations (e.g., the Wellcome Trust) and are making the data available so everybody can get access to them. The scientific community wanted policies to ensure that the high-throughput sequencing centers did not get unfair advantage because of sole access. My reading is that the rapid disclosure policy was not so much a reaction against commercial practices but rather a concern about hoarding data. The idea was that the yeast and nematode genetics science communities were healthier because of the way they shared data. They are more spontaneous. They are more creative and do better science, in part because they share their data at an earlier stage. Out of that movement grew some very concrete policies, including the so-called Bermuda Rules or Bermuda Principles, which grew out of a meeting that the Wellcome Trust and other funders held in Bermuda. The policy was an agreement to dump data quickly into public sequence databases. As such, the high throughput sequencing centers that were funded as part of big human genome projects agreed to provide their data very quickly, usually on a 24-hour basis, a remarkable forcing out of information into the public domain. That is the open extreme of public-domain policy making, defining the opposite pole from HGS.

However, within the academic sector, many intermediate points on the continuum are represented. Many university laboratories have "reasonable delays" for publication of some research. There are thousands and thousands of labs that do sequencing. Most of them do not behave according to the Bermuda Principles. They release data in "publishable units" after a gene has been fully sequenced and partially characterized and once the sequence data are verified. Researchers in academic genetics, that is, who identify themselves as geneticists, report withholding data to honor an agreement with a commercial sponsor or to protect the commercial value of the data. However, I will point out that the big reason for nonsharing or withholding of data is the effort required. So in many university molecular biology laboratories, data and materials may be unpublished for protracted periods.

Yet in private industry, there may be rapid open disclosure of data. Here is the topsy-turvy part of the Human Genome Project. Pharmaceuticals are very patent dependent; it is famous in business schools for being an information-intensive, secrecy-intensive business. However, in 1994, Merck, a private firm, funded the human expressed sequence tag and cDNA sequencing effort at Washington University, the results of which were added to the public domain. Merck did not get to see the data before anybody else. Why did a private firm pay for information to be generated and dumped into the public domain? I believe it is because of what Incyte, HGS, and other genomics startups were doing. Merck pursued one scientific strategy to counter the appropriation of DNA sequence data by funding research to put data into the public domain to defeat the strategies being pursued by some start-up genomics firms.

I have already talked about the Bermuda Principles. They were adopted by the major genome sequencing centers. The SNP Consortium is another interesting model. A "SNP" is a single nucleotide polymorphism. There is a difference in the A, G, T, or C at one place in the genome, which differs among individuals. If you can find a sequence difference, it is a SNP; polymorphism merely means "difference" to geneticists. SNPs are useful as markers on the chromosomes of humans or any other organism. SNPs are particularly valuable for looking for genes of unknown function and location, or for studying the whole genome at once. Some companies were established to find and patent SNPs. As a result, some academic institutions and 13 private firms formed a consortium to make sure that this stays in the public domain. However, the SNP Consortium did not just dump the data. They filed patent applications and then characterized the SNP markers enough so that they could be sure that nobody else could patent them.[3] At that point, they would abandon the patent. It is a very sophisticated intellectual property strategy that in the end was intended to bolster the public domain. It requires coordination, lots of paperwork, and it costs money to file and process applications, but it appears to be an effective defensive patenting strategy.

The publicly funded Mammalian Gene Collection and its parallel program, the Cancer Genome Anatomy Program, pursue policies to promote rapid data disclosure into the public domain. Under Mammalian Gene Collection government contracts, groups do the same thing that Incyte and HGS were doing, which was to sequence identified genes—the juicy bits of the genome that are translated into proteins. In this case, NIH had to go to the Department of Commerce to declare "exceptional circumstances" under the Bayh-Dole Act. As a condition under those contracts, the government gets to keep all the patent rights. Because the government is not filing patent applications, it is basically a de facto nonpatenting strategy. That is the only case that I know of where the exceptional circumstance clause of the Bayh-Dole Act has been invoked, although I know others at NIH are discussing other possible uses.

Remember that when the genome project started out, it was supposed to be a public-domain infrastructure project so we could all do human genetics faster and at less cost. Those of us who thought about the Human Genome Project in the early years were not thinking primarily of commercial potential, but look at what happened to the funding streams in the year 2000. It looks like over $1.5 billion of public and nonprofit funding went into genomics in the year 2000 (see Figure 11-2). The aggregate R&D spending of "genomics" firms is in the

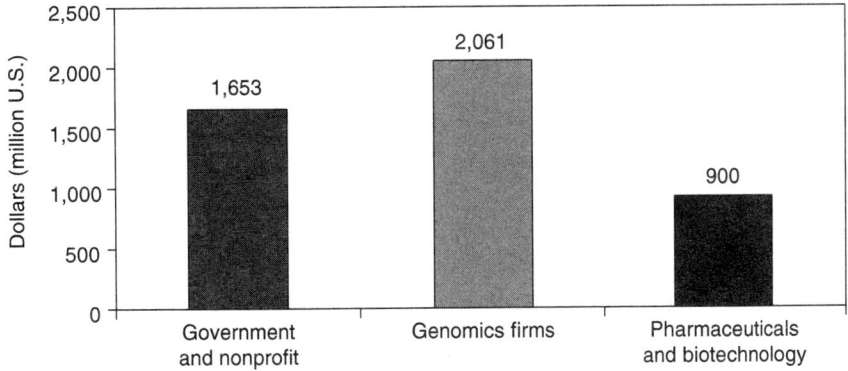

FIGURE 11-2 Private and nonprofit genomics funding, 2000. **Source:** World Survey of Funding for Genomics Research Stanford in Washington Program at http://www.stanford.edu/class/siw198q/websites/genomics/.

---

[3]For more information, see Chapter 28 of these *Proceedings*, "The Single Nucleotide Polymorphism Consortium," by Michael Morgan.

neighborhood of $2 billion (genomics firms devote some substantial part of their R&D to genomics). Major pharmaceutical firms spent another $800 million to $1 billion on genomics in 2000, based on an estimate of 3 to 5 percent of R&D of members of the Pharmaceutical Research and Manufacturers' Association. That does not include the privately held genomics firms that tend to be smaller but are much more numerous. These figures are quite uncertain, but if they can be used as a rough indicator, about a third of genomics funding comes from government and nonprofit organizations, and two-thirds of the funding, at least in 2000, was spent in private R&D.

Figure 11-3 shows government and nonprofit genomics research funding for 2000 by country. The United States, not surprisingly, is number one; the genome project in many ways originated here, even if the science did not. If you normalize for gross domestic product, you see that many countries are investing more in genomics as a fraction of their R&D and as a fraction of their economy than the United States. Estonia, the United Kingdom, Sweden, the Netherlands, Japan, and Germany are all higher than the United States. This happened very fast. In 1994 there were eight publicly traded genomics companies; in 2000 there were 73 publicly traded firms. In 2000, commercial genomics grew to about $94 billion or $96 billion capitalization (before declining precipitously in 2001 and 2002).

What was going on under the surface? Let me return to this private-public mutualism. The main target that everybody was aiming at was that $200 billion market for therapeutic pharmaceuticals, expected to grow much larger in future years. Genomics was thought of as a way to develop those products faster at the front end of the discovery process. Pharmaceutical firms were interested but they came late to the game.

How do I know that? Well, let us look at the patent holdings. Figure 11-4 is only a thousand of the DNA-based patents that were issued from 1980 to 1999. Initially, we read every patent and coded them to be able to get these data, and then we went back to figure out how many patents were owned by whom through 1993. We then used the same patent search algorithm to identify DNA-based U.S. patents and augmented the database through 1999, the basis for the bar charts on which institutions own DNA-based U.S. patents (see Figure 11-4). The U.S. government is the number one patent holder followed by the University of California. Incyte, a genomics company, was number three. Chiron, which is a first-generation biotech firm, was number four, and most of their patents actually

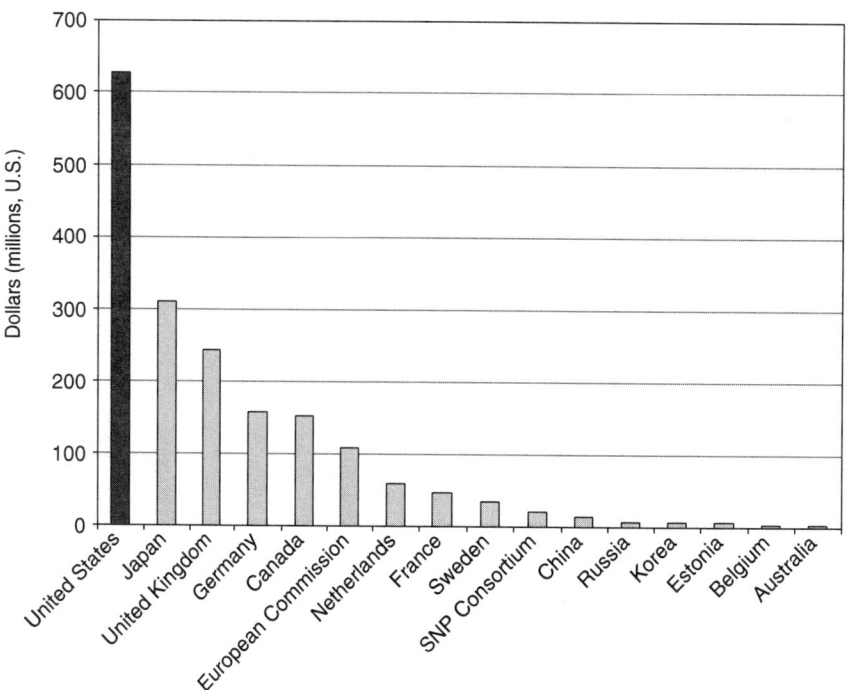

FIGURE 11-3 Government and nonprofit genomics research funding, 2000.

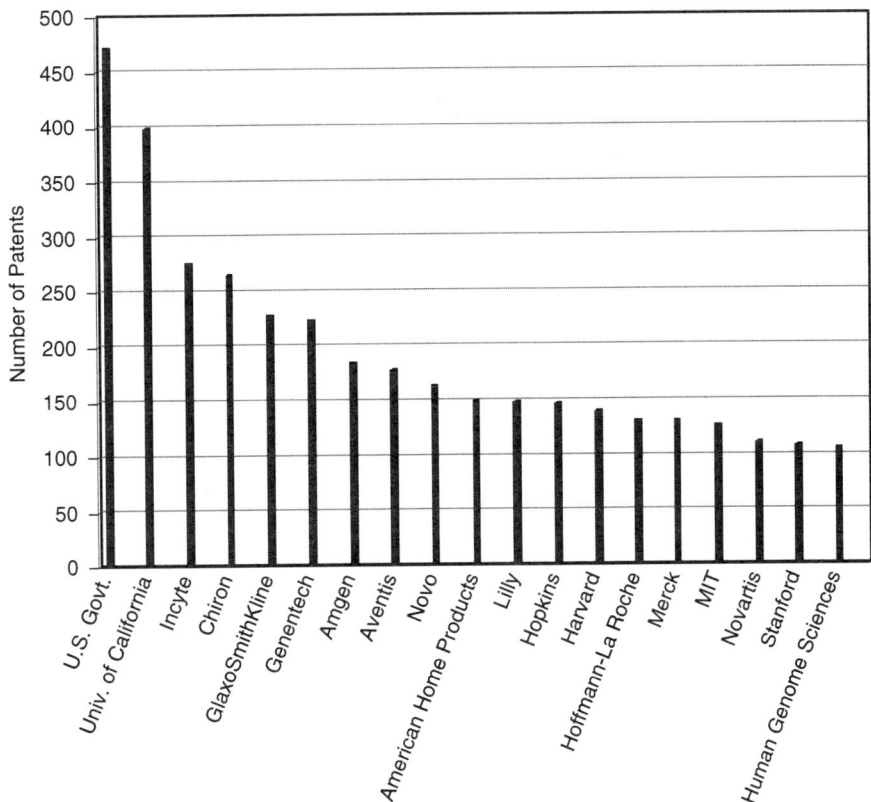

FIGURE 11-4 Number of patents in the DNA patent database, 1980-1999. **NOTE:** Data through the end of December 1999.

came from a firm called Cetus, which they acquired. You have to get to number five before you get to a company that is in the business of creating the end products that everybody is aiming at: GlaxoSmithKline. This represents five different kinds of players that own the intellectual property here.

There were no DNA-based patents in 1970. There were some DNA and RNA-based patents issued before 1980, but not many. A DNA sequencing method was patented in 1973; I do not think anybody uses it, but it is patented and it is in the DNA Patent Database. Starting in the mid-1990s, there was the beginning of an exponential rise. It kept that way through 1996 or 1997. The growth dipped in 1999, which may be a policy dip, because the patent office that year began to change the rules for DNA sequence patents. They raised the threshold basically, changing to a higher utility standard for examining DNA sequence patents—requiring applicants to show a "credible, substantial, and specific" utility in the patent. The U.S. Patent and Trademark Office also demanded a higher standard of written description of the invention in gene-based patents. The growth increases again in 2001 and it looks like we will be below the 2001 number for 2002, although there are a few months of patents we have not examined. The years of exponential growth may be over, but the bottom-line message here is there are 25,000 pieces of intellectual property that have already been created.

A pattern of ownership can also be observed. Of those patents that were issued between 1980 and 1993 that we read through and coded, companies only own about half of these patents. If you add up the others, which are mainly academic research institutions, almost half of the patents are owned by "public" organizations such as universities, nonprofit research centers, and government. That is a very unusual ownership pattern. Overall, academic institutions own only 3 percent of U.S. patents. Figure 11-5 illustrates ownership of 1,078 DNA-based patents from 1980 through 1993.

What is going on here? Molecular genetics appears to be a field in which private-sector actors step in to bolster the public domain. The public domain or the public funders create intellectual property subject to the Bayh-Dole

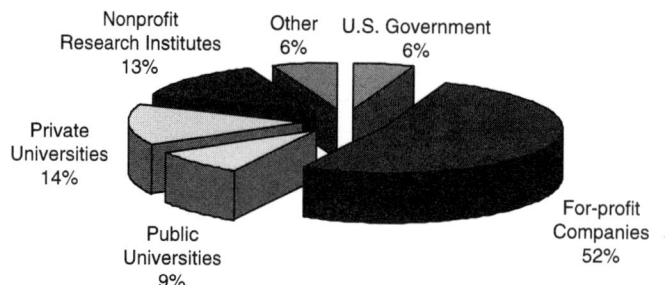

FIGURE 11-5 Ownership of DNA patents, 1980-1999.

Act, which gets protected by intellectual property rights. Private firms draw on public science and also pursue private R&D, which produces more intellectual property owned by private firms.

A couple of generalizations do flow from that. One interpretation is that, the more money you make, the more you are going to plow into R&D. That is the argument that is made by the pharmaceutical and the biotech firms, and it is probably true. If firms make more money, they are going to spend more on R&D, particularly if R&D is the way they believe they made their initial profit. As a result, there will be more innovation, there will be more of the products and services that we like, and, of course, we are going to have to pay more for them. We have been doing that for two decades. One thing to consider is whether double-digit rates of growth are sustainable. We may be beginning to encounter resistance to growth in drug and device expenditures in the early part of the twenty-first century. It is very clear that at least for these genomic startups, private firms believed that their patent portfolio mattered. Their intellectual property mattered and that partially drove this high level of private investment in genomics and in biotechnology more generally.

One of the really interesting things about genomics as a case study is that, in fact, we have a scenario that feels like a race between the public sector and the private sector. I am not sure competition is the right word because you have one group of people that are dumping data into the public domain and another group that are developing data to make a profit. But they are doing the same things in their labs. In the private sector, they have done it in a very capital-intensive way that tends to be fairly centralized.

Public and private sectors are pursuing similar lines of research, and we have a natural experiment that has been going on now for almost a decade, the outcome of which we do not really know. We can say that the academic sector has been a more important part of this story than it has been in most other technologies, such as informatics and computing, although universities were important there too. But it is not possible to compare whether more public good will come from the HGS and Incyte's private sequencing, Celera's quasi-public sequencing, or the strong public-domain policies under the Human Genome Project and Mammalian Gene Collection. We may never know, as products, services, and discoveries are apt to draw from many streams.

The big fight over the publication in February of the 2001 sequence data is how much is going to be put into the public domain where everybody can use it without restriction, how much of it will be publicly accessible with some restrictions, and how much of it is kept behind closed doors. That is not a resolved debate, but it is very rich. In 15 years we may be able to make more educated guesses about what our policy should have been over the past four or five years.

We have races for money, but we also have races for credit, and they are inextricably intertwined. It is very clear that part of Celera's business strategy was to be well known and famous, as well as to sell database subscriptions. Therefore, is not just about money and it is not just about credit, it is about both. That is true on both sides of the academic and industrial divide.

12

Legal Pressures in Intellectual Property Law

Justin Hughes[1]

I wondered whether this audience should be addressed as a political constituency or as a science community. I think that is important to think about because it is a dilemma implicit in this symposium and a lot of the scientific community's study of intellectual property (IP) issues: do you look at intellectual property policy as a subject of study or do you address IP policy as a battleground in which the science community will, whether you like it or not, either be warriors or cannon fodder.

The proof of a major change in policy should either be a good empirical case or a very good theoretical case. The reason I adopt that position is because whether you are a law professor, who is supposed to be committed to the truth, or you are a scientist, who is supposed to be committed to the truth, it is a position that you can take into the activist battlefield and you can do it with principles. You can discuss these policy issues and be a policy activist on what the law will be, my position is that you should not do a major change in IP law, unless you can show me a good empirical case to make the change or a very good theoretical case.

I wrote a paper entitled "Political Economies of Harmonization, Database Protection and Information Patents"[1] for a conference this summer in Paris, cosponsored by the University of Maryland, and Professor Brian Kahin was one of the organizers. I talk about that position in relationship to the fight we have been having in the United States for years over extra copyright protection of databases and also in relationship to software patents. The name of that paper really should be "Everything You Wanted to Know about Database Politics, But You Didn't Know Who to Ask Who Would be Stupid Enough to be Honest With You." I have often wondered if I should edit a lot of the things I report on. I wrote this paper not only at the behest of Professor Kahin, who was a great influence, but also Professor Reichman, who does a great job about writing about database politics and database issues, but he always seems to write something like "there was intense lobbying." I decided I would say a little more than there was intense lobbying.

I wanted to briefly talk about three IP areas and, with each of these IP areas, I want to return to two themes. The first theme, which we have heard in this symposium, is it is not just about what the law is, it is about the tools of law and how they are used, misused, or not used by private players. This is very important in the scientific community's assessment of what kind of IP laws are supportable or acceptable to the community. Second, I wanted to talk about science's interest versus broader public-domain arguments. We were invited to

---

[1] See J. Hughes. 2002. "Political Economics and Harmonization: Database Protection and Information Patents," Conference on Frontiers of Ownership in the Digital Economy, June 10-11, Paris. See http://cip.umd.edu/hughesifri.doc.

say controversial things. On that topic, I am going to go against Jamie Boyle's support of what I call the togetherness principles.

Let me talk about the three IP areas: patents, copyrights, and database protection. In the area of patents, probably unquestionably in the past few years, the most cataclysmic development has been the policy following State Street Bank of allowing patents over business methods. The business method patent issue is not something the scientific community should dwell on at great length because it is not likely to affect it a great deal. Business method patents are probably not a place where the scientific community could have a great deal of influence in turning the tide. IBM, which is consistently the number one patenting entity in the United States, is against business method patents. Of course, the reason is because IBM has been patenting everything else, and they did not think to patent business methods. They suddenly found themselves outflanked by Dell Computer, who forced them to take several patents on Dell business methods.

In the patent area, the important issues that the scientific community has rightfully focused on are patenting of research tools and of express sequence tags. The scientific community needs to approach these issues and ask whether there is a problem because there has been a lot of good legal scholarship that identifies problems in the area of research tools and express sequence tags, an anticommons problem of too much propertization, too much overlapping of property rights. Recently, there was a study done by Wes Cohen of Carnegie Mellon and others that, admittedly with a small sample, concluded that there was not yet any breakdown, any anticommons problem because of IP. There are undoubtedly some increased transaction costs and there are clearly financial transfers going on, but there is not yet the kind of breakdown that Professor Reichman talked about. Why not?

One reason is there is not much enforcement, which is very important because law professors tend to focus on what the law is; whereas as activists you need to be concerned about how the law is enforced. It appears to be the case that the holders of research tool patents to date do not seem to go after universities or nonprofit research. Another issue is to what degree this patenting of research tools has benefited the university community. When you look at the statistics, it is interesting because when you take the U.S. patent pool as a whole, fairly consistently year in and year out, universities—both private and public institutions—command about 2 percent of the patents. However, in certain biomedical categories of patenting, the university numbers are much higher. In the mid-1970s in three of those categories, universities had about 8 percent of the patents. In those same categories by the mid-1990s, universities had 25 percent of the patents. So, if patent fees and licensing fees are being paid, an increasing percentage of them are being paid to universities. That does not necessarily mean more money for research. We have to be honest about that. That could just be transaction costs and that could just mean that universities are hiring more technical management officers and more patent lawyers.

There is a real issue of studying and looking at what is happening in these areas. I said that the standard that you have to take into the battlefield, if you want to do this in a principled way, is there should be no major change in the law until you can show a good empirical or theoretical case. I think that is true for major changes in the law. For smaller changes in the law, it is all right to advocate reform on the basis of your intuitions and on the basis of what you believe is the right way to tinker with the system. As James Boyle said earlier, we are experimenting with a massively important entity, operation, and aspect of our society. If you propose a major change, you might sink the whole ship, but if you propose tinkering and small amendments, you have a different way of organizing opposition procedures of the U.S. Patent and Trademark Office (PTO) or you have a slightly different way of organizing exceptions for encryption or reverse engineering. That is something that I think you should go out and advocate. On that count, I think it was very good that people forced the issue with the PTO about patenting of express sequence tags, because that forced the PTO to reevaluate the issue and in a sense back down. As a result, there is substantially less patenting and stricter patenting in this area of biomedical technology.

Let me turn to copyright issues. Here science has to decide, as a community, its place in the broader battlefield. I think that the togetherness principle might be the right strategy, but it might also be the wrong strategy. Paul Uhlir and Jerry Reichman presented about the possibility of an express research commons, separate from our general understanding of the public domain.

Let me spin out a few things here. Professor Boyle observed earlier that younger scientists may be accustomed to paying for data, just as they pay for reagents.[2] That strikes me as right. Doesn't that seem ironic to you? A younger generation of scientists is now accustomed to paying for data, just as a younger generation of college students is unaccustomed to paying for music.

Professor Boyle also raised an interesting question that I want to try to answer. He said we do not know whether fair uses are part of the public domain. I think that the principled answer, and the answer that the scientific community needs to take up, is that some fair uses are part of the public domain. The fair uses that are part of the public domain are those that are needed for a robust, democratic, civil society, and there are elements of the Berne Convention, which is the multilateral treaty that establishes legal norms for a copyright, that suggest that. Fair uses that are part of the public domain are also those uses that directly advance progress in science and the useful arts. Note that I have torn the public domain apart. There are many uses that are not part of the public domain. Making the fifth copy of the Madonna music track is not necessarily part of the public domain, and it is certainly not part of the public domain that the scientific community needs to care about.

The parts of fair uses that are critical to the public domain are those that are transformative uses—reuses of information, ideas, and expressions that advance the civic or scientific dialogue. I think that most intellectual property scholars are not troubled by traditional IP rights. They are troubled by the digital locks that are being deployed by private players and the laws that make it illegal to tamper with the digital locks.

Professor Cohen is going to talk later about digital rights management technologies.[3] The one place where the scientific community has come together against these digital rights management technologies and the Digital Millennium Copyright Act (DMCA) is the issue of encryption technology. The argument that many computer scientists have made is that the DMCA frustrates their scientific advances by creating a dark and ominous shadow over much of what they want to do in computer security research and encryption research.

I wanted to make some observations about that. The DMCA has some exceptions written into it for security testing and for encryption research. I do not know if these exceptions are adequate, but neither does the computer scientist community and neither do the most shrill voices who are saying the DMCA should be repealed. Just because someone gets a cease and desist letter from an overactive lawyer or an overstaffed corporate legal department does not mean that it is the proper interpretation of the DMCA—that they cannot engage in the particular activity they are engaging in.

Brian Kahin and I were involved at different points in drafting the DMCA. When it came to encryption research, there were some people at the table who we thought knew a great deal about it. They were the spooks. If you ever have to draft legislation with the National Security Agency and the Central Intelligence Agency, good luck. Negotiating with them was kind of like a hot or colder game. I would give them wording and they would say colder, colder. I would give them different wording and they would say warmer, warmer. But this is how the encryption exceptions received a certain amount of scrutiny when they were drafted by people who presumably know what legitimate encryption researchers need. Now there have been some bad stories, such as the attempt by the recording industry to go after Professor Felten. That is one of those stories that I call "anecdata"—these horror stories over which we try to construct theories about how something is or is not working in IP law and policy.

The problem of studying the effects of the DMCA on computer science is different because it has become so politicized. I am not sure how you do empirical research, how you find out from computer scientists whether the DMCA has impacted their activities. We recently held a convocation of IP professors, and one of them wanted to do empirical research. His first question was to ask the computer scientists how the DMCA has affected their lives. Given the politics, the polemics, and the rhetoric, I can tell you what that kind of survey is going to get you. What is needed is a very careful survey that, without mentioning IP, tries to determine if there has been a shift in research activities in the encryption community, a shift in the normal activity and the functioning of the computer security

---

[2]See Chapter 2 of these *Proceedings*, "The Genius of Intellectual Property and the Need for the Public Domain," by James Boyle and Jennifer Jenkins.

[3]See Chapter 15 of these *Proceedings*, "The Challenge of Digital Rights Management Technologies," by Julie Cohen.

community in response to fear about the DMCA. That is a situation where, as scientists or as researchers, you need to keep on your hat of neutrality and not engage in the rhetoric that the DMCA is bad.

Let me talk a little bit about database protection, which is a serious issue. Paul Uhlir, Jerry Reichman, Brian Kahin, and I, along with many others in this room, have worked on it for years. I was surprised to hear Professor Reichman say that university technology managers are imposing conditions on databases built with federal funding. If that is true, we have a problem. However, the problem is not IP law. We need to galvanize the administration so that the National Science Foundation, the National Institutes of Health, the Department of Energy, and all those federal agencies that fund these databases go after their funding and say thou shalt not impose these kinds of conditions on the data that I have paid for.

Let me talk about the European Union Database Directive. The good news is that it is not doing anything literally. Stephen Maurer, Bernt Hugenholtz, and Harlan Onsrud have done some work looking at whether there has been any growth in the commercial database industry in Europe following the implementation of the Database Directive. It is not working to date.

Now, keep my words carefully in mind. I said to date it is not working. When you ask the European Commission if the directive is serving as an incentive for the generation of commercial databases, they say "yes." Their evidence appears to be that they go around Barnes and Noble occasionally and look at database products or they count the number of litigations that they can find. The litigation all involves databases that existed before the directive came out. As such, it is not doing much on incentive structure. Scariest of all, the European Commission commissioned a survey on the effects of the Database Directive. They invited me to fill it out. Two important questions are not on the survey; have you introduced any new database products since the promulgation of the directive? Was the directive instrumental in influencing your decision to introduce any new database products?

As far as we know, the Database Directive is not doing what they thought it would do. Having said that, keep in mind your neutral position. We should not expect it to have done anything yet. It has been promulgated only for a couple of years. We know that the business community is not the most attuned to how IP law works. All you have to do is read *Rembrandts in the Attic* to understand how clueless they generally are about patent and copyright law, which has been around for centuries.

If the European Database Directive is going to have an incentive effect, it will not do so in just a couple years. So, although I am pleased that there are no results yet, I do not want to say that the directive is a failure. All we can say now is that the directive is not working. Now, the other piece of good news that I want to mention about the directive is that none of the cases has involved anyone in the scientific community. All of the cases appear to be corporate entities in the European Union slugging it out with each other. Now, that is important. I am not saying this is not a great concern to science, but it is interesting to look at, and it also makes it more viable for science to request an exception because clearly no one is concerned about what we have been doing.

In the United States, as Professor Reichman said, there has been equipoise between political forces for several years, and we are not going to get any international movement until the United States is somewhere. Both Professor Reichman and I probably agree that we need to have some modest database protection law in the United States to serve as a counterpoise to the European Union. It is just that we do not have a convincing case, empirically or in theory, for anything nearly as strong as what they have done. That is really what the scientific community ultimately should worry about.

I have preached that you can go into battle with this position—there should not be big changes in the law, unless there is a great empirical case or a great theoretical case, but I want to add that it is fine to go into battle on amending the law with your intuitions. If you are thinking about introducing something into the system that might kill off a lot of things, then you ought to hold back and take a principled position and insist that the other people who want the change, whether the change is more IP or the change is less IP, ought to prove it.

13

Legal Pressures on the Public Domain: Licensing Practices

Susan R. Poulter

Imagine, as suggested in a recent article in the *National Law Journal*,[1] that you enter a bookstore (or a library) and you find some books, unlike the ones you typically find, that have locks or snaps on the cover. There is a notice on the cover that states, "As a condition of opening (or purchasing) this book, you agree to the terms contained inside." Since your curiosity is engaged, you open the book, only to find additional terms and conditions that state, among other things,

> You agree that you will not copy any portion of this work or disseminate any information contained herein without the express written permission of the publisher and that you will not publicly criticize this work or the authors thereof. You further agree that in the event you make any commercial use of this work or develop any commercial enterprise or product from the use of this work, you will negotiate an agreement for such use with the publisher, terms of which may include but are not limited to a reasonable royalty.

"Ridiculous," you say? In the paper world this would be a startling development. In the world of electronic publishing and databases, however, such terms can already be found.

Licensing practices are indeed a significant pressure on the public domain; they have the potential to become the vehicle for restricting access to and use of public information that trumps other existing and proposed forms of protection, such as copyright and database protection.

The term "licensing" in the context of electronic publishing means contracting for various activities involving the access and use of articles or the information in the articles. It includes licensing an article for publication (which can also occur through the conveyance of copyright, as has been typical in scientific publishing) and licensing the subscriber to access and read (and print) a copy. By way of comparison, paper copies of journals, including the accompanying scientific data, are typically sold outright; the electronic forms of journals more typically are licensed, however.

There are pressures on the public domain at the level of access to articles and the underlying data and on the uses that can be made of scientific and technical information once accessed. Access to articles (and the data they contain) requires subscription, and few would question the need for a mechanism to recoup the costs of electronic publishing. The primary issues are reasonableness of the subscription price and the scope and duration of the license. A more significant issue, in my view, is whether the subscriber (user) has the right to use the data; that is,

---

[1] McManis, Charles R. "Do Not Support 'Privatizing' of Copyright Law," *The National Law Journal* 24 (Oct. 13, 1997) (online archive).

the right to extract, reanalyze, repeat, build upon, and extend the work. These rights typically have been available to any reader of paper publications, subject only to the limits of patent law. As other speakers have noted, however, the online environment facilitates technologically enforceable restrictions on access and acquiescence to contract terms that could restrict further use of the data. Moreover, these terms usually will not be negotiated in the usual sense of the word because of the unequal bargaining power of subscriber and publisher.

Will licensing provisions that restrict further use of data be upheld? Will such contracts become commonplace? For the remainder of this discussion, I will talk about the applicable legal principles and then will highlight some developments in scientific publishing that may indicate which way the wind is blowing.

THE LEGAL ENVIRONMENT

The primary legal principle applicable to licensing provisions, or at least the starting point for any analysis, is "freedom of contract." Freedom of contract is not unlimited, however, and traditionally has been subject to limitations imposed by various other legal doctrines and policies, such as preemption by federal law (including copyright and patent law), competition policy, the doctrine of unconscionability in contracts, and perhaps even First Amendment principles. Lemley, Reichman, and others have written extensively about these possibilities in limiting objectionable terms in electronic licenses.[2]

With regard to restrictions on the use of scientific data, preemption under copyright law seems the most promising because the U.S. Supreme Court held in *Feist Publishing v. Rural Telephone* that copyright does not protect facts, and protects compilations of facts only to the extent that the selection and arrangement are original.[3] Thus, copyright does not prohibit the extraction of isolated facts from a compilation, even significant portions of them, so long as the selection and arrangement are not copied (assuming it is original enough to be protected in the first place). The Supreme Court has not had occasion to decide, however, whether *Feist's* declaration that facts are free for all to use can be trumped by contract. That question has been addressed in a number of lower court decisions considering "shrinkwrap" and "click-wrap" licenses that typically accompany commercial software and other digital or electronic products.

For a time, most courts held, on contract formation grounds, that the shrinkwrap licenses were invalid and that their terms did not bind the purchaser. Because the consumer was not aware of the terms of the license before making the contract (i.e., agreeing to purchase for the stated price), the license was not part of the contract between buyer and seller. (This problem need not be an obstacle in the online publishing environment, in any event, because the subscription to an online journal or database can be conditioned on acceptance of terms made known to the subscriber at the time of the subscription, before the contract is made.)

The trend toward invalidating shrinkwrap licenses began to wane in later cases, however. In *ProCD v. Zeidenberg*, the Seventh Circuit upheld a shrinkwrap license that accompanied a CD-ROM containing a nationwide telephone directory, overcoming the contract formation issue.[4] More significantly for our purposes, Judge Easterbrook held that the license's prohibition on further copying and use of the data on the CD-ROM was enforceable, rejecting the argument that the Copyright Act preempted terms that were in conflict with the policy of copyright as expressed in *Feist*.

The *ProCD* court appeared to be concerned about the impact of the defendant's complete copying of the database on the plaintiff's investment of about $10 million and on the incentives for others to invest in electronic databases of public information. The problem with the decision, however, is that the court's reasoning suggests that other kinds of restrictions, such as a prohibition of extraction of any part of the database or reach-through provisions attempting to capture royalties on downstream uses and developments, would be upheld. The Supreme

---

[2] See, e.g., Lemley, Mark A. 1999. Beyond Preemption: The Law and Policy of Intellectual Property Licensing, 87 Cal. L. Rev. 111; and Reichman, J. H. and Jonathan A. Franklin. 1999. "Privately Legislated Intellectual Property Rights: Reconciling Freedom of Contract with Public Good Uses of Information," 147 *U. Penn. Law Rev.* 875.

[3] See *Feist Publications, Inc. v. Rural Telephone Service Co.*, 499 U.S. 340 (1991).

[4] See *ProCD v. Zeidenberg*, 86 F.3d 1447 (7th Cir. 1996).

Court has not addressed the preemption issue in the context of contractual limitations on copying and use, but in the meantime, the lower courts are tending to uphold online contracts and find that preemption does not apply.

It is unclear whether there is room for a more nuanced preemption analysis of license terms limiting further use of electronic information. The *ProCD* court applied the express preemption of Section 301 of the Copyright Act. This might be considered a narrow preemption analysis, because Section 301 requires a close comparison of the right asserted and the subject matter protected by the contract, with the subject matter and rights of copyright law. Another court might consider "Supremacy Clause" preemption, which depends on whether the state right in question interferes with the policy of the copyright law. Under this kind of preemption, the court might conclude that, because a strong policy underlying copyright law is that facts should be freely available for all to use, contracts cannot restrict their use once they have been made public (or available to the public). This kind of preemption analysis might also allow a database creator to prevent wholesale copying of a database but prohibit a restriction that would prevent extraction and use of more limited amounts of data.

Several proposed changes to the law will also impact licensing rights, generally strengthening the hand of content providers. Database protection has been considered in Congress since the mid-1990s, at least partly in response to the European Commission Database Directive. A database protection statute will likely strengthen contract claims because a federal law would expressly sanction protection of databases. Moreover, database protection is not the only legislative initiative that will strengthen the hand of electronic database providers and publishers. In 1999 the National Conference of Commissioners on Uniform State Laws proposed a new uniform law, intended for consideration by state legislatures, that would generally validate shrinkwrap or clickwrap licenses. The Uniform Computer Information Transactions Act (UCITA), although instigated by the software industry, is not limited to software licenses, but would cover online publishing.

UCITA contains a number of provisions unfriendly to consumers, but for our purposes perhaps the issue of most concern is that the law appears to endorse the concept that, even where information is marketed to a large group or to the general public (i.e., a "mass-market" license), the publisher can restrict the further dissemination and use of information provided pursuant to the license, whether or not that information qualifies for copyright protection. In other words, the commentary prepared by the drafters endorses the kind of approach endorsed in *ProCD*.

At present, UCITA has been adopted in only two states, Virginia and Maryland, and three other states have enacted legislation to prohibit the enforcement of UCITA against their citizens. UCITA also has been revised, but not to address the problem identified above. Renewed efforts to enact UCITA in other states seem likely.

ELECTRONIC PUBLICATIONS AND DATABASES IN SCIENCE

The remainder of this discussion will provide a sampling of practices and events in scientific publishing that are essentially anecdotal, although in some instances they are very significant anecdotes. I also will mention some countervailing developments, even trends, although I will leave most of that to the speakers later in the program.

At the level of access, high prices (particularly for commercial journals) are a widespread concern. But more broadly, a typical one-year subscription permits online access only during the year of the subscription. A subscription may provide access to several years of back issues; beyond that period, back issues usually are charged separately. Clearly, making back issues available online is a costly activity—the issues here are primarily ones of how electronic publishing products are packaged. It is worth noting, however, that the limited period of access means that even a long-term subscriber has no access once her subscription ends, rather than being able to retain access to those issues to which she subscribed.

A number of practices also limit the usefulness of online publications, especially for the typical researcher today who does most literature research online. In one instance, a leading scientific journal required an individual subscription for full-text access online, even where the institution had an institutional subscription. (This practice has apparently been discontinued, however, by the journal in question.) Commonly, linking to references within an online article requires subscription to the referenced journal as well. Another worrisome trend is that proprietary databases are the only source of some information developed with public funds (see the Commercial Space Act of 1998).

At the level of restrictions on further use of data, in other fields, online database providers assert the right to restrict the use of data extracted from their databases, even when that information is available elsewhere and is not separately entitled to any form of intellectual property protection. An analogous area involves the use of Material Transfer Agreements (MTAs) for the sharing of unpatented specimens or samples. MTAs commonly prohibit commercial use of the materials, or include "reach-through" provisions, requiring negotiation of licenses for downstream commercial uses.

Traditionally, professional society journals have not imposed limits on use of data. Recently, however, *Science* magazine contracted with Celera Genomics for the publication of the Venter et al. paper on the sequencing of the human genome, without making the sequence data freely available to all subscribers. Contrary to the usual practice, Celera (rather than *Science*) maintains the database of sequence data supporting the article, and readers must access the data through the Celera Web site. Access for academic, noncommercial users is free, and such users can download one megabase per week. They may not commercialize the data, but there are no reach-through provisions on developments from searching the database. Access for commercial users requires the execution of a MTA, again prohibiting commercial use, or negotiation of a separate agreement with Celera. Despite the expectation that Celera data would make their way into GenBank, the free public repository of genetic sequence data, an article that appeared as recently as August 2002 indicated that no Celera data had been deposited.

The justification for *Science's* arrangement with Celera is that it made important data available that otherwise would have been accessible only through commercial arrangements, if at all. The arrangement is precedent setting, however, and is contrary to scientific norms in most fields in that it permits the authors to "have their cake and eat it too"—garnering scientific recognition while retaining the ability to exploit the underlying data by limiting access.

The influence of the *Science*-Celera deal remains to be seen. It does, however, provide a precedent for a scientific publication to require less than full disclosure of the underlying data. It further is a precedent for agreements prohibiting commercial use of data, perhaps requiring reach-through provisions, and for restricting extraction and further publication of data. How will journals evaluate other cases where authors want similar arrangements, perhaps even academic authors who want recognition while maintaining a head start much like the situation with x-ray crystallographers in past years that engendered a great deal of criticism? Will journal editors be able to determine when the benefits of such an arrangement outweigh the negatives by making available data that otherwise would not be available? Or will such arrangements allow restrictions on data that otherwise would be published without restriction? Ultimately, will restrictions and reach-through provisions delay and inhibit worthwhile downstream uses of science and technological information?

Members of scientific professional organizations can be expected generally to be opposed to proprietary rights in data or publication with less than full disclosure, although some, especially those who perceive commercial value to the data they generate, may seek to make such arrangements themselves. The structure of scientific publishing may also facilitate such arrangements. Many journals now use supplemental, online repositories for detailed experimental procedures and data, which could easily be subjected to licenses that include use restrictions. These supplemental data repositories are not always published in the paper edition of the journal, although many journals now make them available without an online subscription. Notwithstanding their current availability, online restrictions could completely control access to and use of data.

Not all developments, however, point in the direction of greater restrictions. Authors of some journals have retained the right to use their scientific papers for their own research and educational purposes, including, in some cases, the right to post their articles online. These movements have the potential to facilitate new kinds of highly specialized, interlinked information products. A recent report from the American Association for the Advancement of Science (AAAS) supports these efforts.[5] But author-sponsored movements will likely have limited utility if the goal is widespread access. A major function of journals is to vet scientific publications for quality; unless these "entrepreneurial" databases establish similar vetting systems, they are likely to be incomplete.

---

[5]See American Association for the Advancement of Science (AAAS), 2002. *Seizing the Moment: Scientists' Authorship Rights in the Digital Age*. AAAS, Washington, D.C.

Scientific authors have also tried to influence the other side of the equation through various efforts to persuade journals to allow free access online after a period of subscription-only access, say 6 months to a year. This movement has precedent—the American Astronomical Society has all journals online, with free access after 3 years. Other initiatives include the Public Library of Science, PubMed Central, and the Budapest Open Archives Initiative. A movement by life scientists to boycott journals unwilling to agree to this concept apparently has fizzled, but the movement to unrestricted access seems to be gaining momentum, with the decision of *Science* (the leading publication of the AAAS) to allow unrestricted access after 1 year.

In summary, digital lockup is a realistic possibility. Technological and legal tools support the kinds of restrictions that can deplete the public domain. Science today generates many large data sets (e.g., gene sequences) that can be managed only in digital form. Moreover, the legal framework does not currently set limits on licensing provisions needed to preserve public domain. Member interests may restrain professional society publications and databases from imposing more onerous restrictions, although it is unlikely that there will be unanimity on this issue. Directions are even less clear for proprietary and industry-sponsored databases.

Trends are not uniformly pointing toward greater restrictions, however. Authors' rights movements and the movement toward free online access are welcome developments. The real question is how the system of technological tools and legal rules can be adjusted to strike the right balance—maximizing the data and information in the public domain to "further the progress of science and the useful arts."

14

Legal Pressures in National Security Restrictions

David Heyman

I will address the new national security restrictions and the pressures that they may impose on the public domain. The key questions to address today are what information can we create and share with the world and what information cannot be openly shared, in light of the increased threat environment? I want to start by reading to you from a government directive that attempts to answer this question in part. It was issued in December 2001 and provides regulations on publications. It says that no publication is allowed to have the following, contents that: "insult or slander other people"; "publicize pornography, gambling and violence, or instigate crimes"; "leak state secrets, endanger national security, or damage national interests"; "endanger social ethics and outstanding national cultural traditions"; "disrupt public order and undermine social stability"; or "publicize cults and superstitions."[1]

This document is from the People's Republic of China's "Regulations on Management of Publications," issued December 31, 2001. I bring it to our attention because clearly the question as to what information can and should reasonably be controlled is really open for interpretation—some of the regulations in China seem quite reasonable, some are questionable, and some are beyond the pale. What is important here is that there is a line, which at some point we cross, as you probably did while listening to me. It is a line that contemplates legitimate restrictions on traditional public-domain information. I believe if our policies go to one extreme of that line, we risk national and economic security, and if our policies go to the other extreme, we risk freedom. Our job today and in the future is to clarify that line and shine a light on it so that reasonable people can make reasonable policy.

I am going to focus on federal investments in science and technology and the contributions these investments make to creating and disseminating public-domain information and, ultimately, to improving the security and well-being of our nation.

Let us first look at the new pressures that are being put on the public domain as a result of the attacks on September 11. I think what was so shocking about the terrorist attacks was the realization that the terrorists lived among us and used our open society against us. Our experience turned out to be a bit of a perversion of Pogo's famous quote, "we have met the enemy and it is us." We felt victimized after September 11 in part because the terrorists exploited the very aspects of American society that make our country strong: its openness, easy access to information, freedom of association, ease of mobility, and right of privacy. The terrorists lived among us and used our freedoms

---

[1] See "PRC Regulations on Management of Publications," issued on December 31, 2001. Translated from the Foreign Broadcast Information Service and available online at http://www.fas.org/irp/news/2002/05/xin123101.html.

against us. The recognition that we were vulnerable to this, and the fear that this vulnerability caused, generated a wave of response by the government to limit those very attributes of an open society that were used against us—in particular, access to public-domain information. I will discuss some specific examples later in my remarks.

As a result of this experience, we are now witnessing two key components of public-domain information being constricted due to national security pressures: first, the creation of public-domain information, and second, its availability for others to use.

Let us start by looking at the creation of public-domain information. Science can provide us with the capability to acquire information about the nature of the physical world, as well as the technological alternatives that we do not presently possess. This information, in the long run, is vital to the future of the U.S. economy, our national defense, and general well-being. Science can only do this through sustained investments over the long term and through the continuous development of a talented workforce to perform research and development (R&D). Today, the United States is investing more in R&D than it ever has in the past, even with adjustments for inflation. Nonetheless, the U.S. share of total world R&D is decreasing. More research capabilities are becoming available outside the United States. Over the past 50 years, we have seen the United States go from performing more than 70 percent of the world's total R&D in dollars spent to a point where today the rest of the world performs approximately twice as much R&D as the United States.

The changing investment patterns are having an impact on where research results are being produced and where they are found. For example, since World War II, U.S. scientists have led the world in authoring scientific publications, a measure of where scientific discoveries are being made. Recently, as a result of the increased quality and volume of scientific activity in many countries, more and more articles are submitted to journals from scientists outside of the United States. In the physical sciences, in the early 1980s, U.S. publications accounted for nearly 70 percent of all articles. Today, U.S. publications have fallen to approximately 25 percent of the world's total.

The changes in the U.S. science and engineering workforce are more interesting, particularly in the context of national security. As I noted, there has been an increasing amount of research and technical resources available outside the United States and a growing internationalization of science and technology. As a result, U.S. scientists and engineers comprise a diminishing share of the total global technical workforce. Fewer U.S. scientists are pursuing physical sciences and other comparable hard sciences, and those that are pursuing them prefer academia and industry to a government career. Last, more foreign-born students are being trained as scientists and engineers in the United States, and approximately 50 percent remain in the United States to become part of the workforce. At the same time, domestically, the increased investments in the private sector and the employment opportunities that emerged as a result of the technology boom in the 1990s created a honey pot in the private sector and competed with the government to fill the valuable R&D positions.

To summarize, the creation of science-based public-domain information, in a simplified world, is a product of investment and effort. As more of the U.S. share in total world R&D diminishes, so likely does its share of the creation of the world's total public-domain information. As investments shift from civilian to military priorities, as can be anticipated with a war on terrorism, we might expect to see some crowding-out of investments that lead to public-domain information as well. And lastly, as the U.S. workforce becomes increasingly reliant on foreign scientists and engineers, the challenges for controlling transfer of technology will become greater, limitations on interactions with foreign scientists may increase, and we will likely see more restrictions on access to public-domain information. This is, in fact, what we are seeing.

The availability of public-domain information has been squeezed significantly more by recent national security developments. There is a multitude of ways in which public-domain information is made available. They include publications, Web sites, conferences, and presentations, as well as through working collaborations. Security professionals are not only concerned with what information is being published, but how it is transferred to the public and, in particular, how interactions among scientists and engineers serve as a mechanism for exchanging information.

The post-September 11/post-anthrax attacks security environment has raised concerns regarding the possible malicious use of scientific and technical information and puts greater pressures on scientific institutions to strengthen security to prevent the unintended transfer of technology to those who would harm us. We felt this pressure when we realized that the perpetrators of the September 11th attacks lived secretly in our neighborhoods, and operated freely within our open society. Some of the September 11 terrorists entered the United States on student visas, but never matriculated to the school to which the visa applied; some received pilot training in the United States. These examples raised concerns that terrorists or other enemies of the United States may seek to gain entry to the United States under acceptable and rather innocent guises, but may, in reality, seek to enter the country to acquire knowledge, skills, or technologies to mount future attacks against the United States.

At about the same time, we also saw the publication of three research papers that have generated significant alarm. First, a study published in the *Journal of Virology* described an experiment by Australian scientists to re-engineer a relative of smallpox, called mousepox, in a way that made the virus far more deadly.[2] Some have argued that if the same technique were applied successfully to smallpox, the consequences to society could be devastating. Similarly, the *Proceedings of the National Academy of Sciences* published a study by scientists at the University of Pennsylvania that provided details about how smallpox uses a protein to evade the human immune system.[3] Again, such information, critics suggest, could be quite harmful if misapplied. Lastly, in July 2002, *Science* magazine published a paper in which scientists at the State University of New York at Stony Brook described how to make poliovirus from mail order DNA.[4] The publication of that study spurred Rep. Dave Weldon, a Florida Republican, to introduce a resolution criticizing *Science* for publishing "a blueprint that could conceivably enable terrorists to inexpensively create human pathogens for release on the people of the United States."[5]

As a result of these and other developments, there are a number of growing efforts today to protect and limit access to scientific information, including efforts to restrict the activities of foreign nationals and the interactions of U.S. nationals with foreign nationals.

The Bush Administration, the U.S. Congress, and some scientific communities have adopted or are considering implementing new security measures that could dramatically shrink the availability of public-domain information. These include increased foreign student monitoring,[6] restricted access to certain technical materials or tools, expanding export controls, tightening visa requirements,[7] and limiting publications.[8] Security reforms also include efforts to limit information historically provided to or already in the public domain;[9] to expand the use of a

---

[2] See Jackson, R.J., Ramsay, A.J., Christensen, C.D., Beaton, S., Hall, D.F., Ramshaw, I.A. 2001. "Expression of mouse interleukin-4 by a recombinant ectromelia virus suppresses cytolytic lymphocyte responses and overcomes genetic resistance to mousepox," *Journal of Virology*, 75(3):1205-10, Feb.

[3] See Rosengard, A.M., Liu, Y., Nie, Z., Jimenez, R. 2002. "Variola virus immune evasion design: expression of a highly efficient inhibitor of human complement," *Proceedings of the National Academy of Sciences*, (13):8808-13, June 25.

[4] See Cello, J., Paul, A.V., Wimmer, E. 2002. "Chemical synthesis of poliovirus cDNA: generation of infectious virus in the absence of natural template," *Science*, 297(5583):1016-8, Aug. 9.

[5] See House Resolution 514. 2002. 107th Congress. Introduced by Dave Weldon, July 26.

[6] The USA Patriot Act of 2001 (Public Law No: 107-56), requires universities and "other approved educational institutions [including] any air flight school, language training school, or vocational school" to build and maintain a sizable database on its students and transmit that information to the Department of Justice, the Immigration and Naturalization Service (INS), and the Office of Homeland Security. The database system, called the Student and Exchange Visitor Information System, would automatically notify the INS of a student's failure to register or when anything goes wrong in the student's stay. Further, failure of a university to provide the information may result in the suspension of its allowance to receive foreign students (the ability to issue I-20s or visa-eligibility forms).

[7] The Patriot Act also allows for the U.S. Attorney General to detain immigrants, including legal permanent residents, for seven days merely on suspicion of being engaged in terrorism. The bill denies detained persons a trial or hearing, where the government would be required to prove that the person is, in fact, engaged in terrorist activity. Further, in September 2002, the INS implemented the initial phase of the National Security Entry-Exit System (NSEERS) at selected ports of entry. Under the NSEERS program, certain individuals will be interviewed, fingerprinted and photographed upon entry into the United States, and their fingerprints will be checked against a database of known criminals and terrorists. These individuals also must periodically confirm where they are living and what they are doing in the United States, as well as confirm their departure from the United States.

[8] Taking the first significant step toward self-regulation in this area, in February 2002 the publishers of some of the most prominent science journals in the country issued a pledge to consider the restriction of certain scientific publications in the name of security. The statement outlined the unique responsibility of authors and editors to protect the integrity of the scientific process, while acknowledging the possibility that "the potential harm of publication [of certain research may outweigh] the potential societal benefits."

[9] In October 2001, Attorney General Ashcroft revised the federal government's policy on releasing documents under the Freedom of Information Act, urging agencies to pay more heed to "institutional, commercial, and personal privacy interests." The administration wants the new Department of Homeland Security exempted from many requirements of the Freedom of Information Act. In March 2002, the President's Chief of Staff issued a memo to executive agencies requesting that they safeguard information that could reasonably be expected to assist in the development or use of Weapons of Mass Destruction, including information about current locations on nuclear materials. As a consequence, Federal agencies have removed a range of information from their websites and other public access points.

category of classification known as sensitive, unclassified information; to broaden the enforcement of a concept called deemed exports, which is the oral transfer of technology between people; broaden classification authority in the executive branch;[10] and to impose new restrictions on fundamental research.

A concern raised by all of these developments is the impact that they may have on the scientific community and, consequently, on the scientific enterprise. Additional security requirements may wittingly or unwittingly diminish the amount of scientific and technical data available in the public domain. Furthermore, requirements may slow the production of new knowledge or reduce the ability for some to publish or present their findings because of new classification concerns, which, in turn, will diminish peer recognition, career advancement, and, ultimately, morale. At that point, students may choose not to matriculate to U.S. universities, and scientists may choose to leave their government positions or reject government funding, rather than endure the environment in which they must operate. And who will replace them and who will do their work?

This scenario is not fabricated, and it is not without precedent. In fact, it is exactly what we witnessed at the Department of Energy (DOE) a few years ago. Between 1998 and 2000, the United States faced three national security crises involving the potential loss of scientific and technical information. First, a high-level congressional investigation determined that China had stolen advanced missile technology from the United States, from U.S. corporations, as well as plans for the W88, one of the nation's most sophisticated nuclear weapons. Second, a scientist at one of the DOE's premier national security laboratories was accused of giving sensitive nuclear information to China. This is the Wen Ho Lee case. Last, less than a year after the first two issues surfaced, two computer hard-drives containing classified nuclear weapons information disappeared from a DOE laboratory for over a month. These incidents spurred dramatic reforms from both the legislative and the executive branches, including the institution of numerous new security measures at DOE to protect scientific and technical information and to prevent access of foreign nationals to the labs in certain circumstances.

Concerned about the consequences of these new reforms, then Secretary of Energy William Richardson established a high-level commission led by former Deputy Secretary of Defense John Hamre to assess the new challenges that DOE faces in operating premier science institutions in the twenty-first century, while protecting and enhancing national security. An analysis by the commission, which included former FBI Director William Webster, former Deputy Director Robert Bryant, and numerous Nobel scientists, reveals that, although most reforms were well intentioned, many security reforms were misguided or misapplied and only exacerbated existing tensions between scientists and the security community, contributing to a decline in morale and, in some instances, productivity. I recommend the report, which goes into much more detail on this topic.[11]

In the end, the commission found "that DOE's policies and practices risk undermining its security and compromising its science and technology programs." Relevant to today's discussion, the commission found management dysfunction that impairs DOE's ability to fulfill its missions. In other words, good policy could be undermined by poor management.

In the area of information security, the commission found that the process for classifying information was, in fact, disciplined and explicit. However, the same could not be said for the category of sensitive, unclassified information, for which there is no usable definition at the department, no common understanding of how to control it, no meaningful way to control it that is consistent with its various levels of sensitivity, and no agreement on what significance this category has for U.S. national security. Consequently, security professionals found it difficult to design clear standards for protection, and scientists felt vulnerable to violating rules on categories that are ill defined. As a consequence, scientists and engineers began opting out of conferences and in some cases opting out of publishing.

We have to understand that heightened security is, in fact, appropriate and necessary after September 11. But we should also be deliberate and learn from the DOE's experience, or, like the DOE, we will risk undermining the very security we seek and diminish the scientific programs vital to our national security and our economy.

---

[10] Through Executive Orders issued in December 2001, and May and September 2002, the Secretary of Health and Human Services, the Administrator of the Environmental Protection Agency, and the Secretary of Agriculture were respectively granted the authority to classify information originally as "secret."

[11] See Center for Strategic and International Studies (CSIS). 2002. *Science and Security in the 21st Century: A Report to the Secretary of Energy on the Department of Energy Laboratories,* CSIS, Washington, D.C.

In conclusion, I would like to offer a few principles from the DOE experience that may be instructive on the question of limitations for public-domain information.

The first principle focuses on security. Any policies that we seek to derive or practices we wish to employ regarding safeguarding scientific and technical information must be determined by collaboration between the scientific and technical community and the intelligence and security communities. Scientists cannot be expected to be aware of all the risks they face from hostile governments and agents. At the same time, security professionals can only understand what is at stake by working with scientists. In fact, these two communities must depend on each other to do their shared job successfully.

Second, we must know what we want to protect. What is secret? We cannot make judgments on the architecture of security without first understanding the nature and conduct of science and the scientific environment in which we operate. For example, it is crucial to understand that today classified work has come to be dependent on classified science and technology, and unclassified science, in turn, has become more international and connected by digital communications. There are consequences to this in terms of whom U.S. scientists and engineers seek to collaborate with and whom they seek to employ. There are also costs if we choose to impose limitations on this methodology of work.

Third, we must know what threats we face. We must understand that we exist in a changing world of dynamic threats and national security interests. Technological advances not only serve our social goals, but they also may enable our adversaries not only to exploit our critical systems but also our key personnel. There are legitimate security concerns that we must confront.

Fourth, we must understand that absolute protection is impossible. Security is a balance of resources, which are limited, and risks, which can never be eliminated. We will have to make choices that will mitigate the risks. That means that we must understand the value of what we seek to protect, the consequences of it being compromised, and the cost of protecting it.

Fifth, security processes should minimize disruptions to scientific activity. Security procedures must strike a balance. They must be unobtrusive enough to permit scientific inquiry, but effective enough to maintain strong security.

Sixth, we should control information where there are no other cost-effective alternatives to ensuring national security. I think this is similar to what Justin Hughes proposed earlier. Finally, if information security is required, use understandable, meaningful, and workable classification systems to protect information. I think these principles represent hard decisions, which we must make to manage our growing information society and vital scientific enterprise today.

Misguided or misapplied limitations on scientific activities motivated by security concerns pose a clear threat to science and society today. Information, in the end, is the oxygen that feeds science, our economic system, and our democracy. Consequently, we must be deliberate in defining the line between what should be in the public domain and what should be restricted.

15

The Challenge of Digital Rights Management Technologies

Julie Cohen

I am going to discuss the challenge of digital rights management (DRM) technologies for the public domain. First I will address the technologies themselves, some of the functions that they can be used to implement, and some of the implications of those functions. Then, I will address the Digital Millennium Copyright Act (DMCA), a law designed to provide DRM technologies with an extra layer of legal protection, and discuss the implications of that extra layer of legal protection.

I should preface my remarks by noting that I disagree with Justin Hughes about how bad the DMCA is.[1] I come from a family of scientists and I like data, but sometimes if you wait for a lot of evidence, you have waited too long. One thing that we have seen quite clearly in this symposium is that the scientific enterprise is a very complex system that exhibits enormous path dependencies. That is in large part the message of the other speakers who have argued that making major changes in a complex system that one does not fully understand can be fraught with peril. This can be true even for what could be characterized as "tinkering" with the system if one does not understand all of the path dependencies. When I discuss DRM technologies, the DMCA, and their long-term implications for scientific research and the public domain, I will lay out some of what I think are the worst-case consequences. Not everything that I am about to describe is happening yet. Many of the technologies that I will mention are in experimental use somewhere, usually in markets for video games or digital music files rather than in markets for scientific databases. In my view, the best way to prevent the worst case consequences of DRM technologies and the DMCA is to make sure that everybody sees them coming.

DIGITAL RIGHTS MANAGEMENT TECHNOLOGIES

DRM technologies include, first, technologies that can be used to impose direct functionality restrictions on digital content. A simple example is encryption technology that restricts access to a database to those individuals or devices having the appropriate password or key, but DRM technologies also can impose more complex restrictions. For example, they can be designed to prevent users from taking particular actions with the data, or to regulate the manner in which they make take those actions. Thus, DRM technologies can prevent or limit the acts

---

[1] See Chapter 12 of these *Proceedings*, "Legal Pressures in Intellectual Property Law," by Justin Hughes.

of copying, extracting some of the data, or transferring some of the data to a different document or to a different device, such as another computer or a personal digital assistant.

Second, DRM technologies can be used to effectuate "click-wrap" contractual restrictions. It is possible to use a combination of direct functionality restrictions and click-wrap contract restrictions to produce a fairly broad range of regulation of the behavior of database users. Click-wrap restrictions might be used to implement a pay-per-use scheme that allows metered access, and possibly some copying, for a fractional fee. Alternatively, they might be used to impose narrower restrictions, such as a prohibition against disclosure of the data to the public, or against use of the data for a commercial purpose, or against use of the data to reverse engineer a computer program. One recent case brought by the New York State attorney general involved a click-wrap contract restriction prohibiting the publication of a critical review of a software package.[2]

Finally, DRM technologies can be designed to effectuate what I will call self-help, such as disabling access to the database or to some portion of the database if the system detects an attempt to engage in some sort of impermissible action, or detects unauthorized files residing on the user's computer. For example, if a copyright holder can detect unauthorized MP3 files somewhere on an individual's system, it might use that fact to disable access to a lawful subscription service. A lot of myth and legend surround the potential capabilities of self-help technologies, and I have not heard anything (yet) to suggest that these capabilities are being implemented in the scientific database realm, but certainly they are the subject of experimentation elsewhere in the market for DRM.

IMPLICATIONS OF DRM TECHNOLOGIES FOR SCIENTIFIC RESEARCH

What are some implications of these technologies for access to and use of public-domain information? First of all, direct functionality restrictions will have some obvious implications for access to and use of unprotected, uncopyrightable information. Authentication restrictions can inhibit initial access to the information, allowing access based on the user's device or domain or on possession of a valid subscriber identification. In cases involving collaborative research, this can generate added transaction costs because researchers will have to make sure that everyone with whom they want to collaborate is coming from an authorized device, domain, or subscriber identification. If the DRM restrictions prevent excerpting or extraction of the data, this will hinder research efforts that require extraction and manipulation. If the DRM software or hardware is designed to require proprietary file formats for any data that are extracted, these restrictions may cause other kinds of problems. Papers intended for publication may be subject to limits imposed by the demands of the DRM system. Direct functionality restrictions also raise the risk of loss of access to data, either because a subscription has expired or because the system has invoked self-help functionality to disable itself. Loss of access in turn raises the possibility of damage to other files or programs that may reside on the researcher's system.

Pay-per-use provisions and other click-wrap restrictions also have some important long-term implications for research. First, the pricing of some of these subscriptions can represent a significant cost. In addition, the use of click-wraps to restrict subsequent use and disclosure raises concerns about secrecy and freedom to publish.

It is worth separately highlighting some of the potential effects of DRM technologies on libraries. Libraries will have the headache of managing all of the authentication restrictions. They also will need to worry about loss of access to back issues of journals and databases when subscriptions expire. That is not how their print and microfiche collections have worked, and it is a fairly significant concern for obvious reasons. Libraries also need to be concerned about loss of control over the formats for archival storage, search, and retrieval of data. Search tools have to be able to interact with the file structure of the databases they are designed to search. If the file structures are proprietary for reasons related to the imposition of DRM, then a search engine capable of interacting with that proprietary wrapper may also be considered a proprietary tool. This in turn raises questions about who will be permitted to develop those tools, and what restrictions will be placed on their use. All of these issues are critical to libraries' mission of facilitating access to information by their user communities.

---

[2]*People v. Network Associates, Inc.*, No. 400590 (N.Y. Sup. Ct. Jan. 14, 2003).

THE DIGITAL MILLENNIUM COPYRIGHT ACT

One might respond to this "parade of horribles" by noting that nothing can be built that cannot be hacked. So far, that has been true, but Congress and the content industries have fought back. In 1998, Congress enacted the DMCA, which has four main types of provisions.

First, the DMCA has an anti-circumvention provision, which is relatively simple. It prohibits circumvention of a technological measure that effectively controls access to a protected work. Note what this provision does not say. It does not say: "Thou shalt not circumvent a technological measure that effectively protects the right of the copyright owner in a protected work, such as the right to copy the protected work." Circumvention of such copy-control measures is not prohibited. The statute prohibits only circumvention of access-control measures. But consider the different ways that a DRM system can be designed to work. Access to a database may be provided via a click-wrap system that treats every act of opening up the database and using it as a separate act of access to the database. If so, then arguably the anti-circumvention provision applies to each instance of use, even by an authorized user. So too if the DRM measure requires the application of some sort of password or key every time one wants to use the database. One cannot circumvent that password for any reason, because it is a technological measure that effectively controls access to a protected work.

Second, the DMCA contains what I will call anti-device provisions. These provisions are somewhat more complex. They prohibit manufacturing, distributing, or trafficking in a technology that meets any one of three criteria: (1) it is primarily designed or produced for circumvention of a technological protection measure; (2) it has only limited commercially-significant purpose or use other than to circumvent; or (3) it has been knowingly marketed for use in circumvention. The anti-device provisions apply both to devices for circumventing access-control measures and to devices for circumventing DRM measures that effectively protect rights of the copyright owner, such as the right to copy. Think back to the wording of anti-circumvention provision. One is not allowed to circumvent to unlawfully gain access to a work. In theory at least, nothing prohibits circumvention to get around a technology that protects a right of the copyright owner. But where are users going to get the tools that would allow them to undertake lawful acts of circumvention? The anti-device provisions exist to ensure that such tools cannot be offered on the market.

Technically, the anti-device provisions protect only DRM measures that are applied to copyrighted works. Some types of databases may not be covered by copyright. But database providers still may apply DRM measures to these databases, and if they use DRM standards that are also widely used for copyrighted works, then the anti-device provisions probably will prevent the sale of circumvention technologies anyway.

Third, the DMCA includes some exceptions to the anti-circumvention and anti-device provisions. An exception for nonprofit libraries allows some circumvention of access control technologies, but only to decide whether or not to make an acquisition of a work, not subsequent to the acquisition decision. Libraries are not exempted from the anti-device prohibitions so they cannot develop circumvention tools in any case.

As one might predict, there is a reasonably broad exemption for law enforcement, intelligence agencies, and the like to circumvent DRM measures, and to develop circumvention tools.

Another exception allows circumvention and the development of circumvention tools for the purpose of reverse engineering computer software to create interoperable software, subject to some conditions: (1) the copy of the software must have been lawfully obtained; (2) the interoperability information must not previously have been readily available; (3) the circumvention and tools are for the sole purpose of reverse engineering; and (4) the information and tools can be shared with other people only for that purpose.

The DMCA also includes an exception for "good faith encryption research," with some fairly stringent conditions on who can qualify. The research has to satisfy a criterion of necessity, and the researcher must have made a good-faith effort to obtain authorization from the copyright owner to undertake the circumvention. There are "manner" limits on the dissemination of information and circumvention tools that are similar to the ones that apply to the reverse engineering exception. The researcher can disseminate the information gained from the circumvention only in a manner that is calculated to further the research, and not in a manner likely to facilitate infringement. The "good faith encryption research" exception also includes some credentialing requirements for the researchers themselves. One thing that helps the court determine whether someone qualifies to claim the exemption is whether that person is engaged in a legitimate course of training, study, or research in computer science and therefore (by necessary implication) not just some hacker somewhere. Finally, the information that is gained from the research must be shared with the copyright owner.

There is also an exception for computer security testing, which is subject to conditions similar to those that apply to the reverse engineering and good faith encryption research exceptions. The circumvention must satisfy a sole purpose criterion and there are manner limitations on subsequent dissemination of the information gained.

Finally, the DMCA contains some provisions identifying other rights that it supposedly does not affect. These include limitations or defenses to copyright infringement, such as fair use, rights of free speech or the press, and principles governing vicarious or contributory liability for the design of electronics, computing, or telecommunications equipment.

RECENT DMCA LITIGATION

What do all of these provisions actually mean, and what are their implications for the use of public-domain and scientific and technical (S&T) data and for collaborative research? Three recent high-profile cases shed some light on these questions.

The first case, *Universal City Studios v. Reimerdes*,[3] is often referred to as the "DeCSS case." A 15-year-old Norwegian, Jon Johansen, developed a technology called DeCSS, which circumvents the Content Scrambling System (CSS) that protects the movies encoded on DVDs. According to his later court testimony, he did this to create a Linux-based DVD player; i.e., an open-source DVD player. The initial decryption program, however, was a Windows-based program, because Johansen was working from a Windows-based DVD player. Johansen shared this program, DeCSS, fairly widely. One of the organizations that ended up with DeCSS was a hacker magazine based in New York City called *2600.com*, which put the program on its Web site. *2600.com* and its principals were promptly sued by members of the Motion Picture Association of America, and were permanently enjoined from posting the program. The court also enjoined the defendants from knowingly providing links to any other Web site that provided DeCSS. Both parts of that injunction were upheld on appeal.

The second case involved Professor Ed Felten, who took the "SDMI challenge." The Strategic Digital Music Initiative (SDMI) was a project to develop a secure technology for protecting digital music files. The Recording Industry Association of America (RIAA) challenged researchers to try and crack the prototype SDMI technology. Professor Felten and his team at Princeton succeeded. Rather than submit their results confidentially to the technology provider and claim a cash prize, Felten decided to publish them and arranged to present the paper initially at a computer security conference. The RIAA notified the conference organizers and Princeton University's legal counsel of the possibility of a lawsuit if the paper was presented. Felten withdrew the paper from that conference and publicized the circumstances, generating considerable uproar within the scientific community. The RIAA immediately issued a press release stating that it did not intend to sue. Felten filed a declaratory judgment suit against the RIAA and the technology provider, challenging the lawfulness of what he claimed was a threatened suit or possible prosecution for violation of the DMCA. He also arranged to present the paper at a different conference, and did so. Subsequently, the court granted the RIAA's motion to dismiss the suit. It ruled that there was no credible threat of suit or prosecution after the other parties disclaimed intent to sue.[4]

The final case, *United States v. Elcom*,[5] was the first criminal prosecution under the DMCA. It involved a Moscow-based software firm, Elcom, which developed a technology that disabled certain DRM features of Adobe's eBook Reader software so that one could, for example, make a copy of an eBook to back it up or to transfer it to a different device. Elcom distributed its software via a Web site that was accessible in the United States. Shortly thereafter, one of its leading programmers, Dmitry Sklyarov, came to the United States to attend a software conference. Sklyarov was arrested at the airport, extradited to the Northern District of California, and

---

[3] See *Universal City Studios, Inc. v. Reimerdes*, 111 F. Supp. 2d 294 (S.D.N.Y. 2000), *aff'd sub nom. Universal City Studios, Inc. v. Corley*, 273 F.3d 429 (2d Cir. 2001).
[4] See *Felten v. Recording Industry Association of America*, No. 01-CV-2669 (D.N.J.).
[5] See *United States v. Elcom, Ltd.*, 203 F. Supp. 2d 1111 (N.D. Cal. 2002).

arraigned under the criminal provisions of the DMCA. He subsequently cut a deal securing his release in exchange for agreement to testify in the government's prosecution of Elcom, his employer. The court denied defense motions to dismiss the case on constitutional grounds and also on the ground that the court lacked personal jurisdiction over Elcom for actions taken in Russia. Ultimately, however, the jury acquitted Elcom of all charges, finding that even if Elcom had violated U.S. law, it had not done so willfully.

What do these cases tell us? First, they tell us some things about the general scope of the DMCA's prohibitions. In the DeCSS case, defendants argued that the CSS for DVDs could not possibly be the kind of technological measure protected by the statute because it was so easy to hack. Specifically, they noted that the statute refers only to "effective" technological measures, and argued that CSS was relatively ineffective. It probably will not surprise you that the court ruled this cannot possibly be what the DMCA means, because otherwise the statute would not protect very much. To be protected under the DMCA, an "effective" technological measure does not have to be hack-proof. This, though, means that given the broad language of the statute, virtually anything could qualify as the kind of technological measure that is protected by the DMCA. The statute protects any measure that requires the application of authorized information or an authorized process to gain access to the work, or that prevents or restricts the exercise of a right of the copyright owner. A simple password requirement that one could get around quite easily might qualify. The DMCA, therefore, potentially covers many kinds of DRM gateways.

Also under the heading of general scope, the cases establish that one can be liable for knowingly linking to another site that offers a circumvention tool. According to the *Reimerdes* court, the requirement of knowledge is intended to avoid First Amendment problems, and parallels the requirements for defamation liability. Here it is important to remember that knowledge can be established based on notice, and that the copyright industries are fairly diligent about sending out such notices. Even if the server hosting a circumvention tool is located outside the United States, then, a copyright owner can use notices to ensure the disruption of links that might lead U.S.-based users to the circumvention tool.

A final thing that we know about the general scope of the DMCA is prosecutors can rely on it to arrest people from other countries when they get off the plane in the United States, and that courts will uphold personal jurisdiction over those arrested. *Elcom* suggests, however, that the harshness of this rule may be mitigated in practice by the difficulty of establishing willfulness directed specifically toward U.S. law.

Second, the cases tell us some things about the relationship of the DMCA to the doctrines of fair use and contributory copyright infringement, both of which are designed to avoid overly broad infringement liability that might threaten other important public policies. The courts have concluded that there is no general fair use defense available under the DMCA. As already noted, the DMCA does have a provision stating that defenses to copyright infringement, including fair use, are not affected. The courts have reasoned, however, that a cause of action under the DMCA is not a cause of action for copyright infringement. Instead, it is a separate and distinct cause of action for circumventing DRM measures or for manufacturing or distributing a circumvention technology. Nowhere in the DMCA did Congress provide a fair-use defense to either of those causes of action. Therefore, the DMCA contains no open-ended safety valve, comparable to the fair-use doctrine, designed to avoid overly broad anti-circumvention liability.

Contributory copyright infringement is a doctrine that protects technology providers in certain circumstances. One can sue a technology provider for providing a technology that facilitates copyright infringement, but there will be no liability if the technology is capable of substantial non-infringing use. In the *Sony Betamax* case,[6] the Supreme Court held that the VCR was capable of many substantial non-infringing uses; therefore, Sony could not be held liable simply because people could also use VCRs to engage in unlawful copying. In contrast, the courts have read the DMCA's anti-device provisions to say that there is no "substantial non-infringing use" defense to a charge of manufacturing or distributing circumvention tools. Recall that the anti-device provisions cover technologies that are primarily designed for circumvention, that have only limited commercially significant use other than to circumvent, or that are knowingly marketed for circumvention. Those are very different standards than whether the technology has "substantial non-infringing" uses. Therefore, the court in the DeCSS case concluded that the

---

[6]*Sony Corp. of Am. v. Universal City Studios, Inc.*, 464 U.S. 417 (1984).

contributory infringement doctrine has been overruled by Congress in the circumvention context to the extent of any inconsistency with the new statute.

Third, the first wave of DMCA litigation highlights the narrowness of the reverse engineering and encryption research exceptions. The defendants in the DeCSS case argued that they could claim both of those exceptions. They noted that DeCSS was developed in the course of reverse engineering and for encryption research intended to produce a Linux-compatible DVD player. The court rejected this argument on the ground that the defendants, as third parties not involved in the reverse engineering process, lacked standing to invoke either exception. The clear import of the court's ruling is that disseminating information to the general public to solicit participation in a process of reverse engineering or encryption research will not shield the recipients of that information.

Fourth, the courts have uniformly rejected constitutional challenges to the DMCA. In all three cases, the challengers argued that the statute was facially overbroad and therefore violated the First Amendment because it regulated far more speech (in the form of computer code) than was necessary to achieve the legitimate purpose of preventing infringement. In *Reimerdes* and *Elcom*, the courts reasoned that defendants lacked standing to argue overbreadth on the basis that others might use the disputed technologies to make fair uses of copyrighted works. Since defendants themselves had not done so, whether the statutory prohibitions might be unconstitutional as applied to somebody else was irrelevant. This issue remains unresolved, and it is difficult to predict how the courts might rule on it. The *Felten* case, however, suggests a way for courts to avoid doing so.

The only party who clearly was making a fair use, Felten, had to surmount the initial threshold of demonstrating that there was some reasonable likelihood of suit or prosecution based on his activities. As already noted, the court did not think that any chilling effect on Felten's speech existed once the RIAA issued its press release stating that it would not sue. That seems inconsistent with some other First Amendment case law about chilling effects. For the present, however, it seems that in DMCA cases, courts will require more than a threat hastily retracted following bad publicity to demonstrate a chilling effect. To a very real extent, this ruling insulates the copyright industries from suit, and the DMCA from constitutional threat, by bona fide fair users.

The *Reimerdes* and *Elcom* courts also rejected arguments that under Article I of the Constitution, Congress lacked the power to enact the DMCA in the first place. Again, they reasoned that since defendants themselves did not seem to be making any fair uses of protected works, there was no need to consider whether there would be constitutional problems if the statute prevented others from making fair uses. The courts also opined that "horse and buggy" fair use would save the statute in any event. By this, I mean the kinds of fair uses that one can make without direct copying. For example, one can point a video camera at a DVD playing on one's computer screen, or directly transcribe by hand the contents of a protected eBook. The courts opined that the fair-use privilege does not give users the right to make the best technological kind of fair use that they could possibly want to make.

Finally, the early DMCA cases shed some light on the ways that the DMCA is likely to be deployed in the future. Whatever the differences between these three cases, the bottom line is that all of these cases have involved the anti-device provisions. They have all involved people who were charged with or, in Felten's case, threatened with charges for creating technologies that could be used by other people to circumvent access-control or copy-control measures. As I noted earlier, the main purpose of the statute seems to be making sure that actual or potential circumvention tools do not get created or disseminated. In holding that technology providers have no standing to invoke possible fair uses by others to support constitutional defenses, courts reinforce this message, and ensure that would-be fair users who are not technologically savvy are out of luck. Even though users technically retain circumvention privileges in cases involving copy controls (as opposed to access controls), they have to be able to develop the tools themselves.

IMPLICATIONS OF THE DRM/DMCA REGIME

If these first few DMCA cases hold up, what are the long-term implications of DRM plus the DMCA for research and innovation? I am going to talk briefly about four issues.

First, the new regime of DRM controls backed by law intensifies the impediments to information sharing and collaboration previously discussed. It is helpful to keep in mind that we are talking about two different kinds of information: (1) ordinary information, i.e., the content that is actually protected by the DRM technology, such as weather or fisheries data; and (2) technical information with circumvention potential, such as cryptographic systems or computer security information. For the former, if DRM restrictions apply, circumvention is either prohibited (for access controls) or impossible as a practical matter for most people. For the latter, there is very limited standing to invoke the reverse engineering, encryption research, and security testing exceptions. It seems that they can be invoked only by the person who actually does the initial act of circumvention, and the scope of dissemination of that information is quite limited. Further, some technical investigations will require permission of or sole use by the rights holder.

Second, does *Felten* mean that there is nonetheless a zone of safety for academic computer science researchers? Some have argued that Felten was never really in any danger of being sued, and therefore that other academic researchers are in no danger. I certainly agree with Justin Hughes' comment that the RIAA's initial threat to sue Felten was monumentally stupid. It is not completely clear to me, though, that no future threats remain. Academic researchers still must determine when an academic research paper containing technical information with circumvention potential is also a prohibited technology. If a research paper contains computer code, as many such papers do, I do not think that the language of the statute would clearly exempt it. For that reason, I think there is certainly a residual chill that applies to those researchers who want to put code in their papers. Some notable foreign computer scientists filed declarations in the *Elcom* case saying that they would no longer attend conferences in the United States because they were afraid of being arrested. Some of that was theater. Evaluation of threats, however, is partly subjective. If people say they are afraid, the fact that a well-trained copyright lawyer might conclude that their fears are boundless may in some cases be beside the point.

Third, it is worth noting separately that the DMCA is profoundly hostile to the open-source software community. Focusing on the language of the reverse engineering, encryption research, and computer security testing exceptions makes that crystal clear. The way that collaboration works within open-source communities is that researchers put information they have learned on the Web or on discussion lists and invite anybody who is interested to help with the project at hand. Such widespread sharing of information does not seem to fall within the kind of behavior contemplated by the exceptions. The exceptions direct courts to consider whether the information is disseminated to others in a manner calculated to further research or, alternatively, in a manner calculated to further infringement. I think that it is going to be very hard to argue that sharing information about cryptography research or reverse engineering with the open-source community as a whole, which includes anybody who wants to join it, satisfies that condition. In addition, the encryption research exception includes a credentialing provision, which directs the court to consider whether the person claiming the exception is employed or engaged in a legitimate source of study in computer science. The computer science research community is far broader than that, and this wording seems clearly to privilege a certain kind of scientific elite over others who might want to tinker with and improve software and enhance understanding of programming techniques. I think we all understand that is not the way that scientific progress historically has worked in this country. Our scientific tradition includes many people who invented pretty cool stuff in their garages. The encryption research exception seems to contemplate a very different sort of regime.

Finally, it is important to consider the potential network effects of DRM systems and standards. Recall, once again, that the universe of scientific and technical innovation is a very complex system that exhibits a lot of path dependencies. One cannot just drop DRM technology into the world of computer software and networking and expect nothing else to happen. We need to consider the effects these standards are going to have, both as initially implemented in discrete areas of the system and as they start to migrate deeper into the network.

An initial class of network effects relates to the modification of other standards and technologies to increase their interoperability with DRM systems and increase the efficacy of the larger DRM/DMCA system. All consumer electronics equipment and blank media will have to be made interoperable with these technologies. Manufacturers who decline to comply with DRM standards may be shut out of content markets; manufacturers who design their equipment and media to override or ignore DRM standards may be vulnerable to charges that they have created circumvention tools. This has implications for researchers even if you think that the most egregious instantiations of DRM will apply only to such things as music files and video games. Quite possibly, the same blank media that researchers want to use to write their research papers will be encoded with DRM protection at the behest of copyright owners who simply want to make it more difficult to copy music. In other words, the standards do not have to penetrate all the way into markets for S&T data in order to have an effect on the conduct of research and innovation.

What happens if DRM standards do gradually extend into S&T markets, or if they migrate even deeper into the various technical layers of computers and computing networks? Some people who are developing DRM systems have realized that they are relatively easy to circumvent if they are implemented in particular applications or peripheral devices, and that they would be harder to circumvent if embedded in computer operating systems and even harder to circumvent if one could wire them into hardware or embed them in the basic Internet protocol.

More widespread extension of DRM regimes will reshape the ways in which information storage, retrieval, and exchange are handled. Earlier, I raised the question of who will develop search tools that can interact with individual DRM systems. We now can extend this point to network searches more generally. If one needs a license to develop a search engine, what kinds of consequences will that have for the development of innovative search technologies? If archiving and storage become proprietary activities, will the risks of format obsolescence increase? Maybe we do not have enough data to answer these questions. It is certainly a change from the way the development of search tools has worked so far.

If DRM functionality continues to migrate deeper into the computing layer, we also may see decreased penetration of open-source systems simply because it is going to be difficult legally to create the kinds of interoperability that are necessary for open-source systems to attain greater market share. In the pre-DMCA world, if consumers wanted their DVD players or their word processing program to behave in a certain way, the open-source community could do that relatively easily. If members of that community wanted to make it happen, they would. But if the information about how to make these systems interoperate with other components of the computing platform is protected under the DMCA, achieving interoperability will be much more difficult.

CONCLUSION

It is usually easy to convince academics and researchers that the worst-case potential consequences of a phenomenon are worth studying more closely. Yet the worst-case consequences of DRM regimes and the protection given them under the DMCA are worth more than further study. The culture of scientific research is in some ways extraordinarily robust, but in other ways it is extraordinarily fragile. In particular, it is premised on a series of assumptions about the public domain, and about access to and use and sharing of information, that may soon warrant serious revision. In my view, waiting for these worst-case consequences to materialize would be a terrible mistake.

SESSION 3: POTENTIAL EFFECTS OF A DIMINISHING PUBLIC DOMAIN

16

Discussion Framework

Paul Uhlir

In the first session of this symposium, we described some of the potentially limitless possibilities for research and innovation that might ensue from using digital technologies to exploit scientific data available from the public domain as it was traditionally constituted. However, these prospects dim the moment we consider the ramifications for science from the economic, legal, and technological assaults on the public domain that are described in the Session 2 presentations. Here, we explore some of the likely negative implications of these trends for science and innovation unless science policy directly addresses these risks.

In the interests of clarity, I outline the effects of present trends on a sectoral basis, in keeping with the functional map of public-domain data flows presented previously.[1] I begin with the government's role as primary producer of such data and then consider the implications of present trends for academia and for our broader innovation system.

If a basic trend is to shift more data production and dissemination activities from government to the private sector, one should recognize at the outset that the social benefits can exceed the costs under the right set of circumstances. In principle, private database producers may operate more efficiently and attain qualitatively better results than government agencies. Positive results are especially likely when markets have formed; competition occurs; and the public interest, including the needs of the research community that was previously served by the government activity, continues to be met.

There are also numerous drawbacks associated with this trend, however, that require careful consideration. To begin with, the private data supplier will seldom be in a position to produce the same quantity and range of data as a government agency and still make a profit while charging prices that users can afford. In other words, the government agency has typically taken on the task of data production and dissemination as a public good precisely because the social need outweighs the market opportunities. Social costs begin to rise if the profit motive induces the private supplier to reduce the quantity and range of data to be produced or made available. For example, a private data producer typically markets highly refined data products to end users in relatively small quantities, whereas basic research, particularly in the observational sciences, generally requires raw or less commercially refined data in voluminous quantities. On the whole, overzealous privatization of the government's data produc-

---

[1] For information on these data flows, please refer to Chapter 1 of these *Proceedings*, "Session 1 Discussion Framework," by Paul Uhlir.

tion capabilities poses real risks for both science and innovation, because the private sector simply cannot or will not duplicate the government's public-good functions and still make a profit, not to say extract maximum rents.

Moreover, unless the private sector can demonstrably produce and distribute much the same data more effectively and with higher-quality standards than a government agency, privatization may become little more than a sham transaction. On this scenario, the would-be entrepreneur merely captures a government function and then licenses data back to a captive market at much higher prices and greatly increased restrictions on access and use. In the absence of market-induced competition, there is a very high risk of trading one monopolist with favorable policies toward science and the broader society—the government—for another monopolist driven entirely by the profit motive and the restrictions that makes necessary.

Absent a sham transaction, one cannot say a priori that any given privatization project necessarily results in a net social loss. The outcome will depend on the contracts the agency stipulates and on the steps it is willing to take to ensure continued access to data for research purposes on reasonable terms and conditions. In contrast to buying data collection services, the licensing of data and information products from the private sector raises serious questions about the types of controls the private sector places on the redistribution and uses of such data and information that the government can subsequently undertake. If the terms of the license are onerous to the government and access, use, and redistribution are substantially restricted, as they almost always are, neither the agency nor the taxpayer is well served. This is particularly true in those cases where the data that need to be collected are for a basic research function or serve a key statutory mission of the agency.

A classic example of what can go wrong was the privatization of the Landsat earth remote sensing program in the mid-1980s. Following the legislatively mandated transfer of this program to the Earth Observation Satellite (EOSAT) Company, the price per scene rose more than 1,000 percent, and significant restrictions were imposed even on nonprofit research uses. Use by both government and academic scientists plummeted, and subsequent studies showed the extent to which both basic and applied research in environmental remote sensing was set back. This experiment also failed in commercial terms, as EOSAT became unable to continue operations after a few years.

The legal and technological pressures identified in this symposium will also affect the uses that are made of government-funded data in academic and other nonprofit institutions. They will intensify the tensions that already exist between the sharing norms of science and the need to restrict access to data in pursuit of increased commercial opportunities.

Although the enhanced opportunities for commercial exploitation that new intellectual property rights (IPRs) and related developments make possible are clear, they will affect the normative behavior of the scientific community gradually and unevenly. Academics are already conflicted in this emerging new environment, and these conflicts are likely to grow. As researchers in public science, they need continued access to a scientific commons on acceptable terms, and they are expected to contribute to it in return. As members of academic institutions, however, they are increasingly under pressure to transfer research results to the private sector for gain, and they themselves may want to profit from the new commercial opportunities.

The government itself fuels these conflicts by the potentially contradictory policies that underlie its funding of research. One message reminds scientists of their duties to share and disclose data, in keeping with the traditional norms of science. The other, more recent, message delivered by the Bayh-Dole Act urges them to transfer the fruits of their research to the private sector or to otherwise exploit the intellectual property protection their research may attract.

At the moment, these conflicts are strongest where the line between basic and applied science has collapsed, and where commercial opportunities are inherent in most projects. Obvious examples are biotechnology and computer science. In the future, the enactment of a powerful IPR in collections of data might be expected to push these tensions into other areas where the lines between basic and applied research remain somewhat clearer and the pressures to commercialize research results have been less noticeable thus far. In exploring the implications of these developments for academic research, we continue to focus attention on the two distinct, but overlapping, research domains we previously characterized as "formal" and "informal."

In what we term the formal sector, science is conducted within structured research programs that establish guidelines for the production and dissemination of data. Typically, data are released to the public in connection with the publication of research results. Data may also be disclosed in connection with patent applications and supporting documentation. One should recall that, even without regard to the mounting legal and technological pressures, there are strong economic pressures that already limit the amount of data investigators are inclined to release at publication or in patent applications, there are growing delays in releasing those data as researchers consider commercialization options, and more of the data that are released come with various restrictions.

The enactment of a hybrid IPR in collections of data such as the E.C. Database Directive would introduce a disruptive new element into an already troubled academic environment. To some extent, this development would tend to erase some of the previous distinctions between the "formal" and "informal" domains. In both domains, access to data might nonetheless have to be secured by means of brokered, negotiated transactions, and this outcome is rife with implications. For present purposes, it seems clear that any database protection law, coupled with the other legal and technological measures discussed previously, will further undermine the sharing ethos and encourage the formation of a strategic trading mentality, based on self-interest, that already predominates in the informal domain.

We also predict that these pressures will necessarily tend to blur and dilute the importance of publication as the line of demarcation between a period of exclusive use in relative secrecy and ultimate dedication of data to the public. Suddenly, such a right would make it possible to publish academic research for credit and reputation while retaining ownership and control of the underlying data, which would no longer automatically lapse into the public domain. Once databases attract an exclusive property right valid against the world, the legal duty of scientists publishing research results to disclose the underlying data would depend on codified exceptions permitting use for verification and for certain "reasonable" nonprofit research and educational purposes. We recognize that this new proprietary default rule must ultimately be reconciled in practice with the disclosure obligations of the federal funding agencies. Our point is that the new default rule nonetheless places even published data outside of the public domain, and we note further that much academic research is not federally funded or is not funded in ways that waive such disclosure requirements.

Moreover, the role of academic journal publishers in this new legal environment bears consideration. At present, scientists tend to assign their copyrights to such publishers on an exclusive basis, and many of these journals now produce electronic versions, sometimes exclusive of a print version. This already complicates matters because, as discussed in Session 2, the data that traditional copyright law puts into the public domain may be fenced to a still unknown extent by the technological measures that the Digital Millennium Copyright Act reinforces. If, in addition, a database law is enacted, any data that the scientist assigns to the publisher with the article will become subject to the statutory regime. The publisher would then be in a position to control subsequent uses of the data and to make them available online under a licensed subscription or pay-per-use basis, and with additional restrictions on extraction or reuse.

Even if individual scientists are willing and able to resist the demands for exclusive assignments of both their copyrights and any new database rights, the fact remains that publication of the article in a journal will no longer automatically release the data into the public domain as before. On the contrary, unless the scientist waives the new restrictive default rule, even the data, revealed in the publication itself, will remain subject to the scientist's exclusive right of extraction and reuse, at least as formulated under the E.C. database protection model.

With or without a new statutory database right in the United States, scientists in public research also appear certain to come under increasing pressure to retain data for commercial exploitation. The research universities are already deeply committed to maximizing income under Bayh-Dole, with varying degrees of success, and they will logically extend these practices and procedures to the commercialization of databases as valuable research tools. A key question is whether they will make the commercialized data available for academic research on reasonable terms and conditions.

As with government-generated data, university efforts to commercially exploit their databases could produce net social gains under the right set of circumstances. In addition to the incentives to generate new and more refined data products that an IPR may promote, greater efforts may be made to enhance the quality and utility of selected databases than would otherwise be the case. Absent such incentives, many scientists may not take pains to organize and document their data for easy use by others, particularly outside their immediate discipline, and they may not refine their data beyond the level needed to support their own research need and related publication objectives. Legal incentives may thus stimulate the production of more refined databases, especially where markets for such products have formed.

At the same time, these new commercial opportunities tempt university administrators and academics to attenuate or modify the sharing and open-access norms of science and to circumvent obligations in this regard that the federal agencies have established. Were this to occur on a large-scale basis, the unintended harms to research could greatly exceed those we are accustomed to coping with concerning patented inventions under Bayh-Dole. The licensing of academic databases, reinforced by a codified IPR, would thus limit the quantity and quality of data heretofore available from the public domain.

At present, the primary bulwarks against such a breakdown of the sharing ethos are the formal requirements of the federal funding agencies, which in many cases continue to require that data from the research projects they fund should be transferred at some point to public repositories, or made available upon request. To avoid the negative results we envision, the agencies would have to strengthen these requirements—and their enforcement—and adapt them to the emerging high-protectionist intellectual property environment. We elaborate further on this topic in the next session. The point for now is that, absent express overrides that universities voluntarily adopt or that funding agencies impose in their research grants and contracts, the new restrictive default rules of ownership and control will automatically take effect if Congress enacts a database protection law. Indeed, they could become general practice even without such a law as the result of routine, unregulated database licensing practices.

In the informal zone, researchers are not yet ready to publish, or they are working independently on "small science" projects beyond the formal controls and requirements of a federal research program that requires open access or public deposit. This includes research funded by state governments, foundations, and the universities themselves, which leave more discretion in these matters to researchers, and by private companies, which normally require secrecy.

Much of what has been said about the effects of the new legal and technological pressures on the formal academic zone thus applies with even greater force to the informal zone because the impetus to commercialize data will encounter fewer regulatory constraints. The changing mores likely to undermine disclosure and open access in the formal zone will make it ever harder to organize cooperative networks in the less structured and more unruly informal domain.

These tendencies would predictably become more pronounced over time as more scientists became aware of the new possibilities to retain ownership and control of data, even after publication of research results. Indeed, one would logically expect that strategic behavior in the informal zone would increasingly be geared to efforts to maximize advantages from postpublication opportunities. Should this occur, academics themselves would exert pressure on the federal system that defends open access and on their universities to fall in line with the needs of commercial partners.

One can thus project a kind of cascading effect if a strong database protection right is enacted and the scientific community fails to take steps to preserve and reinforce the research commons. On this view, today's formal zone built around release of data into the public domain at publication would begin to resemble the informal zone, while that same informal zone would look more and more like the private sector. Under these circumstances, one cannot necessarily assume that the open-access policies currently supporting the formal sector would continue in force, in which case, even basic research could be adversely affected, as occurred in the United Kingdom in the 1980s.

What the new equilibrium that will result from the conflict between these privatizing and commercializing pressures on the one hand, and the traditional norms of public science on the other, will look like cannot be predicted with any degree of certainty. In previous articles, however, we have outlined the cumulative negative effects that such tendencies likely would have on scientific endeavor. For the sake of brevity, we recall them here in summary form:

• less effective domestic and international scientific collaboration, with serious impediments to the use, reuse, and transformation of factual data that are the building blocks of research;

• increased transaction costs driven by the need to enforce the new legal restrictions on data obtained from different sources, by the implementation of new administrative guidelines concerning institutional acquisitions and uses of databases, and associated legal fees;

• monopoly pricing of data and anticompetitive practices by entities that acquire market power, or by first entrants into niche markets that predominate in many research areas; and
• less data-intensive research and lost opportunity costs.

What could well be the greatest casualty are the new opportunities that digital networks provide to create virtual information commons within and across discipline-specific communities that are built around optimal access to and exchange of scientific data. To the extent that public science becomes dominated by brokered intellectual property transactions, then the resulting combination of high transaction costs, unbridled self-interest, and anticommons effects would defeat the fragile cooperative arrangements needed to create and maintain such virtual information commons and the distributed research opportunities they make possible.

Finally, to see why some critics in the United States harbor deep concerns about the long-term consequences of the E.U.'s approach, it suffices to grasp how radical a change it would introduce into the domestic system of innovation and to consider how great the risks of such change really are. Traditionally, U.S. intellectual property law has not protected investment as such, a tradition that still has constitutional underpinnings. At the same time, the national system of innovation depends on enormous flows of mostly government-generated or government-funded scientific and technical data and information upstream, which everyone is free to use, and on free competition with respect to downstream information goods.

The domestic intellectual property laws protected downstream bundles of information in two situations only: copyrightable works of art and literature and patentable inventions. However, the following conditions apply in both cases:

• these regimes require relatively large creative contributions based on free inputs of information and ideas;
• they presuppose a flow of unprotected information and data upstream; and
• they presuppose free competition with regard to the products of mere investment that are neither copyrightable nor patentable.

As previously observed, the E.C.'s Database Directive changes this approach, as would the last parallel proposal, H.R. 354, to enact strong database rights in the United States. Specifically, these *sui generis* regimes confer a strong and, in the European Union, potentially perpetual exclusive property right on the fruits of mere investment, without requiring any creative contribution. They also convert data and information—the previously unprotectible raw materials and basic inputs of the modern information economy—into the subject matter of this new exclusive property right.

The *sui generis* database regimes would thus effectuate a radical change in the economic nature and role of IPRs. Until now, the economic function of IPRs was to make markets possible where previously there existed a risk of market failure due to the public-good nature of intangible creations. Exclusive rights make embodiments of intangible public goods artificially appropriable, they create markets for those embodiments, and they make it possible to exchange payment for access to these creations.

In contrast, an exclusive IPR in the contents of databases *breaks* existing markets for downstream aggregates of information, which were formed around inputs of information largely available from the public domain. In effect, the *sui generis* database regimes create new and potentially serious barriers to entry to all existing markets for intellectual goods owing to the multiplicity of new owners of upstream information in whom they vest exclusive rights, any one of whom can hold out and all of whom can impose onerous transaction costs analogous to the problem of multimedia transactions under copyright law. This thicket of rights fosters anticommons effects, and the database laws appear to be ideal generators of this phenomenon.

Under the new *sui generis* database regime, in short, there is a built-in risk that too many owners of information inputs will impose too many costs and conditions on all the information processes we now take for granted in the information economy. At best, the costs of research and development activities might be expected to rise across the entire economy, well in excess of benefits, owing to the potential stranglehold of data suppliers on the raw materials. This stranglehold will increase with market power if databases are owned by sole-source providers. Over time, the comparative advantage from owning a large, complex database will tend progressively to elevate these barriers to entry.

Supporters of strong database protection laws and of strong contractual regimes to reinforce them believe that the benefits of private property rights are without limit, and that more is always better. They expect that these powerful legal incentives will attract huge resources into the production of electronic databases and information goods. In contrast, critics fear that an exclusive property right in noncopyrightable collections of data, coupled with the proprietors' unlimited power to impose adhesion contracts in the course of online delivery, will compromise the operations of the national system of innovation, which depend on the relatively free flow of upstream data and information. In place of the explosive production of new databases that proponents envision, opponents of a strong database right predict a steep rise in the costs across the information economy and a progressive balkanization of that economy, in which fewer knowledge goods may be produced as more tithes have to be paid to more information rent seekers along the way.

17

Fundamental Research and Education

R. Stephen Berry

In this presentation I will emphasize fundamental research and focus less on education, but I will comment on the impacts on education. First, I am going to look at this issue from the viewpoint of a scientist.

Two fundamental characteristics govern the way scientists carry out their activities. First, they depend on open access to information, because that information continually is expected to be used and to be challenged. One of the most important ways in which that information is used is in sustaining the verifiability that makes science different from virtually any other subject. It is the verifiability, which is the second characteristic, that makes scientific knowledge a firmer kind of knowledge that anything else we have. This information includes not only data in databases, but also the information found in journals and textbooks, the interpretation of data, and the concepts that underlie these.

I want to address almost exclusively information that is generated by either governments or not-for-profit institutions; I will not address proprietary information. In Session 2, we heard that information generated by science supported this way constitutes a public good. The justification for the support of that research is the production of the public good that comes from the science. A public good is one that does not diminish with use and has virtually no marginal costs for all of the users after the first user. But there is a special characteristic to scientific public goods: not only does the value of the scientific information not diminish, but it increases with its use. To satisfy the intent of the supporter of the research, society has to use that information and maximize its use, if possible, to achieve the values of the public good.

Historically, the scientific community and the publishing community in the broad sense—that is, the private publishers, the professional societies, and government through its own publications—always had a symbiotic relationship, as long as paper publishing was the sole outlet for the distribution and archiving of this information. That all changed with the Internet, which provided a faster, cheaper, and more efficient way for the scientific community to distribute and share its information. I think probably the sharpest example of that is the online e-print archive that Paul Ginsparg started in the area of high-energy physics.[1]

When that technological development happened, the relationship changed. It was no longer that comfortable symbiotic relationship; many scientists wanted to make use of the new medium. Many publishers, including some professional societies, did not want to use the Internet as a principal mode of distribution. In fact, many publishers

---

[1] See the arXiv.org e-print archive Web site at http://arxiv.org for additional information.

saw the use of electronic distribution not as a new way to provide a different kind of added value, but as a threat to the way that they make their living.

We have to keep one point in mind, which is very difficult for people outside the scientific world to realize. This issue became apparent to me when we were carrying out the *Bits of Power* study.[2] The study committee consisted of scientists, technologists, economists, and lawyers. During the first two meetings, the scientists and technologists and the lawyers and the economists were making no contact. They were talking as if they were in different worlds, but there was a key step that was a breakthrough. That was the articulation of the realization that for scientists the motivation is not the same as it is for the author of a novel. It is not making money from publication. For the scientist, the primary motivation, the currency if you will, is the propagation of ideas. This is the reason why scientists want to publish the results of their work. The scientist's primary goal is to distribute ideas and influence the thinking of others. If you use that as the basis of a value system, then an economist can slightly recast traditional economics with this other currency.

With this realization, the scientists and the lawyers and the economists on the committee were able to talk to each other in a very productive way. We simply had to find a way to establish the bridge to allow the economists to use their tools with the analog of what they normally use as the basis of evaluation. The financial monetary basis and the idea distribution basis were compatible when there was only one way to distribute the information in a storable, preservable way. In addition, the existing social and legal structure made open access via copyright and its exceptions. Protective, or restrictive, approaches changed that, or at least raised the specter. Those restrictions basically created an incompatibility, or threatened to create an incompatibility, between the way that the scientists operate and the way that the publishers operate. That incompatibility has been very difficult to explain, because people outside the scientific world usually do not understand the motivations of scientists.

The federal agencies that support the research have an interest in maintaining the distribution and archiving of the scientific information. And, of course, when a body, a law, or an activity acts to inhibit the distribution of that scientific information, then it is acting against the interests of the funding agency and acting against the interests of the national goals that justify the funding agency. That inhibition diminishes the public-good value of that information.

In extreme terms, which apply more to the case of the European Union Database Directive than to anything we have enforced in the United States right now, this thwarting of the distribution of information created in the national interest can be thought of as a theft of government property. The privatization, the inhibition of distribution, is in effect stealing from the government and putting into private hands the information that the government created for the purpose of public distribution.

People argue that scientists withhold information. However, the socially acceptable withholding of information in the scientific community is basically to allow scientists to (a) verify and establish the validity of what they are doing and (b) to be able to study, capture, and exploit their own research. So, for example, when crystallographers keep coordinates for one year, it is a way that the researcher with two graduate students in a small department can take the results of his own measurements and study them for that year. If the coordinates were published earlier, then a group of 30 could very easily do the studies much faster and publish in a few weeks something that would take the group of one faculty member and two graduate students several months to do. This is a kind of courtesy within the scientific community that is well accepted. It is a recognition, call it a soft spot or a weakness in the system, in which scientists compete with each other, and it is accepted.

There are some journals that require that data be deposited in publicly available databases. This is counter to that acceptance of the temporary withholding of information. By and large in the scientific community withholding data is really a bad thing socially. Scientists are very much looked down upon or scorned for withholding data. There is an ethic in this community that most of the time works pretty well.

Let us turn to the question of whether there could be a sound stable market for scientific information of the kind that would be captured by the legislation that has been proposed in the United States or by the European Union Database Directive. The value of small bits of scientific information is uncertain. The uncertainty of the

---

[2]See National Research Council. 1997. *Bits of Power: Issues in Global Access to Scientific Data*, National Academy Press, Washington, D.C.

value of scientific data diminishes as it is aggregated more and more. For example, all the data generated by the research supported by the mathematical and physical sciences division of the National Science Foundation have a high value. But what part of that body of data contributes the high value is very unpredictable. The value actually may not be achieved for a number of years.

One of the difficulties we have to contend with is that, although the aggregate data have a very high value, the way we decide what research to do or the way we fund the production of that information is at a much more desegregated basis. So the value of the data produced by the programs supported by one program officer at the National Science Foundation is very uncertain. The value of the data produced by one division of the National Science Foundation is somewhat less uncertain, because it is aggregated. But the funding is not done on the aggregated basis, the funding is done on a very disaggregated basis. Consequently, because of the high risk associated with each decision in the funding process, the result is that we cannot establish the value of the information produced by any one research project or even one program officer's set of projects. This is one of the ways that Congress justifies the relatively low support for the kinds of research that the United States supports, the National Institutes of Health currently notwithstanding. Real venture research is particularly unlikely to be adequately funded until it has proved itself.

What will privatizing do in this pinched market? It essentially will price the basic science community out of buying the information it needs. Or, alternatively, the scientific community may very well find its own ways to sustain itself, with its own new ways to distribute information outside the commercial market. The scientist does not have to publish in the existing journals, he does not have to deposit his data on a privatized basis. He can find his own pathways to do it.

To see how this is a plausible course, we can just look at the fact that scientists do have this other motivation—to maximize the distribution of information. The basic science community may find its own way to provide the information in a public domain, or some other open-access mechanism, outside the commercial publishing community. We have existing models for pathways that the scientific community can create for itself, such as the ArXiv.org and the Protein Data Base.

Professional societies represent a range of models that go all the way from astronomers, mathematicians, and physicists that have moved very much toward open access and public domain all the way to the other pole, to the American Chemical Society, which basically sees its publications as the principal source of its own support and therefore is very protective of its publications. One thing that we have not seen yet, and I think we will, is professional societies examining other models to support their publishing activities. The organizations that have considered it necessary to support themselves through publication have not yet started to look at other possible ways of doing this, but I think that we can expect to see that in the next four or five years.

There is a question now of who will pay for the distribution of scientific information. We have heard in this symposium that if a truly competitive market that would establish suitable pricing would do it, then that would be fine. If basic scientific information cannot be managed in a stable way by a competitive market, then society faces a choice: Which is more important to the society, the sustenance of scientific enterprise or the sustenance of the private information business? We have evidence from the fact that the federal government is a very important, key supporter of basic research and that we place a high value on the maintenance of science.

What is the responsibility of the private sector? If we look back at the publishing business during the 1960s and 1970s, when there was a lot of money for science and a lot of money for highly specialized journals, libraries were able to pay for the subscriptions on specialized journals. Publishing all kinds of scientific journals was a reasonable and profitable thing for publishers. However, a responsible publisher must monitor the profitability of every one of its ventures.

We have heard about the number of subscriptions that are being dropped by the university libraries that provide the principal sustenance for scientific publishers. As a result of the decreasing subscriptions, publishers must determine whether to discontinue these journals. Publishers have been very reluctant to take the responsibility to decide whether continuing to publish high-priced scientific journals is profitable. I challenge the publishers to examine one by one the specialized journals that they publish and decide whether they should continue to do so. I think it is a business decision that is hard to face up to, because it has been very profitable. But it is not clear that it is going to continue to be profitable. If the publishers decide it is not, then they should drop the journals. The scientific community simply will have to find other alternatives to distribute its information, and we have some models for that to happen.

The public funder of research has some responsibilities. We talked about the costs of publication, but these costs along with those of collecting and distributing the information are far lower than the costs of the research. When we think of the observational sciences, I include the gathering of meteorological and astronomical data as part of the research, rather than as part of the publication process. This research done for public good is valueless unless the results are distributed. As such, the supporter of the research carries the responsibility to see that there is some mechanism to distribute the information. If the market mechanism does not do it, then the publisher of the information must be some institution or some mechanism supported by the supporter of the research. That small added cost for getting the information out has to be included.

Let us turn now to education. Education has thrived on access to scientific information through fair use for many years, and we will count on that in the future. But there is a problem that I will not discuss in much detail about the use of online and distance education, and the vehicles that are used for this. Are these going to become captured, privatized, and turned into the kinds of instruments that are not available for fair use? This is one of the new problems that education faces. As you know, there are open-source materials available, as well as commercially marketed counterparts.

One effect on education is already apparent, which is the impact of the nondisclosure constraints in some university-industry collaborations. Some of these collaborations have nondisclosure restraints that literally prohibit graduate students from one research group talking to the graduate students in another group about their work. This is an erosion of the environment in which we want our graduate students to be trained. This is a very serious surrender of principles of education to essentially gain a fast buck.

I think it is very disturbing that in much of our discussion even at this symposium, we have talked about universities as though their primary function is turning out commercially useful research. The primary purpose of a university, the primary product of a university, is educated students, and we must never lose sight of that. We must never surrender the mechanisms that produce truly educated students for secondary purposes such as commercially productive research. This is a very important perspective that we have to retain.

Let me go back now to the online education issue, which will lead me to a final perspective. In the case of online education where we have both models, open source and commercial, why not let them compete? Let us do the experiment and see whether the commercial products are the ones that people want to use, or the open-source ones, or both. We may very well have two kinds of users in the long run.

What are the next steps? In education, in scientific data, we are not at a stage where we have a clear-cut course ahead of us; we are going through a period of adaptation. We do not know what will be best. The only sensible thing for us to do is to try the different alternatives and see what works where. The worst thing would be to follow a restrictive course through legislation. The most productive course we could take is a permissive one to allow the different modes of activity to compete with each other. The Digital Millennium Copyright Act in this sense is going in exactly the wrong direction, because it is an inhibiting, rather than a permissive legislation. We need legislation that encourages the competition between different methods and allows us to try different options and see what works where.

I hope that we can recognize and adjust to that before we reach a crisis in which, for example, the scientific community strangles. My own personal hope, optimistic and naive as it may be, is that if we do face these restrictive forms of legislation, then the scientific community will be inventive enough to find its own way to solve its problems and sustain itself independent of those who insist on capturing the real estate and listing databases at the cost of whatever the scientific community might have to pay.

18

Conflicting International Public Sector Information Policies and their Effects on the Public Domain and the Economy

Peter Weiss[1]

It is regrettably a well kept secret in Washington that the open and unrestricted policies regarding dissemination of U.S. federal government-produced, taxpayer-funded information are not based merely on abstract notions of government transparency or support for the scientific endeavor, but are based on a fundamental economic concept—information and data paid for by the taxpayers are an important input to our overall economy. The large quantities and varieties of taxpayer-funded information have been demonstrated to be important inputs in a number of industries. Economic research is somewhat sparse on this, but it is clear that the information retrieval and database industries are highly dependent on the open and unrestricted availability of government information. For example, the industry grew from $4 billion in 1994 to an expected $10 billion in 2002, and the number of database vendors grew from 900 in 1991 to 2,400 vendors in 1999. These numbers show that the Internet revolution and the ability to use government information as an input to value-added commercial products has been a significant economic boon.

Focusing on one particular sector of information, meteorological and related environmental information, we know that the weather and climate impact about a third of the gross domestic product of the United States, about $3 trillion. Many industries are weather sensitive including construction, agriculture, energy, and tourism. That has resulted in two interesting phenomena. The United States has a large and growing commercial meteorological industry. You have probably heard of the Weather Channel, which is globally unique in its size and scope. The Weather Channel can exist only because of the federal government's policy of open and unrestricted access to the taxpayer-funded meteorological information, model outputs, satellite data, surface observations, oceanic observations, and so on. There is no Weather Channel in Europe, and there are reasons for that which we will explore later.

In addition, the financial community has learned that it can assist weather-sensitive industries to hedge their risks. So, for example, if you are a natural gas marketer in the Midwest, you make more money in a cold winter and less money in a warm winter. If you run a resort on the Florida coast, you make more money when it is warm and sunny. If you are a ski resort operator, you make more money when it is cold and snowy. Your fortunes vary, depending on the conditions that occur in any given year. The financial markets now can help you hedge those risks through financial instruments commonly called derivatives, which act as insurance policies. That industry has boomed in just five years and is now a nearly $12 billion industry. The reason it can prosper is because of the

---

[1] The views expressed are those of the author and do not necessarily represent those of the National Oceanic and Atmospheric Administration/National Weather Service.

fact that all the meteorological information, current and historic, gathered and generated by the U.S. government is openly and unrestrictedly available to the financial community to provide these specialized services.

In the United States we hold as self-evident the truth that taxpayer-funded information belongs to the taxpayers. However, that truth is not broadly accepted worldwide. Indeed, the United States stands close to alone in following and advocating an open and unrestricted data policy. Some notable exceptions are Japan, Australia, New Zealand, and the Netherlands. Rather, much of the world, for example Great Britain, France, and Germany, treats its government information not as a public good, but rather as a private revenue-generating mechanism to supplement or offset agency appropriations.

It has only been in the past few years that researchers and economists have been starting to think about the economic effects of open-access policies. One seminal study was funded by the European Commission.[2] For the purposes of this discussion, remember that the European Union economy and the American economy are about the same size. The European Commission study found that the United States spends twice as much money in creating public-sector information than the Europeans do in total, but the economic value to society in terms of job creation, wealth creation, and taxes is a factor of five larger in the United States than it is in Europe.

The United States follows a policy that encourages, even sometimes forces, federal agencies actively to disseminate that information to all comers, so that we can have, for example, database industries, more robust publishing industries, commercial meteorology firms like the Weather Channel, or weather risk management firms that are selling new forms of insurance. European government restrictions on dissemination and use of public-sector information are thwarting the kind of economic development in these sectors that we have already seen in the United States.

For example, in the United States the commercial meteorology sector totals approximately half a billion dollars annually. It includes about 400 companies employing about 4,000 people. In Europe, the commercial meteorology sector is a factor of ten smaller. Again, the European economy and the American economy are of approximately equal size.

Why does this phenomenon exist? I claim it exists because European government agencies often restrict taxpayer-funded information for shortsighted reasons. An even more telling economic statistic illustrates this issue. The value of contracts issued by the weather risk management industry over the four years ending in 2001 was over $7 billion; it is now nearly $12 billion, adding nearly $5 billion in notional value in one year alone. By contrast, according to research done by the Weather Risk Management Association and PriceWaterhouse Cooper, the European weather risk management market is $720 million over the same five-year period.[3] Again, the reason for this is the difference in the public information policies of the United States versus those of Europe. European government agencies often assert copyright as well as the *sui generis* database protection right on taxpayer-funded information. In the United States, they do neither.

A specific example is illustrative of this phenomenon. A particular firm requested the entire historic record of meteorological observations in the United States, from 1948 on, from the National Oceanic and Atmospheric Administration's (NOAA) National Climate Data Center (NCDC). Following the policies of the Paperwork Reduction Act of 1995 and Office of Management and Budget (OMB) Circular A-130, NCDC burned a stack of CDs for them, 15 gigabytes of data, and charged a little over $4,000, which covered dissemination costs, including time, effort, labor, burning the CDs, postage, etc. This same research firm requested analogous data from the German government, their entire postwar meteorological record. The German government quoted $1.5 million. The volume of the data is significantly smaller, because it is one country in Europe versus the entire United States. They also quoted 4,000 German marks, which now would be $2,500, for the historical record of only one observing station in Germany. The United States has well over a thousand, the German government fewer than 200. The interesting thing about this example is that, because this firm could not afford the German data, it did

---

[2]PIRA International. 2000. *Commercial Exploitation of Europe's Public Sector Information*. Final Report for the European Commission, Directorate General for the Information Society, PIRA International, LTD, University of East Anglia and KnowledgeView, Ltd.

[3]See also Mr. Weiss' powerpoint presentation from the symposium at http://www7.nationalacademies.org/biso/STIsymposium_Weiss.ppt.

without. So not only did the German weather service quote an astronomical price for their data, but they did not make a pfennig off the data because the research firm could not afford them.

What is wrong with this picture? An example relating to basic scientific research paints it starkly. There is a team at the India Institute of Technology in New Delhi attempting to develop a method to predict monsoons by comparing a long series of climate model output data with the record of actual observations of the monsoons over the same time period. Whether the onset, strength, and duration of the monsoons can be predicted with any skill is a fundamental research question. The team has access to all of the U.S. model reanalysis data for a 30-year period, essentially for free from our NCDC. The European Center for Medium-Range Weather Forecasts, which is the equivalent to what we do here at NOAA, quoted the researchers a price they could not afford. The research team asked if they could get free access to use the data for this basic scientific research purpose potentially affecting the lives of over a billion people annually. They were refused.

There is now an emerging realization in Europe about the benefits of open access to public-sector information. The European Commission commissioned that seminal study previously described. They have recently released two very interesting documents, one from the Directorate General for the Information Society on a more open public-sector information policy and one from the Directorate General on the Environment espousing an open policy for environmental data.[4]

A draft directive on the "Re-use and commercial exploitation of public sector documents" is quite broad, encompassing most public-sector data and information not otherwise protected due to privacy or security considerations. It urges more transparency in the pricing practices of member states' agencies, but does not tackle the issue of dissemination cost pricing versus cost recovery pricing[5] or the question of the propriety of restrictive terms intended to control downstream uses of the information. Although weak in these areas, the draft directive does seem to have significant support both at the commission level and, perhaps more significantly, in the European Parliament.

A draft directive on public access to environmental information could have a more significant short-term impact on European agency practices. It contains a strong definition of covered environmental information, which extends to most information about the environment, including meteorological data quite specifically. The definition is significantly more specific than the 1990 environmental information directive it is intended to replace, which European meteorological services have construed as being limited to information relevant to environmental regulatory enforcement and not to meteorological, climatological, or other data that merely describe the state of the environment. Most importantly, it sets a cost of dissemination standard for the pricing of this information, which would preclude cost recovery pricing for data. It too is essentially silent regarding restrictions on downstream use. This draft environmental information directive is also garnering significant support in the European Parliament. Because it would replace an existing directive, it may be adopted more promptly than the draft public-sector information directive, which is new.

It is unclear, given the political realities and institutional interests in Europe, what the practical effects these documents will have, should they be adopted. But they certainly are a step in a forward-looking direction.

At the national level, the Netherlands stands out as having adopted a policy for taxpayer-funded public-sector information that is very similar to the U.S. Paperwork Reduction Act and OMB Circular A-130. Indeed, much of the recent growth in the European weather risk management sector is attributable to a large group of contracts issued in the Netherlands to ensure their construction industry from weather-related risk. The reason it exists is because the Dutch government has adopted an open, unrestricted policy with regard to its historic meteorological

---

[4]Commission of the European Communities. 2000. "Proposal for a Directive of the European Parliament and of the Council on the re-use and commercial exploitation of public sector documents." COM207. Brussels, July 5. Council of the European Union. 2002. "Common position adopted by the Council on 28 January 2002 with a view to the adoption of a Directive of the European Parliament and of the Council on public access to environmental information and repealing Council Directive 90/313/EEC." 11878/1/01 Rev 1. Brussels, January 29.

[5]Office of Management and Budget Circular A130 Section 8(a)(1) defines cost pricing versus cost recovery pricing: "Agencies should . . . (c) Set user charges for information dissemination products at a level sufficient to recover the cost of dissemination but no higher. They shall exclude from calculation of the charges costs associated with original collection and processing of the information." OMB. 1996. Circular No. A-130, "Management of Federal Information Resources," 61 Federal Register 6428, February 20 at http://www.whitehouse.gov/OMB/circulars/a130/a130.html.

record and current observations. So it should not be surprising that the largest commercial meteorology firm in Europe also happens to be Dutch. By contrast, the weather risk management market and the commercial meteorological sector are relatively much smaller in Great Britain, France, and Germany.

Recent economic research, most of it European, reviewed in my paper "Borders in Cyberspace,"[6] leads to some general conclusions. First, cost recovery is not the best approach to maximizing the economic value of public-sector information to society as a whole, not even from the viewpoint of government finances. Again, for example, the German weather service did not make $1.5 million by selling its historic record because the research firm could not afford it. Second, prosperity effects are maximized when data are sold at marginal cost. Direct government funding and free provision to all are favored with their contribution to national welfare maximized at the point where marginal benefits equal marginal costs. That may sound like economists' rhetoric, but the recent research suggests it is true.

In the area of atmospheric sciences, as I said, there is relatively little commercial meteorology or weather risk management activity in Europe because most European governments do not have open-access policies, resulting in data not being readily, economically, and efficiently available. Because the size of the European and U.S. economies are approximately the same, there is no reason for the European market not to grow to U.S. size with the accompanying revenue generation and job growth. A significant contributor to these disparities is a difference in information policies between the United States and Europe.

Luckily there is a slowly emerging recognition in Europe that open access to government information is critical to the information society, environmental protection, and economic growth. However, the slowly growing trend toward more liberal policies faces opposition from government agencies themselves. For example, the German Parliament recently rejected a modest Freedom of Information Act. The political argument on which it was rejected was that the public has no particular right to know about the internal workings of the government. Great Britain enacted its first Freedom of Information Act in 2000. According to my colleagues in the British press, it has many loopholes, but at least they are moving in a positive direction.

This concept of government commercialization and the idea of the "entrepreneurial bureaucrat"—which I claim is an oxymoron—do not succeed in the face of economic realities and under open competition policies. My paper documents a number of instances of anticompetitive practices by European government agencies.

In sum, the research to date strongly suggests that open government information policies foster significant but not easily quantifiable economic benefits to society. Hence, the necessary impetus for adopting open information policies worldwide may turn on further economic research to better quantify the benefits of open and unrestricted public-sector data. That economic research should prove relevant not only to the question of governmental policies, but also to the larger questions about the value of the so-called "public domain" to society over all.

---

[6]See Peter Weiss. 2002. "Borders in Cyberspace: Conflicting Public Sector Information Policies and their Economic Impacts-Summary Report," at http://weather.gov/sp/Borders_report.pdf.

19

Potential Effects of a Diminishing Public Domain in Biomedical Research Data

Stephen Hilgartner

In Session 2, my colleague Sherry Brandt-Rauf presented some of the findings of our studies of data access practices among biomedical researchers.[1] I envy her ability to talk about data that we actually collected, as opposed to my charge, which is to talk speculatively about what might happen if the public domain were to diminish in the biomedical area. I can only present guesses and conjectures about what might in fact occur.

The central question that I want to examine is what might happen to research systems if the public domain diminishes. I will confine my attention to "small science" biomedical research, which does not include areas like large clinical trials, which arguably are a form of big science, given their many collaborators and complex organizational structures (e.g., GUSTO III Investigators, 1997). I will focus on areas such as molecular biology, crystallography, structural biology, and cell biology. In these fields, academic research is usually conducted on a benchtop scale by small research groups, and there are many independent laboratories working and often directly competing. In these particular scientific cultures, it is understood that the scientist who runs the laboratory—the "lab head" as he or she is called—is the only person who can speak for the lab. Other people, such as postdocs, can only speak with special authorization from that person (Knorr Cetina, 1999). So this kind of science features a culture of autonomous, independent, highly entrepreneurial operators who work to build a research enterprise, produce findings, get grants, keep a lab going, and so forth. I want to speculate about what kinds of changes you might expect in this cultural setting in science, based on what we know from ethnographic studies about the social practices that regulate access to data in these areas of science.

Although Sherry Brandt-Rauf described some of our work on how scientists control access to data and resources, there are several points I want to underline about data access practices. First, it is very important to recognize that these practices are specific to particular research communities. Many scientists tend to talk about all science as if it were uniform in the ways it handles data access, without recognizing the diversity of scientific cultures. Thus, it is common to observe that "all scientists want to publish," as if this were a universal truth about scientists. Indeed, there is no doubt that this statement is true at a very general level in all areas of academic science, as well as in some industrial contexts (Hicks, 1995). But the details of how publication is managed—what constitutes "enough" for a paper, which data are "ready" to be published when, who decides, and how strategic concerns about competition are addressed—vary tremendously across different scientific fields. Scientists do not

---

[1] See Chapter 9 of these *Proceedings*, "The Role, Value, and Limits of S&T Data and Information in the Public Domain on Biomedical Research," by Sherry Brandt-Rauf.

simply publish everything that they produce; they engage in strategic maneuvers about who is going to get access to what data and materials under what terms and conditions. Publication is only one move (albeit an extremely important one) in an extremely complex process. The practices used to regulate access to data are quite different in molecular biology, as opposed to high-energy physics, as opposed to a large clinical trial. There are different expectations and rules about control over the flow of data in those settings. As a result, any analysis of how changes in the public domain might affect science must focus on particular research communities, not science as a whole. When I refer to scientists, I am referring to researchers working in molecular biology and other benchtop biomedical fields that exhibit similar cultures.

DATA ACCESS PRACTICES

Brandt-Rauf and I set out to create a theoretical framework and analytic method for comparing data access practices across diverse scientific fields. We concluded that such a framework must treat the category "data" as problematic; that is, one cannot focus on what the scientists themselves in a particular area regard as "data," as if their notion of data were unambiguous and universal to all fields, but instead to consider the full range of forms of data and heterogeneous resources that researchers produce and use (Hilgartner and Brandt-Rauf, 1994).

In molecular biology, these data and resources include all sorts of written inscriptions (such as sequence data) and biomaterials. They also include instruments, software, techniques, and a variety of "intermediate results." In the laboratory, these entities are woven together into complicated assemblages. An isolated, single biological material sitting alone in a test tube is a useless thing; to be scientifically meaningful, it must be linked using labels and other inscriptions to the source of the sample and its particular characteristics. Moreover, to use the material, one needs a laboratory equipped with an appropriate configuration of people, techniques, instruments, and so forth. As scientific work proceeds, materials and inscriptions are processed and reprocessed, so these assemblages continuously evolve, producing new data and materials (Latour and Woolgar, 1979). Many of the items found in a laboratory can be found in any laboratory, but some of the items—especially those toward the "leading edges" of these evolving assemblages—are available only in a few places, or perhaps only in one place. These scarce and unique items can convey a significant competitive edge. For example, the laboratory that first develops a useful new technique, the researcher who collects a particularly interesting set of DNA samples, and the creator of a powerful new algorithm all end up controlling strategically important resources. They can enter into negotiations about collaborations and other exchanges from a strong position, owing to the value and scarcity of the resource.

In small-scale biomedical research, with its many independent operators, a dynamic, invisible economy exists below the radar screen of what looking at the published literature reveals. There is a huge range of transactions going on all the time. Scientists have to decide whether to publish a result immediately or delay publication until an even better result is achieved. In many areas, such as gene hunting, several research groups may be racing to reach the same goal, and an early publication from one group may help competing groups to catch up (Hilgartner, 1997). Given such strategic considerations, scientists have to decide whether to publish right away, or to delay publication, or to provide information on a limited basis to specially targeted audiences. Often, they work to negotiate agreements with the heads of other academic laboratories, or perhaps with commercial organizations. Many of these exchanges entail at least temporary restrictions on publication. As scientists work to build collaborations, they seek to avoid arrangements that will cause them to become merely the provider of a "service," as molecular biologists put it, to another lab without benefiting themselves (Knorr Cetina, 1999). Sometimes researchers provide these services expecting a quid pro quo later. Sometimes they provide them out of the goodness of their hearts. Sometimes they provide them because funding agencies or other policy makers encourage them to do so (Hilgartner, 1998). But complex negotiations, replete with strategic gamesmanship and uncertainty, are routine in small-scale biomedical research.

PUBLIC DOMAINS

Having briefly characterized the strategic role of data and the wide variety of transactions that surround data and resources, it is finally time to turn to my main question: What would happen to this area of science if the public domain were to diminish? To address this question, it is first necessary to consider how the public domain fits into this research area. The public domain is a complex concept, and it is important to recognize that scientists may not think about this concept in precisely the same way that lawyers and legal scholars do. Legal categories permeate scientists' consciousness, but not in the systematic, formalized ways that one might find in a law review article. For this reason, when scientists talk about "the public domain," the concept that they have in mind may not neatly map onto a formal, legal definition. When a scientist asks whether a resource is in the public domain—or, more colloquially, "is that public?"—what they mean is something like "Can I get it? Can I use it? What do I have to do to get it and what encumbrances will restrict my use of it?" In other words, the central issues are usually availability and the terms of access.

Legal ownership is only one of many things that constitute availability and shape the terms of access. So scientists deal less with the public domain—if we construe that as a legal category produced in court decisions, statutes, briefs, and formal legal negotiations—than with resources that are more, and sometimes less, "available." Shifting from a legal concept of the public domain to this more pragmatic concept centered on access directs our attention not only to formal ownership, but also to the practical difficulties of obtaining data and resources. When molecular biologists refer to some data and resources as "public" they typically mean that they are readily available to any scientist. I will refer to data that are public in this sense as "public resources." Important public resources are found in many domains: from scientific literatures, to Internet-accessible databases, to biomaterials repositories, to stock centers that house strains of organisms (Fujimura, 1996; Kohler, 1994).

Scientists also may regard instruments as public resources if they are available at reasonable prices on open markets. In contrast, some instruments are not public resources. For example, access to beam time on a synchrotron may be allocated by peer review (White-DePace et al., 1992). Similarly, an instrument that is not yet commercially available might be offered to selected scientists for beta testing through a special arrangement that provides early access: "We have this new instrument," says the firm, "and you want to try it out. Well, you get to use it first, but let us know how you like it, and if you like it, tell your friends and cite our product in your published work." The point is that so long as the instrument is for sale (at a reasonable price), scientists often describe it as "public," even though the instrument in fact is probably someone's intellectual property.

As the above discussion suggests, scientists do not deal with an abstract public domain; they interact with diverse public domains, including open literatures, open databases, open materials repositories, and open markets. The plural term—public domains—is important here, both to emphasize their diversity and to underline how these public domains are not coterminous with abstract definitions of the public domain. These public domains are what we should consider when thinking about the effects on scientific research of increasing privatization and restriction of domains that were once public.

EFFECTS OF DIMINISHING PUBLIC DOMAINS

Before launching into a speculative discussion of the possible effects of diminishing public domains, one must ask a crucial question: Can public domains really diminish? One might be forgiven for suspecting that they cannot. After all, the scientific literature continues to expand rapidly, and biomedical science is experiencing an unprecedented deluge of biomolecular data (Lenoir, 1999). For example, the volume of DNA sequence information available in public databases has been increasing exponentially, and it probably will continue to grow rapidly for awhile. (Of course, at the same time, we know that the amount of sequence data available in private databases is also growing, although it is harder to estimate how rapidly, because such information is private.)

Even given an expanding literature and an explosion of data, public domains clearly can diminish in at least six ways. First, absolute reductions in public domains occur when particular items are removed from them for various reasons. Second, items that were previously conceived of as things to be shared openly among scientists can be redefined as proprietary (Mackenzie et al., 1990). Third, there can be delays of the release of information into public domains. We know that such delays occur with some regularity in such high-impact fields as genetics, and this constitutes at least a short-term limitation of the scope of such public domains as the scientific literature (Hilgartner, 1997; Campbell et al., 2002). Fourth, items in public domains can have new encumbrances attached to them. These encumbrances might include publication restrictions, reach-through licenses and similar mechanisms, or just the transaction costs of negotiating access. Fifth, some items available for purchase can be acquired only at high prices. Finally, the relative size of public domains versus private domains can diminish, even when both public and private domains are growing. Such reductions in the relative size of public domains arguably constitute a form of diminishment. In short, there are a variety of ways that public domains can remain "public" in the sense that I have described, but at the same time diminish.

There are three different orders of effects that you might expect if public domains diminish. The first order includes direct effects on the transactions that drive this fast-moving world of exchanges among biomedical scientists. Second-order effects involve changes in research communities and cultures and how they manage data access, such as a shift toward more restrictive practices. Third-order effects include the ways that diminishing public domains might alter the position of science in the wider polity.

Let us turn initially to direct effects on transactions. As a starting point for considering these effects, imagine a small academic laboratory that engages in several kinds of transactions. It obtains some inputs for its research from public domains; it releases some of its outputs to public domains, such as the literature; and it gets some inputs from (and deploys some outputs in) restricted-access transactions. "I give you this, you give me that, maybe this deal is more to my advantage than to yours, but there will be another exchange later and that one will work out the other way." Importantly, whenever this laboratory acquires an entity from a public domain, it immediately begins to process it, manipulate it, and combine it with other data and materials. Through this processing the laboratory reprivatizes the entity—or, more precisely, produces new entities that end up under its exclusive control. Put otherwise, laboratories not only release material into public domains, but they also continually incorporate entities from public domains into their own private domains. Viewed in this light, laboratories emerge not only as mechanisms for creating new knowledge, but also as devices for redrawing the contours of the public-private boundary.

What effects might diminishing public domains have on such a laboratory's transactions? If public domains diminish, then there will be less material in them, at least in relative terms. Thus, we might expect that perhaps our imaginary laboratory would acquire fewer inputs from public domains. If fewer inputs are obtained from public domains, then the laboratory must get them from some other source (or do without). Most likely, it will acquire a higher percentage of its inputs from restricted-access transactions, and this will lead the laboratory into negotiations with people who hold resources privately.

Of course, these restricted-access transactions will most likely entail quid pro quos, such as confidentiality agreements or rights to prepublication review. As a result, one might expect the laboratory to release fewer of its outputs into the public domain, or at least to do so later. More generally, if many laboratories became increasingly entangled in proprietary agreements, this might drag down the quality of public domains on a wider scale, for example, by limiting their content or causing delays in the introduction of new information. Indeed, one can imagine a synergistic process that would increasingly lead laboratories to rely on restricted-access transactions, producing a progressive impoverishment of the public domain that would, in turn, encourage further reliance on restricted-access transactions. If things were to go very badly, such effects could reduce the vitality and creativity of biomedical science for some of the same reasons that the lack of a strong public domain restricts the creative use of European weather data.[2]

I want to move on to possible second-order effects. What kinds of effects might diminishing public domains have on research communities and research cultures? Research communities play a key role in constituting public domains in science. If you think of public domains not as an abstract legal category, but instead as material entities produced actively through social action, then research communities are central players in building them. It takes a tremendous amount of work to make science and scientific information "public" (Callon, 1994). Research communities accomplish this in part by building institutional arrangements that lead individual laboratory scientists to put things into public domains. The published literature itself is the solidified sediment of a huge set of institutional arrangements that give academic scientists incentives to publish information. There were not always scientific journals; their history goes back to the Enlightenment, and over time, they have grown into a central scientific institution. Today, a complex set of institutional arrangements—from the tenure system to the research funding

---

[2]See Chapter 18 of these *Proceedings*, "Potential Effects of a Diminishing Public Domain in Environmental Information," by Peter Weiss.

system to a socialization process that makes publication important to a scientist's identity—encourages researchers to publish.

Building such institutions entails creating informal expectations and formal rules, and these expectations and rules are historical achievements, not timeless, stable features of science. Social and technological change creates openings for institutional innovations that can influence the contours of public domains. The emergence of DNA sequence databases, a new kind of public domain in science, provides a good example. DNA sequencing began in a small way in the 1970s, and a visionary group of scientists conceived of the Los Alamos Sequence Library (LASL) at the end of the decade. LASL gathered previously published sequences, which at that point were published in print in scientific journals, and prepared them in machine-readable form to permit mathematical analysis. In this way, these scientists created a new kind of public domain—the sequence database—for biology (Cinkosky et al., 1991; Hilgartner, 1995).

LASL later evolved into GenBank. Early in the history of GenBank, sequence data began accumulating so fast that journals became reluctant to publish it. GenBank decided to ask scientists to submit sequences directly to the database, but the incentive structures did not encourage them to do so. Sending in sequence data took time and required effort, and there was little payoff in terms of scientific credit for submitting sequences. Only later did GenBank, with help from the relevant scientific journals, negotiate a new deal to compel scientists to submit sequence data to the public databases (Hilgartner, 1995, p. 253). A policy-making journal publication contingent on database submission was first implemented by *Nucleic Acids Research* in 1988, and many other journals followed suit (*Nucleic Acids Research*, 1987; McCain, 1995). This example illustrates how new institutional arrangements, combined with technological developments such as DNA sequencing and the Internet, can be deployed to constitute important new public domains in science.

However, the ability to seize such opportunities depends on the existence of a scientific culture conducive to creating collective resources. Excessive concern with the protection of intellectual property can erect barriers to establishing new public domains. To illustrate this point, consider a counterfactual example. Imagine that you were trying to set up the first sequence database today. One proposed plan (which follows closely the model of LASL) might be to copy all the DNA sequences from the published literature, draw them together in machine-readable form, and provide access to the entire collection on the Internet. But in this post-Bayh-Dole era, with more than two decades of increasing commercialization of biology, would such a proposal be taken seriously? Perhaps not. And if it were, there is little doubt that one would need to convene a small army of university technology transfer officials, lawyers, and technology licensing specialists to negotiate about ownership of the database.

Of course, even given the increasing importance of proprietary regimes in biomedical science, commercial entities may at times decide to create public domains with unrestricted access. In Session 2, Robert Cook-Deegan mentioned the example of dbEST—the expressed sequence tag (EST) database funded by Merck.[3] Michael Morgan will discuss the Single Nucleotide Polymorphism Consortium,[4] which is a good example of a situation in which large pharmaceutical companies funded the development of a public-domain resource (in part to prevent other companies from creating monopolies over that resource). Clearly, public domains can still be constituted in a commercialized culture, but the question then becomes, how often will this happen? Can the scientific community safely assume that large corporations will create public domains in the future whenever they are needed? I think not.

I want to close by briefly mentioning possible third-order effects. Given the centrality of scientific knowledge and science advice to many critical public issues, it is worth considering how changes in public domains might affect the position of science in the wider polity. Arguably, the rapid commercialization and privatization of science has the potential to undermine the Enlightenment notion of science as a special form of knowledge, open to public scrutiny and collective verification (Shapin and Schaffer, 1985). If fundamental data pertaining to

---

[3]See Chapter 11 of these *Proceedings*, "The Urge to Commercialize: Interactions between Public and Private Research and Development," by Robert Cook-Deegan.

[4]See Chapter 28 of these *Proceedings*, "New Paradigms in Industry: The Single Nucleotide Polymorphism (SNP) Consortium," by Michael Morgan.

important public issues get caught up in proprietary arrangements that make it difficult for people to access them, reanalyze them, criticize them, or incorporate them into critiques of things going on in the world, then the notion that science is public knowledge would be seriously threatened.

Perhaps such effects are the hardest to predict and the hardest to be certain about. But it is clearly worth asking how far science can move in the direction of privatization before people stop perceiving it as a credible and disinterested source of public knowledge, and instead begin to think of science as just another private interest—one that cannot be scrutinized and cannot be counted on to speak the truth.

REFERENCES

Callon, Michel. 1994. "Is science a public good?" *Science, Technology, and Human Values*, 19(4): 395-424.

Campbell, Eric G., Brian R. Clarridge, Manjusha Gokhale, Lauren Birenbaum, Stephen Hilgartner, Neil A. Holtzman, and David Blumenthal. 2002. "Data withholding in academic genetics: evidence from a national survey," *Journal of the American Medical Association*, Vol. 287(4), January 23/30, pp. 473-480.

Cinkosky, M. J., J. W. Fickett, P. Gilna, and C. Burks. 1991. "Electronic data publishing and GenBank," *Science*, Vol. 252, pp. 1273-1277.

Fujimura, Joan H. 1996. *Crafting Science*. Harvard University Press, Cambridge, MA.

GUSTO III Investigators. 1997. "A comparison of reteplase with alteplase for acute myocardial infarction," *New England Journal of Medicine*, 337(18): 1118-1123.

Hicks, Diana. 1995. "Published papers, tacit competencies and corporate manage of the public/private character of knowledge," *Industrial and Corporate Changes*, Vol. 4(2), pp. 401-424.

Hilgartner, Stephen. 1995. "Biomolecular databases: new communication regimes for biology?" *Science Communication*, Vol. 17(2), pp. 240-63.

Hilgartner, Stephen. 1997. "Access to data and intellectual property: scientific exchange in genome research," in National Research Council's *Intellectual Property and Research Tools in Molecular Biology: Report of a Workshop*. National Academy Press, Washington, D.C.

Hilgartner, Stephen. 1998. "Data access policy in genome research." pp. 202-18 in Arnold Thackray, ed., *Private Science: Biotechnology and the Rise of the Molecular Sciences*, University of Pennsylvania Press, Philadelphia.

Hilgartner, Stephen and Sherry-Brandt-Rauf. 1994. "Data access, ownership, and control: toward empirical studies of access practices," *Knowledge: Creation, Diffusion, Utilization*, Vol. 15(4), pp. 355-72.

Knorr Cetina, Karin. 1999. *Epistemic Cultures*, Harvard University Press, Cambridge, MA.

Kohler, Robert E. 1994. *Lords of the Fly: Drosophila Genetics and the Experimental Life*. University of Chicago Press, Chicago, IL.

Latour, Bruno and Steve Woolgar. 1979. *Laboratory Life*. Sage Publications, Beverly Hills, CA.

Lenoir, Timothy. 1999. "Shaping biomedicine as an information science." In *Proceedings of the 1998 Conference on the History and Heritage of Science Information Systems*, Mary Ellen Bowden, Trudi Bellardo Hahn, and Robert V. Williams, eds., ASIS Monograph Series, Information Today, Inc., Medford, NJ, pp. 27-45.

Mackenzie, Michael, Peter Keating, and Alberto Cambrosio. 1990. "Patents and free scientific information in biotechnology: making monoclonal antibodies proprietary," *Science, Technology, and Human Values*, Vol. 15(1), pp. 65-83.

McCain, K.W. 1995. "Mandating sharing: journal policies in the natural sciences," *Science Communication*, Vol. 16, pp. 403-436.

Nucleic Acids Research. 1987. "Deposition of nucleotide sequence data in the data banks," *Nucleic Acids Research*, Vol. 15(18), front matter.

Shapin, Steven and Simon Schaffer. 1985. *Leviathan and the Air-Pump*, Princeton University Press, Princeton, NJ.

White-DePace, Susan, Nicholas F. Gmur, Jean Jordan-Sweet, Lydia Lever, Steven Kemp, Barry Karlin, Andrew Ackerman, and Jack Presses, eds. 1992. *National Synchrotron Light Source: Experimenter's Handbook*, National Technical Information Service, Springfield, VA.

SESSION 4: RESPONSES BY THE RESEARCH AND EDUCATION COMMUNITIES IN PRESERVING THE PUBLIC DOMAIN AND PROMOTING OPEN ACCESS

20

Discussion Framework

Jerome Reichman

A CONTRACTUALLY RECONSTRUCTED RESEARCH COMMONS FOR SCIENCE AND INNOVATION

The presentations in Session 2 of this symposium described the growing efforts under way to privatize and commercialize scientific and technical information that was heretofore freely available from the public domain or on an "open-access" basis. If these pressures continue unabated, they will result in the disruption of long-established scientific research practices and in the loss of new research opportunities that digital networks and related technologies make possible. We do not expect these negative synergies to occur all at once, however, but rather to manifest themselves incrementally, and the lost opportunity costs they are certain to engender will be difficult to discern.

Particularly problematic is the uncertainty regarding the specific type of database protection that Congress may enact and any exceptions favoring scientific research and education that such a law might contain. As we have tried to demonstrate, however, the economic pressures to privatize and commercialize upstream data resources will continue to grow in any event. Moreover, legal means of implementing these pressures already exist, regardless of the adoption of a *sui generis* database right. Therefore, given enough economic pressure, that which could be done to promote strategic gains will likely be done by some combination of legal and technical means.

If one accepts this premise, then the enactment of some future database law could make it easier to impose restrictions on access to and use of scientific data than at present, but the absence of a database law or the enactment of a lower protectionist version would not necessarily avoid the imposition of similar restrictions by other means. In such an environment, the existing elements of risk or threat to the sharing norms of public science can only increase, unless the scientific community adopts countervailing measures.

We, accordingly, foresee a transitional period in which the negative trends identified above will challenge the cooperative traditions of science and the public institutions that have reinforced those traditions in the past, with uncertain results. In this period, a new equilibrium will emerge as the scientific community becomes progressively more conflicted between their private interests and their communal needs for data and technical information as a public resource. This transitional period provides a window of opportunity that should be used to analyze the potential effects of a shrinking public domain and to take steps to preserve the functional integrity of the research commons.

The Challenge to Science

The trends described above could elicit one of two types of responses. One is essentially reactive, in which the scientific community adjusts to the pressures as best it can without organizing a response to the increasing encroachment of a commercial ethos upon its upstream data resources. The other would require science policy to address the challenge by formulating a strategy that would enable the scientific community to take charge of its basic data supply and to manage the resulting research commons in ways that preserve its public-good functions without impeding socially beneficial commercial opportunities.

Under the first alternative, the research community can join the enclosure movement and profit from it. Thus, both universities and independent laboratories or investigators that already transfer publicly funded technology to the private sector can also profit from the licensing of databases. In that case, data flows supporting public science will have to be constructed deal by deal with all the transaction costs this entails and with the further risk of bargaining to impasse. The ability of researchers to access and aggregate the information they need to produce discoveries and innovations may be compromised both by the shrinking dimensions of the public domain and by the demise of the sharing ethos in the nonprofit community, as these same universities and research centers increasingly see each other as competitors rather than partners in a common venture. Carried to an extreme, this competition of research entities against one another, conducted by their respective legal offices, could obstruct and disrupt the scientific data commons.

To avoid these outcomes, the other option is for the scientific community to take its own data management problems in hand. The idea is to reinforce and recreate, by voluntary means, a public space in which the traditional sharing ethos can be preserved and insulated from the commoditizing trends identified above. In approaching this option, the community's assets are the formal structures that surround federally funded data and the ability of federal funding agencies to regulate the terms on which data are disseminated and used. The first programmatic response would look to the strengthening of existing institutional, cultural, and contractual mechanisms that already support the research commons, with a view to better addressing the new threats to the public domain identified above. The second logical response is collectively to react to new information laws and related economic and technical pressures by negotiating contractual agreements between stakeholders to preserve and enhance the research commons.

As matters stand, the U.S. government generates a vast public domain for its own data by a creative use of three instruments: intellectual property rights, contracts, and new technologies of communication and delivery. By long tradition, the federal government has used these instruments differently from the rest of the world. It waives its property rights in government-generated information, it contractually mandates that such information should be provided at the marginal cost of dissemination, and it has been a major proponent and user of the Internet to make its information as widely available as possible. In other words, it has deliberately made use of existing intellectual property rights, contracts, and technologies to construct a research commons for the flow of scientific data as a public good. The unique combination of these instruments is a key aspect of the success of our national research enterprise.

Now that the research commons has come under attack, the challenge is not only to strengthen a demonstrably successful system at the governmental level, but also to extend and adapt this methodology to the changing university environment and to the new digitally networked research environment. In other words, universities, not-for-profit research institutes, and academic investigators, all of whom depend on the sharing of data, will have to stipulate their own treaties or contractual arrangements to ensure unimpeded access to, and unrestricted use of, commonly needed raw materials in a public or quasi-public space, even though many such institutions or actors may separately engage in transfers of information for economic gain. This initiative, in turn, will require the federal government as the primary funder—acting through the science agencies—to join with the universities and scientific bodies in an effort to develop suitable contractual templates that could be used to regulate or influence the research commons.

Implementing these proposals would require nuanced solutions tailor-made to the needs of government, academia, and industry in general and to the specific exigencies of different scientific disciplines. The following sections describe our proposals for preserving and promoting the public-domain status of government-generated scientific data and of government-funded and private-sector scientific data, respectively. We do not, however, develop detailed proposals for separate disciplines and subdisciplines here, as these would require additional research and analysis.

THE GOVERNMENT SECTOR

To preserve and maintain the traditional public-domain functions of government-generated data, the United States will have to adjust its existing policies and practices to take account of new information regimes and the growing pressures for privatization. At the same time, government agencies will have to find ways of coping with bilateral data exchanges with other countries that exercise intellectual property rights in their own data collections.

We do not mean to imply a need to totally reinvent or reorganize the existing universe in which scientific data are disseminated and exchanged. The opposite is true. As we have explained, a vast public domain for the diffusion of scientific data—especially government-generated data—exists and continues to operate, and much government-funded data emerging from the academic communities continues to be disseminated through these well-established mechanisms.

Facilities for the curation and distribution of government-generated data are well organized in a number of research areas. They are governed by long-established protocols that maintain the function of a public domain, and in most cases ensure open access (either free or at marginal cost) and unrestricted use of the relevant data collections. These collections are housed in brick-and-mortar data repositories, many of which are operated directly by the government, such as the NASA National Space Science Data Center. Other repositories are funded by the government to carry out similar functions, such as the archives of the Hubble Space Telescope Science Institute at Johns Hopkins University.

Under existing protocols, most government-operated or government-funded data repositories do not allow the conditional deposits that look to commercial exploitation of the data in question. Anyone who uses the data deposited in these holdings can commercially exploit their own versions and applications of them without needing any authorization from the government. However, no such uses, including costly value-adding uses, can remove the original data from the public repositories. In this sense, the value-adding investor obtains no exclusive rights in the original data, but is allowed to protect the creativity and investment in the derived information products.

The ability of these government institutions to make their data holdings broadly available to all potential users, both scientific and other, has been greatly increased by direct online delivery. However, this potential is undermined by a perennial and growing shortage of government funds for such activities; by technical and administrative difficulties that impede long-term preservation of the exponentially increasing amounts of data to be deposited; and by pressures to commoditize data, which are reducing the scope of government activity and tend to discourage academic investigators from making unconditional deposits of even government-funded data to these repositories.

The long-term health of the scientific enterprise depends on the continued operations of these public data repositories and on the reversal of the negative trends identified earlier in this chapter. Here the object is to preserve and enhance the functions that government data repositories have always played, notwithstanding the mounting pressures to commoditize even government-generated data.

Implementing any recommendations concerning government-generated data will, of course, require adequate funding, and this remains a major problem. In most cases, however, it is not the big allocations needed to collect or create data that are lacking; it is the relatively small but crucial amounts to properly manage, disseminate, and archive data already collected that are chronically insufficient. These shortsighted practices deprive taxpayers of the long-term fruits of their investments in the scientific enterprise. Science policy must give higher priority to formulating workable measures to redress this imbalance than it has in the past.

Policymakers should also react to the pressures to privatize government-generated research data by devising objective criteria for ascertaining when and how privatization truly benefits the public interest. At times, privatization will advance the public interest because the private sector can generate particular datasets more efficiently or because other considerations justify this approach. Very often, however, the opposite will be true: especially when the costs of generating the data are high in relation to known, short-term payoffs. Two recent

National Research Council studies have attempted to formulate specific criteria for evaluating proposed privatization initiatives concerning scientific data.[1] The science agencies should make the formulation of such criteria for different areas of research a top agenda item. In so doing, the agencies also need to analyze the results of past privatization initiatives with a view to assessing their relative costs and benefits.

Once the validity of any given privatization proposal has been determined by appropriate evaluative criteria, the next crucial step is to build appropriate, public-interest contractual templates into that deal to ensure the continued operations of a research commons. The public research function is too important to be left as an afterthought. It must figure prominently in the planning stage of every legitimate privatization initiative precisely because the data would previously have been generated at public expense for a public purpose. After all, the process of privatization aims to shift the commercial risks and opportunities of data production or dissemination to private enterprise under specified conditions that promote efficiency and economic growth. However, the process should not pin the functions of the research enterprise to the success of any given commercial venture; and it must not allow such ventures to otherwise compromise these functions by the charging of unreasonable prices or by the imposition of contractual conditions unduly restricting public, scientific uses of the data in question.

There are two situations in which model contractual templates, developed through interagency consultations, could play a critical role. One is where data collection and dissemination activities previously conducted by a government entity are transferred to a private entity. The other is where the government licenses data collected by a private entity for public research purposes. In both cases, the underlying contractual templates should implement the following research-friendly legal guidelines:

(1) a general obligation not to legally or technically hinder access to the data in question for nonprofit scientific research and educational purposes;
(2) a further obligation not to hinder or restrict the reuse of data lawfully obtained in the furtherance of nonprofit scientific research activities; and
(3) an obligation to make data available for nonprofit research and educational purposes on fair and reasonable terms and conditions, subject to impartial review and arbitration of the rates and terms actually applied, to avoid research disasters such as the Landsat deal in the 1980s.

In cases where the public data collection activity is transferred to the private sector, care must be taken to ensure that the private entity exercises any underlying intellectual property rights, especially some future database right, in a manner consistent with the public interest—including the interests of science. To this end, a model contractual template should also include a comprehensive misuse provision like that embodied in Section 106 of H.R. 1858:

SEC. 106. LIMITATIONS ON LIABILITY
(b) MISUSE—A person or entity shall not be liable for a violation of section 102 if the person or entity benefiting from the protection afforded a database under section 102 misuses the protection. In determining whether a person or entity has misused the protection afforded under this title, the following factors, among others, shall be considered:
(1) the extent to which the ability of persons or entities to engage in the permitted acts under this title has been frustrated by contractual arrangements or technological measures;
(2) the extent to which information contained in a database that is the sole source of the information contained therein is made available through licensing or sale on reasonable terms and conditions;
(3) the extent to which the license or sale of information contained in a database protected under this title has been conditioned on the acquisition or license of any other product or service, or on the performance of any action, not directly related to the license or sale;
(4) the extent to which access to information necessary for research, competition, or innovation purposes have been prevented;
(5) the extent to which the manner of asserting rights granted under this title constitutes a barrier to entry into the relevant database market; and

---

[1]See National Research Council. 1997. *Bits of Power: Issues in Global Access to Scientific Data,* National Academy Press, Washington, D.C., and National Research Council. 2001. *Resolving Conflicts Arising from the Privatization of Environmental Data,* National Academy Press, Washington, D.C.

(6) the extent to which the judicially developed doctrines of misuse in other areas of the law may appropriately be extended to the case or controversy.

The larger principle is that, in managing its own public research data activities, the government can and should develop its own database law in a way that promotes science without unduly impeding commerce. This principle is not new; the government already has a workable information regime, as described in the first session of this symposium. However, the government will need to adapt that regime to the pressures arising from the new high-protectionist legal environment and ensure that its agencies are consistently applying rational and harmonized public-interest principles. Otherwise, the traditional public-domain functions of government-generated data could be severely compromised. This in turn would violate the government's fiduciary responsibilities to taxpayers and raise conflicts of interest and questions concerning sham transactions.

THE ACADEMIC SECTOR

In putting forward our proposals concerning the preservation of a research commons for government-funded data, it is useful to follow the distinction between a zone of formal data exchanges and a zone of informal data exchanges previously discussed in Session 1.[2] Consistent with our earlier analysis, we emphasize that the ability of government funding agencies to influence data exchange practices will be much greater in the formal than the informal zone.

The Zone of Formal Data Exchanges

Where no significant proprietary interests come into play, the optimal solution for government-generated data and for data produced by government-funded research is a formally structured, archival data center also supported by government. As discussed, many such data centers have already been formed around large-facility research projects. Building on the opportunities afforded by digital networks, it has now become possible to extend this time-tested model to highly distributed research operations conducted by groups of academics in different countries.

The traditional model entails a "bricks-and-mortar" centralized facility into which researchers deposit their data unconditionally. In addition to academics, contributors may include government and even private-sector scientists, but in all cases the true public-domain status of any data deposited is usually maintained. Examples include the National Center for Biotechnology Information, directly operated by the National Institutes of Health, and the National Center for Atmospheric Research, operated by a university consortium and funded primarily by the National Science Foundation (NSF).

A second, more recent model, enabled by improved Internet capabilities, also envisions a centralized administrative entity, but this entity governs a network of highly distributed smaller data repositories, sometimes referred to as "nodes." Taken together, the nodes constitute a virtual archive whose relatively small central office oversees agreed technical, operational, and legal standards to which all member nodes adhere. Examples of such a decentralized network, which operate on a public-domain basis, are the NASA Distributed Active Archive Centers under the Earth Observing System program and the NSF-funded Long Term Ecological Research Network.

These virtual archives, known as "federated" data management systems, extend the benefits and practices of a centralized brick-and-mortar repository to the outlying districts and suburbs of the scientific enterprise. They help to reconcile practice with theory in the sense that the investigators—most of whom are funded by government anyway—are encouraged to deposit their data in such networked facilities. The very existence of these formally constituted networks thus helps to ensure that the resulting data are effectively made available to the scientific community as a whole, which means that the social benefits of public funding are more perfectly captured and the sharing ethos is more fully implemented.

At the same time, some of the existing "networks of nodes" have already adopted the practice of providing conditional availability of their data: a feature of considerable importance for our proposals. By "conditional availability," we mean that the members of the network have agreed to make their data available for public science

---

[2]See Chapter 1 of these *Proceedings*, "Session 1 Discussion Framework," by Paul Uhlir.

purposes on mutually acceptable terms, but they also permit suppliers to restrict uses of their data for other purposes, typically with a view to preserving their commercial opportunities.[3]

The networked systems thus provide prospective suppliers with a mix of options to accommodate deposits ranging from true public-domain status to fully proprietary data that has been made available subject to rules the member nodes have adopted. The element of flexibility that conditional deposits afford make these federated data management systems particularly responsive to the realities of present-day university research in areas of scientific investigation where commercial opportunities are especially prominent.

Basic Proposals

Our first proposition is that the government funding agencies should encourage unconditional deposits of research data, to the fullest extent possible, into both centralized repositories and decentralized network structures. The obvious principle here is that, because the data in question are government funded, improved methods should be devised for capturing the social benefits of public funding, lest commercial temptations produce a kind of de facto freeride at the taxpayers' expense.

When unconditional deposits occur in a true public-domain environment removed from proprietary concerns, the legal mechanisms to implement these expanded data centers need not be complicated. Single researchers or small research teams could contribute their data to centers serving their specific disciplines, with no strings attached other than measures to ensure attribution and professional recognition. Alternatively, as newly integrated scientific communities organize themselves, they could seek government help in establishing new data centers or nodes that would accept unrestricted deposits on their behalf. Private companies could also contribute to a true public-domain model or organize their own variants of such a model; these practices should be encouraged as a matter of public policy.

If the unrestricted data were deposited in federal government-sponsored repositories, existing federal information law and associated protocols would define the public access rights. The maintenance of public-interest data centers in academia, however, is problematic without government support. These data centers can become partly or fully self-supporting through some appropriate fee structure, but resorting to a fee structure based on payments of more than the marginal cost of delivery quickly begins to defeat the public good and positive externality attributes of the system, even absent further use restrictions.

Leaving aside the funding issue, the deeper question that this first proposal raises is how the universities and other nonprofit research entities will resolve the potential conflict between the pressures to disclose and deposit their government-funded data and the valuable proprietary interests that are increasingly likely to surface in a high-protectionist intellectual property environment. One cannot ignore the risk that the viability and effectiveness of these centers could be undermined to the extent that the beneficiaries of government funding can resist pressures to further implement the sharing ethos and even decline to deposit their research data because of their commercial interests.

Despite their educational missions and nonprofit status, both universities and individual academics are increasingly prone to regard their databases as targets of opportunity for commercialization. This tendency will become more pronounced as more of the financial burden inherent in the generation and management of scientific data is shouldered by the universities themselves or by cooperative research arrangements with the private sector. In this context, the universities are likely to envision split uses of their data and will prefer to make them available on restricted conditions. They will logically distinguish between uses of data for basic research purposes by other nonprofit institutions and purely commercial applicants. Even this apparently clear-cut distinction might break down, moreover, if universities treat databases whose principal user base is other nonprofit research institutions as commercial research tools.

The point is that the universities may not want to deposit data in designated repositories, even with government support, unless the repositories can accommodate these interests, and the repositories could compromise their public research functions if they are held hostage to too many demands of this kind. The same potential

---

[3] An example of an international network that operates on the basis of conditional deposits is the Global Biodiversity Information Network, headquartered in Denmark, which is substantially supported by U.S. government funding. For additional information, see http://www.gbif.org.

situation exists for individual databases made available by universities (as opposed to their contributions to larger, multisource repositories). This state of affairs will accordingly require still more creative initiatives to parry the economic and legal pressures on universities and academic researchers to withhold data.

With these factors in mind, our second major proposal is to establish a zone of conditionally available data to reconstruct and artificially preserve functional equivalents of a public domain. This strategy entails using property rights and contracts to reinforce the sharing norms of science in the nonprofit, transinstitutional dimension without unduly disrupting the commercial interests of those entities that choose to operate in the private dimension.

To this end, the universities and nonprofit research institutions that depend on the sharing ethos, together with the government science funding agencies, should consider stipulating suitable "treaties" and other contractual arrangements to ensure unimpeded access to commonly needed raw materials in a public or quasi-public space. From this perspective, one can envision the accumulation of shared scientific data as a community asset held in a contractually reconstructed research commons to which all researchers have access for purposes of public scientific pursuits.

One can further imagine that this public research commons exists in an ever-expanding "horizontal dimension," as contrasted with the commercial operations of the same data suppliers in what we shall call the "vertical" or private dimension. The object of the exercise would be to persuade the government, as primary funder, to join with universities and scientific bodies in an effort to develop suitable contractual templates that could be used to regulate the research commons. These templates would ensure that data held in the quasi-public or "horizontal" dimension would remain accessible for scientific purposes and could not be removed or otherwise appropriated to the private or "vertical" dimension. At the same time, these contractual arrangements would expressly contemplate the possibilities for commercial exploitation of the same data in the private or vertical dimension, and they would clarify the depositor's rights in that regard and ensure that the exercise of those rights did not impede or disrupt access to the horizontal space for research purposes.

Ancillary Considerations

In fashioning these proposals, we are aware that considerable thought has recently been given to the construction of voluntary social structures to support the production of large, complex information projects. Particularly relevant in this regard are the open-source software movement that has collectively developed and managed the GNU/Linux Operating System and the Creative Commons, which seeks to encourage authors and artists to conditionally dedicate some or all of their exclusive rights to the public domain.[4] In both these pioneering movements, agreed contractual templates have been experimentally developed to reverse or constrain the exclusionary effects of strong intellectual property rights.

Although neither of these models was developed with the needs of public science in mind, both provide helpful examples of how universities, federal funding agencies, and scientific bodies might contractually reconstruct a research commons for scientific data that could withstand the legal, economic, and technological pressures on the public domain identified in this paper. In what follows, we draw on these and other sources to propose the contractual regulation of government-funded data in two specific situations: (1) where government-funded, university-generated data are licensed to the private sector and (2) where such data are made available to other universities for research purposes.

Licensing Government-Funded Data to the Private Sector

In approaching this topic, one must consider that the production of scientific databases in academia is not always dominated by activities funded by the federal government. It may also entail funding by universities themselves, foundations, and the private sector. Although these sources seem likely to grow in the future, especially if Congress adopts a database protection right, the government's role in funding academic data production

---

[4]See Chapter 23 of these *Proceedings*, "New Legal Approaches in the Private Sector," by Jonathan Zittrain, for a description of the Creative Commons initiative.

will nonetheless remain a major factor, at least in the near term (although its role will vary from project to project). As we discussed earlier in this symposium, this presence gives the federal funding agencies unique opportunities to influence the data-sharing policies of its beneficiary institutions.

Ideally, funders and universities would agree on the need to maintain the functions of a public domain to the fullest extent possible, to provide open access for nonprofit research activities, and to encourage efficient technological applications of available data. At the same time, technological applications and other opportunities for commercial exploitation of certain types of databases will push the universities to enter into private contractual transactions that, if left totally unregulated, could adversely affect the availability of the relevant data for public research purposes. The reconciliation of the conflicts between enhancing the public research interests and freedom of contract will require carefully formulated policies and institutional adjustments.

Assuming the existence of sufficient funds, the maximum availability of academic data for research purposes is assured if those data have been deposited in the public data centers. To the extent that agencies successfully encourage academics and their universities to deposit government-funded data into either old or new repositories established for this purpose, the research-friendly policies of these centers should automatically apply. As long as these policies are not themselves watered down by commercial and proprietary considerations, they should generally immunize the research function from conflicts deriving from private transactions.

However, the universities or their researchers may very well balk at depositing commercially valuable data in these repositories unless a relative degree of autonomy is preserved for depositories to negotiate the terms of their private transactions and to impose restrictions on the uses of the data deposited for commercial purposes. This raises two important questions. The first concerns the willingness of data centers themselves—whether of the centralized brick-and-mortar variety or virtual networks—to accept conditional deposits that impose restrictions on use for certain purposes in the first place. The second question, closely tied to the first, concerns the extent to which federal funding agencies should further seek to define and influence the relations between universities and the private sector to protect the public research function—especially when the data in question have not been deposited in an appropriate repository or when they have been so deposited but the repository permits conditional deposits.

Regarding the first of these questions, we previously observed that the emerging "network of nodes" model is more likely to accommodate conditional deposits or availability than are the traditional centralized data centers. Nevertheless, the practice remains controversial in scientific circles in that it deviates from the traditional norm of "full and open access." For present purposes, we shall simply state our view that the possibilities for maximizing access to scientific data for public nonprofit research will not be fully realized in a highly protectionist legal and economic environment unless the scientific community agrees to experiment with suitably regulated conditional deposits.

The second question, concerning the need to regulate the interface between universities and the private sector with regard to government-funded data, acquires important contextual nuances when viewed in the light of the policies and practices that currently surround the Bayh-Dole Act and related legislation. The Bayh-Dole Act encourages universities to transfer the fruits of federally funded research to the private sector by means of the patent system. In a somewhat similar vein, federal research grants and contracts allow researchers to retain copyrights in their published research results. By extension, the same philosophy could apply to databases produced with federal funding, especially if Congress were to adopt a *sui generis* database protection right, with incalculably negative results, unless steps were taken to reconcile the goals of Bayh-Dole with the dual nature of data as both an input and an output of scientific research and of the larger system of technological innovation.

It would also be a mistake for the science policy establishment to wait for the enactment of database legislation before considering the implications of blindly applying the spirit of Bayh-Dole to any database law that Congress may adopt. Because databases differ significantly from either patented inventions or copyrighted research results, policy makers should anticipate the advent of some database legislation and address the problems it may cause for science—particularly in regard to government-funded data. Special consideration must be given to how the power to control uses of scientific data after publication would be exercised once a database protection law was enacted.

We do not mean to question the underlying philosophy or premise of Bayh-Dole, which has produced socially beneficial results. Its very success, however, has generated unintended consequences and raised new questions that require careful consideration. In advocating a program for a contractually reconstructed research commons, one of our explicit goals is, indeed, to ensure that academics and their universities benefit from new opportunities to exploit research data in an industrial context. This goal reflects the policies behind Bayh-Dole. At the same time, it would hardly be consistent with the spirit of Bayh-Dole to allow the commercial partners of academic institutions to dictate the terms on which government-funded data are made available for purposes of nonprofit scientific research.

On the contrary, a real opportunity exists for government funders and universities to develop agreed contractual templates that would apply to commercial users of government-funded data in general. In effect, the public scientific community would thus develop a database protection scheme of its own that would override the less research-friendly provisions of any *sui generis* regime that Congress might adopt. In so doing, the scientific community could also significantly influence the data-licensing policies and practices of the private sector, before that sector ends up influencing the data-licensing practices of university technology transfer offices.

If one takes this proposal seriously, a capital point of departure would be to address the problem of follow-on applications, which has greatly perturbed the debate about database protection in general. The critical role of data as inputs into the information economy weighs heavily against endowing database proprietors with any exclusive right to control follow-on applications. This principle becomes doubly persuasive when the government itself has defrayed the costs of generating the data in question, in which case an exclusive right to control value-added applications takes on a cast of reverse free-riding. The solution is to allow second-comers freely to extract and use data from any given collection for bona fide value-adding purposes in exchange for adequate compensation of the initial investor based on an expressly limited range of royalty options. If the rules developed by universities and funding agencies imposed this kind of "compensatory liability" regime on follow-on applications of government-funded academic data, in lieu of any statutorily created exclusive right, there is reason to believe it would significantly advance both technological development and the larger public interest in access to scientific data.

Universities and funding agencies could also adopt clauses similar to those proposed above in the context of government-generated data, including a general prohibition against legally or technically hindering access to any database built around government-funded data for purposes of nonprofit scientific research. Clauses that obligate private partners not to hinder the reuse of data in the construction of new databases to address new scientific research objectives seem particularly important, as are clauses requiring private partners to license their commercial products on fair and reasonable terms and conditions. Also desirable are clauses forbidding misuse of any underlying intellectual property rights and establishing guidelines that courts should apply in evaluating specific claims of misuse.

Moreover, when considering relations with the private sector, attention should be given to the high costs of managing and archiving data holdings for scientific purposes and to the possibilities of defraying some of these costs through commercial exploitation. Although government support ought to increase, especially as the potential gains from a horizontal e-commons become better understood, the costs of data management will also increase with the success of the system. For this reason, universities may want to levy charges against users in the private sector or the vertical dimension, in order to help defray the costs of administering operations in the horizontal domain and to make this overall approach more economically feasible.

Finally, care must be taken to reduce friction between the scientific data commons as we envision it and universities' patenting practices under the Bayh-Dole Act. For example, any agreed contractual templates might have to allow for deferred release of data, even into repositories operating as a true public domain, at least for the duration of the one-year novelty grace period during which relevant patent applications based on the data could be filed. Other measures to synchronize the operations of the e-commons with the ability of universities to commercialize their holdings under Bayh-Dole would have to be identified and carefully addressed. We also note that there is an interface between our proposals for an e-commons for science and antitrust law, which would at least require consultation with the Federal Trade Commission and might also require enabling legislation.

In sum, to successfully regulate relations between universities and the private sector in the United States, where most of the scientific data in question are government funded (if not government generated), considerable thought must be given to devising suitable contractual templates that universities could use when licensing such data to the private sector. These templates, which should aim to promote the smooth operations of a research commons and to facilitate general research and development uses of data as inputs into technological development, could themselves constitute a model database regime that optimally balances public and private interests in ways any federally enacted law might not. To succeed, however, these templates must be acceptable to the universities, the funding agencies, the broader scientific community, and to the specific disciplinary subcommunities—all of whom must eventually weigh in to ensure that academics themselves observe the norms that they would thus have collectively implemented.

In so doing, the participating institutions could avoid a race to the bottom in which single universities might otherwise trade away more restrictions on open access and research to attract more and better deals from the private sector. Unless science itself takes steps of this kind, there is a serious risk that, under the impetus of Bayh-Dole, the private sector will gradually impose its own database rules on all government-funded data products developed with their university partners.

Inter-University Licensing of Scientific Data

Whatever the merits of our proposals for regulating transfers of scientific data from universities to the private sector, the need for science policy to regulate inter-university transfers of such data seems irrefutable. In this context, most of the data are generated for public scientific purposes and at public expense, and the progress of science depends on continued access to, and further applications of, such data. Not to construct a research commons that could withstand the pressures to privatize government-funded data at the inter-university level would thus amount to an indefensible abdication of the public trust by encumbering nonprofit research with high transaction and exclusion costs. All the same, implementing this task poses very difficult problems that are likely to exacerbate the conflicts of interests between the open and cooperative norms of science and the quest for additional funding sources we previously identified.

One may note at the outset that these conflicts of interest are rooted in the Bayh-Dole approach to the transfer of technology itself. This legislative framework stimulates universities to protect basic research results through intellectual property rights and to license those rights to the private sector for commercial applications. If Congress enacted a strong database protection law, it could extend Bayh-Dole to this new intellectual property right. In such a case, Bayh-Dole would simply pass the relevant exclusive rights to extract and utilize collected data straight through the existing system to the same universities and academic researchers who now patent their research results and who would thus end up owning all the government-funded data they had generated.

Moreover, the Bayh-Dole legislation makes no corresponding provision for beneficiary universities to give differential and more favorable treatment to other universities when licensing patented research products. On the contrary, there is evidence that in transactions concerning patented biotechnology research tools, universities have viewed each others' scientists as a target market. In these transactions, universities have virtually the same commercial interests as private producers of similar tools for scientific research. Such inter-university deals have accordingly been constructed on a case-by-case basis, often with considerable difficulty, by technology transfer offices striving to maximize all their commercial opportunities.

Without any agreed restraints on how universities are to deal with collections of data in which they had acquired statutorily conferred ownership and exclusive exploitation rights, their technology transfer offices could simply treat databases like patented inventions—despite the immensely greater impact this could have on both basic and applied research. In this milieu, reliance on good-faith accommodations hammered out by the respective technology transfer offices would, at best, make inter-university exchanges resemble the complicated transactions that already characterize relations between highly distributed laboratories and research teams in the zone of informal exchanges of scientific data. All the vices of that zone would soon be imparted into the more formal zone of inter-university relationships. At worst, this would precipitate a race to the bottom as universities tried to maximize their returns from these rights, in which case some technology transfer offices could be expected to contractually override any modest research exceptions a future database law might have codified.

At the same time, the Bayh-Dole legislative framework may itself suggest an antidote for resolving these potential conflicts of interest, or, at least, a sound point of departure for addressing them. Section 203 of the Bayh-Dole Act explicitly recognizes that the public interest in certain patented inventions may outweigh the benefits usually anticipated from private exploitation under exclusive property rights. In such cases, it authorizes the government to impose a compulsory license or otherwise to exercise "march in" rights and take control of the invention it has paid to produce. In fact, these public-interest adjustments have never successfully been exercised in practice.

Nevertheless, the principle (if not the actual practice) behind these provisions presents a platform on which universities and federal funding agencies can build their own mutually acceptable arrangements to promote their common interest in full and open access to government-funded collections of data. Our goal, indeed, is to persuade them to address this challenge now, before a database protection law is enacted, by examining how to ensure the smooth and relatively frictionless exchange of scientific data between academic institutions, regardless of any exclusive property rights they may eventually acquire and notwithstanding any other commercial undertakings with the private sector they may pursue. Absent such a proactive approach, we fear a slow unraveling of the traditional sharing norms in the inter-university context and an inevitable race to the bottom.

- *Structuring Inter-University Data Exchanges in a High-Protectionist Legal Environment.* Because the issues under consideration here pertain to uses of government-funded data produced by academics for university-sponsored programs, one looks to "full and open access" as the optimal guiding principle and to the sharing norms of science as the foundation of any arrangement governing inter-university licensing of data. On this approach, the government-funded data collections held by universities would be viewed as a single common resource for inter-university research purposes. The operational goal would be to nurture and extend this common resource within a horizontally linked administrative framework that facilitated every university's public research functions, without unduly disrupting commercial relations with the private sector that some departments of some universities will undertake in the vertical dimension.

To achieve this goal, universities, funding agencies, and interested scientific bodies would have to negotiate an acceptable legal and administrative framework, analogous to a multilateral pact, that would govern the common resource and provide day-to-day logistical support. Ideally, the participating universities or their designated agents would operate as trustees for the horizontally constructed common resource, much as occurs with what the Free Software Foundation does with the GNU system. In this capacity, the trustees would assume responsibility for ensuring access to the holdings on the agreed terms and for restraining deviant uses that violate those terms or otherwise undermine the integrity of the commons. The full weight of the federal granting structure could then be made to support these efforts by mandating compliance with agreed terms and by directly or indirectly imposing sanctions for noncompliance.

Alternatively, a less formal administrative structure could be built around a set of agreed contractual templates regulating access to government-funded data collections for public research purposes. On this approach, the participating universities would retain greater autonomy, there would be less need for a fully fleshed out "multilateral pact," and the monitoring and other transaction costs might be reduced.

In a less than perfect world, however, there are formidable obstacles standing in the way of a negotiated commons project, over and above inertia, that would have to be removed. Initially, the very concept of an e-commons needs to be sold to skeptical elements of the scientific community whose services are indispensable to its development. Academic institutions, science funders, the research community, and other interested parties must then successfully negotiate and stipulate the pacts needed to establish it, as well as the legal framework to implement it. Transaction costs would all need to be monitored closely and, whenever possible, reduced throughout the various development phases.

Once the research universities became wholeheartedly committed to the idea of a regime that guaranteed them universal access to, and shared use of, the government-funded data that they had collectively generated, these organizational problems might seem relatively minor. The difficulties of winning such a commitment, however, cannot be overestimated in a world where university administrators are already conflicted about the efforts of their technology transfer offices to exploit commercially valuable databases in the genomic sciences and other disciplines with significant potential for commercial development. The prospect that Congress will eventually adopt a hybrid intellectual property right in collections of data could make these same administrators reluctant to lock their institutions into a kind of voluntary pool of any resulting exclusive property rights, even for public scientific research purposes.

Conceptually, the problems inherent in organizing a pool of intellectual property rights so as to preserve access to, and use of, a common resource have become much better understood than in the past—owing to the experience gained from both the open-source software movement and the new Creative Commons initiative. These projects demonstrate that there are few, if any, technical obstacles that cannot be overcome by adroitly directing relevant exclusive rights and standard-form contracts to public, rather than private, purposes.

The deeper problem is persuading university administrators that they stand to gain more from open access to each others' databases in a horizontally organized research commons than they stand to lose from licensing data to each other under more restrictive, case-by-case transactions. Although we believe this proposition to be true, following it could amount to an act of faith, albeit one that resonates with the established norms of science, and with the primary mission of universities.

To the extent that the universities may have to be sold on the benefits of an e-commons for data, with a view to rationalizing and modifying their disparate licensing policies, this project would require statesmanship, especially on the part of the leading research universities. It may also require pressure from the major government funders and standard-setting initiatives by scientific subcommunities. Funding agencies, in particular, must be prepared to discipline would-be holdouts and to discourage other forms of deviant strategic behavior that could undermine the cohesiveness of those institutions willing to pool their resources. In this regard, account will have to be taken as well of the universities' patenting interests, which will need to be suitably accommodated.

Assuming a sufficient degree of organizational momentum, there remains the thorny problem of establishing the terms and conditions under which participating universities could contribute their data to a horizontally organized research commons. The bulk of the departments and subdisciplines involved would almost certainly prefer a bright-line rule that required all deposits of government-funded data to be made without conditions and subject to no restrictions on use. This preference follows from the fact that most science departments currently see no realistic prospects for licensing basic research data, even to the private sector, and have not yet experienced the proprietary temptations of exclusive ownership that a *sui generis* intellectual property right in noncopyrightable databases might eventually confer.

At the same time, such a bright-line rule could utterly deter those subdisciplines that already license data on commercial terms to either the private or public sectors, or that contemplate doing so in the near future. These subdisciplines would not readily forego these opportunities and would, on the contrary, insist that any multilateral negotiations to establish a horizontal commons devise contractual templates that protected their commercial interests in the vertical dimension. If, moreover, Congress enacts a de facto exclusive property right in collections of data, it would probably deter other components of the scientific community, who might become unwilling to forego either the prospective commercial opportunities or other strategic advantages such rights might make possible.

In a word, a bright-line rule requiring unconditional deposits in all cases could thus defeat the goal of linking all university generators of government-funded data in a single, horizontally organized research commons. At the same time, the goal of universality could, paradoxically, require negotiators seeking to establish the system to deviate from the norm of full and open access by allowing a second type of conditional deposit of data into the horizontal domain by those disciplines or departments that were unwilling to jeopardize present or future commercial opportunities.

• *Resolving the Paradox of Conditional Deposits.* Science policy in the United States has long disfavored a two-tiered system for the distribution of government-funded data. Such two-tiered systems for government or academic data distribution have been favored and promoted by the scientific community in the European Union, but these initiatives have been strongly opposed by U.S. science agencies and academics. Under such a system, database proprietors envision split (or two-tier) uses of certain data and will only make them available on conditions that govern the different types of uses they have expressly permitted.

Typically, split-level arrangements distinguish between relatively unrestricted uses for basic research purposes by nonprofit entities and more restricted uses for commercial applications by private firms that license data from scientific entities. The latter conditions may range from a simple menu of price-discriminated payment options to more complicated provisions that regulate certain data extractions, seek grant-backs of follow-on applications by second-comers, or impose reach-through clauses seeking legal or equitable rights in subsequent products. In some cases, moreover, the distinction between profit and nonprofit uses of scientific data becomes blurred, and the two categories may overlap, which adds to the costs and complications of administration. For example, universities may treat some databases as commercial research tools and impose a price discrimination policy that provides access to the research community at a lower cost than to for-profit entities. This becomes more likely when there is a private-sector partner.

We recognize that a decision to allow participating universities to make conditional deposits of government-funded data to a collectively managed research consortium represents a second-best solution, one that conflicts with the goal of establishing a true public domain based on the premise of full and open access to all users. The allowance of restrictions on use breaks up the continuity of data flows across the public sector and necessitates administrative measures and transaction costs to monitor and enforce differentiated uses. It also entails measures to prevent unacceptable leakage between the horizontal and vertical planes, and it may result in charges for public-interest uses that exceed the marginal cost of delivery, even in the horizontal plane.

We nonetheless doubt that a drive for totally unconditional deposits of government-funded data could succeed in the face of mounting worldwide pressures to commoditize scientific data, and we fear that excessive reliance on the orthodox position would, in the end, undermine—rather than save—the sharing ethos. Even if one disregards the prospects for strengthened intellectual property protection of noncopyrightable databases, too many universities have already begun to perceive the potential financial benefits they might reap from commercial exploitation of genomic databases in particular and biotechnology-related databases in general. Their reluctance to contribute such data to a research commons that allowed private firms freely to appropriate that same data could not easily be overcome. Adoption of a database protection law would then magnify this reluctance and encourage the respective technology transfer offices to find more ways to commercially exploit more of the government-funded data products that were subsequently invested with proprietary rights.

Even if a consortium of universities were to formally consent to such an unconditional arrangement, their technology transfer offices might soon be demanding an exceptional status for any databases that contained components produced without government funds. They could persuasively argue that private funds for most jointly created data products could decrease or even dry up if both customers and competitors could readily obtain the bulk of the data from the public domain. Once it became clear that an admixture of privately funded data could elicit the right to deposit data in a research commons on conditions that protected commercial exploitation of the databases in question, academics with an eye to cost recovery and profit maximization would logically make persistent efforts to qualify for this treatment. They would thus seek more private investment for this purpose or seek to obtain the university's own funds for the project. Either way, there would be a perverse incentive to privatize more data than ever if the only legitimate way to avoid dedicating it all to the public domain was to show that some of it had been privatized.

In other words, if the quasi-public research space accommodated only unconditional deposits of data, it could foster an insuperable holdout problem as participating universities found ways to detach and isolate their commercially valuable databases from such a system. In these circumstances, a failure to obtain a best-case scenario premised on "full and open access" would quickly degenerate into a worst-case scenario, characterized by growing gaps in the communally accessible collection and an unraveling of the sharing ethos, that would require case-by-case construction of inter-university data flows and could sometimes culminate in bargaining to impasse.

In our estimation, the worst-case scenario is so bad, and the pressures to commoditize could become so great in the presence of a strong database right, that steps must be taken to ensure universal participation in a contractually reconstructed research commons from the outset by judiciously allowing conditional deposits of government-funded data on standard terms and conditions to which all the stakeholders had previously agreed. Indeed, the goal is to develop negotiated contractual templates that clearly reinforce and implement terms and conditions favorable to public research without unduly compromising the ability of the consortium's member universities to undertake commercial relations with the private sector.

At stake in this process is not just a few thousand patentable inventions, but, rather, every government-funded data product that has potential commercial value to other universities as a research tool or educational device. Sound data management policies thus point to a second-best solution that would preserve the integrity of the inter-university commons by disallowing the principal ground on which concerted holdout actions might take root, by ensuring that only research-friendly terms and conditions apply in both the horizontal and vertical dimensions, and by making it too costly for any institution to deviate from the agreed regulatory framework governing the two-tiered regime.

Those who object to this proposal will argue that it unduly undermines the "full- and open-access" principle by tempting more and more university departments or subdisciplines to opt for conditional deposits than would otherwise have been the case. On this view, once a negotiated two-tiered model were set in place, universities would come under intense pressures to avoid the true public domain or open-access option even when there was no need to do so.

However, a universal and functionally effective inter-university research commons simply cannot be constructed with a bright-line, true public-domain rule applied across the board for the reasons we previously set out. A bright-line rule also carries with it the well-recognized difficulty of distinguishing for-profit from not-for-profit research activities when single libraries increasingly engage in both. In contrast, a regime based on conditional deposits overcomes this problem by allowing a scientific entity to contribute to and benefit from the data commons so long as it respects the agreed norms bearing on that arrangement. In this respect, a normative accommodation will have displaced legal distinctions that cannot feasibly be enforced.

Moreover, the very contractual templates that make the construction of such a commons feasible in a two-tiered system should also mitigate its social costs. Even if conditional deposits are allowed, many subdisciplines will continue to have no commercial prospects and no need to invoke the contractual templates that regulate them. When this is the case, peer pressures reinforced by the funding agencies should make it difficult, if not impractical, for members of those communities to opt out of the traditional practice of making data available unconditionally.

When, instead, given communities find themselves forced to deal with serious commercial pressures, the negotiated contractual solutions that enabled them to make the data conditionally available for public research purposes should also tend to preserve and implement the norms of science. In particular, the applicable contractual templates should immunize deposited data from the vagaries of case-by-case transactions under the aegis of universities' technology transfer offices and would also limit the kinds of restrictions private-sector partners might otherwise seek to impose on universities.

At the end of the day, a set of agreed contractual templates permitting conditional deposits in the interests of a horizontally linked research commons would provide a tool universities could use with more or less wisdom. If used wisely, this tool should ensure that more data are made available to a contractually reconstructed research commons than would be possible if member universities could not protect the interests of their commercial partners. This same tool may also provide incentives for the private sector to work with universities in producing better data products than the latter alone could generate with their limited funds.

- *Other hard problems.* Allowing universities to deposit government-funded data into a contractually reconstructed research commons, on conditions designed to protect their commercial relations with the private sector, solves two difficult problems. First, it avoids the risk that large quantities of government-funded data would remain outside the system on the grounds that they had been commingled with privately funded components. Second, it ensures that any negotiated contractual templates the research consortium adopts to govern its horizontal space will apply to all the data holdings subject to its jurisdiction, including databases to which the private sector had contributed. However, it does not automatically determine the precise conditions that the agreed contractual templates should apply to inter-university licensing of data subject to their collective jurisdiction. In the process of defining these conditions, moreover, those who negotiated the "multilateral pact" among universities, federal funders, and scientific bodies needed to launch the consortium would have to resolve a number of contentious issues.

The guiding principle that should apply to inter-university licensing of data available from the quasi-public space is that depositors may not impose any conditions that impede the customary and traditional uses of scientific data for nonprofit research purposes. A logical corollary is that they should affirmatively adopt the measures that may prove necessary to extend and apply this principle to the online environment. Because the data under discussion are government funded for academic purposes to begin with, the "open access" and sharing norms of science should then color any specific implementing templates that regulate access and use.

With regard to access, the customary mode of implementing these norms would be to make data available to other nonprofit institutions at no more than the marginal cost of delivery. In the online environment, these marginal costs are essentially zero. This represents the preferred option whenever the costs of maintaining the data collection are defrayed by public subsidy or by nonexclusive licenses to private firms in the vertical dimension.

If, however, the policy of free or marginally priced access appears unable to sustain the costs of managing a given project at the inter-university level, an incremental pricing structure may become unavoidable. The options for such a pricing structure range from a formula allowing partial incremental cost recovery when a project is partially subsidized to a formula providing full cost recovery when this is necessary to keep the data collection alive. This may be accomplished through a paying inter-university consortium, such as the Inter-University Consortium for Political and Social Science Research at the University of Michigan, or by means of more ad hoc cost-recovery methods. Examples of subcommunities that have found it necessary to rely on the second option are largely in the laboratory physical sciences.

The prices charged other nonprofit users to access data in the research commons should never exceed the full incremental cost of managing the collective holdings. This premise follows from the fact that the initial costs of collecting or creating the data were defrayed by the government or by some combination of sources (including private sources) that normally subscribe to the open-access principle.

However, when private firms have defrayed a substantial part of the costs of generating the database in question, there are few, if any, standard solutions. Occasionally, even a private partner might view the collective holdings as a valuable resource for its own pursuits, to which it agrees to contribute on an eleemosynary basis. In the more typical cases, the private partner is likely to view the research community as the target market for a database it paid to create and from which it must derive its expected profits.

In that event, the collection of additional revenues from private-sector access charges should depend entirely on freedom of contract, although a likely demand that public research users pay access charges that exceeded data management costs would pose a hard question. On the one hand, as beneficiaries of government funding, the universities should forego "profits" from charges levied to access their partly publicly funded databases for public research purposes. On the other hand, a private partner will not readily forego such profits, especially if it had invested in the project precisely because of its potential commercial value as a research tool. If the university shared these "profits" with its private partner, this practice would deviate from the basic principle governing inter-university access generally, and it would encourage other universities to seek private partners for this purpose, which in turn would yield both social costs and benefits.

In these cases, care must be taken to avoid adopting policies that would discourage either public-private partnerships for the development of socially beneficial data products or the inclusion of such products in a horizontal, quasi-public research space. At the same time, there is a potential loophole here that would allow universities to deviate from the general rules applicable to that space if the private partner could impose market-driven access rights for nonprofit research purposes, and its partner university shared in those profits.

We know of no standard formula for resolving this problem. If the database is also of interest to the private sector, price discrimination and product differentiation are the preferred techniques for reducing access charges levied for public research. In any event, the trustees that manage the inter-university system should monitor and evaluate these charges, and their powers to challenge unreasonable or excessive demands would become especially important in the absence of any alternative or competing sources of supply.

This strategy, however, begs the question of whether and to what extent the universities should be allowed to retain their share of the "profits" from access charges levied against public research users. As matters stand, this is an issue that can be addressed only by the relevant discipline communities themselves, in the absence of some general norm that would not pose insuperable administrative burdens to implement.

Once access to databases available to the research commons has legitimately been gained, further restrictions on uses of the relevant data should be kept to a minimum. In principle, contractual restrictions on reuses of publicly funded data for nonprofit research purposes should not be permitted. This principle need not impede the use of conditions that require attribution or credit from researchers who make use of such data, and it can also be reconciled with provisions that defer access by certain users for specified periods of time or that impose restrictions on competing publications for a certain period of time.

This ideal principle runs into trouble, however, when confronted with the difficult problem posed by commercially valuable follow-on applications derived from databases made available to the research commons. It is one thing to posit that the academic beneficiaries of publicly funded research should be limited to the recovery of costs through access charges and should not be entitled to additional claims for follow-on uses by other nonprofit researchers. Quite different situations arise when the funding is public, but a private firm has invested its own resources to develop a follow-on application for commercial pursuits or when the initial data-generating project entailed a mix of private and public funds and the product subsequently gives rise to a commercially valuable follow-on application. These hard cases become even harder if the follow-on product primarily derives its commercial value from being a research tool universities themselves need to acquire.

Assuming, as we do, that a primary objective of any negotiated solution is to avoid gaps in the data made available for public research purposes in the horizontal domain, there is an obvious need for agreed contractual templates that would respect and preserve the commercial interests in the vertical plane identified above. This goal directly conflicts, however, with the most idealistic option set out above, which is to freely allow all follow-on applications based on data made available to the research commons, regardless of the commercial prospects or purposes and without any compensatory obligation beyond access charges (if any).

This option would represent a true public-domain approach to government-funded data, and it would fit within the traditional legal framework applied in the past to collections of data. However, it might be expected to discourage public-private partnerships formed to exploit follow-on applications of publicly funded databases, contrary to the philosophy behind the Bayh-Dole Act, although this risk is tempered by the fact that all would-be competitors who invested in such follow-on applications would find themselves on equal footing in this respect. This option would certainly discourage public-private partnerships formed to produce scientific databases from making them available to the commons if that decision automatically deprived them of any rights to follow-on applications.

A second option is to leave the problem of commercially valuable follow-on applications to freedom of contract, in which case universities and their private partners could license whom they please and exclude the rest. This solution is consistent with proposals to enact a de facto exclusive property right in noncopyrightable databases and with the philosophy behind Bayh-Dole. It would also alleviate disincentives to make databases derived from a mix of public and private funds available to the nonprofit research community.

However, this second option would relegate the problem of follow-on applications to the universities' technology transfer offices once again, which might be tempted routinely to impose the kind of grant-back and reach-through clauses that are already said to generate anticommons effects in biotechnology and that are inconsistent with the dual nature of data as both inputs and outputs of innovation. Just as a true public-domain approach tends unintentionally to impoverish the commons we seek to construct, so too a true laissez-faire approach undermines the effectiveness of that same commons and triggers a race to the bottom, as universities seek private partners solely for the purpose of occupying a privileged position with respect to follow-on applications.

A third option is to allow freely follow-on applications of databases made available to the research commons for commercial purposes while requiring their producers to pay reasonable compensation for such uses under a predetermined menu that fixes a range of royalties for a specified period of time. For maximum effect, a corollary "no holdout" provision should obligate all universities engaged in public-private database initiatives to make the resulting databases available to the research commons under this "compensatory liability" framework.

This approach enables investors in public-private database initiatives to make their data available for public research purposes without depriving them of revenue flows from follow-on applications needed to cover the costs of research and development or of the opportunities to turn a profit. At the same time, it avoids impeding access to the data for either commercial or noncommercial purposes, in which aspect it mimics a true public domain, and it creates no barriers to entry. Moreover, a compensatory liability approach implements the policies behind the Bayh-Dole Act without the overkill that occurs when publicly funded research results are subjected to exclusive property rights that impoverish the public domain and create barriers to entry to boot.

These, or other, options would require further study and analysis as part of the larger process of reconstructing the research commons we propose. It should be clear, moreover, that any solutions adopted at the outset must be viewed as experimental, subject to review in light of actual results.

The Zone of Informal Data Exchanges

The zone of informal data exchanges is populated by single researchers or laboratories or by small teams of associated researchers whose work is typically expected to lead to future publications. Because this zone operates largely in a prepublication environment or outside the ambit of federal granting agencies, the constraints of government funders on uses of data are relatively less prescriptive, and a considerable amount of the data being produced may not be funded by federal agencies at all. If funding is provided by other nonprofit sources or by state governments, the end results still pertain to public science and its ultimate disclosure norms, but the controls are not standardized. To the extent that private-sector funding is also involved, even the norms of public science may not apply.

Quantitatively, the amount of scientific data held in this informal zone appears large. Despite the relative degree of invisibility that prepublication status confers, these holdings are also of immense qualitative importance for cutting-edge research endeavors. Although these data may not be as well prepared as those released for broad, open use in conjunction with a publication, they typically will reflect the most recent findings. Moreover, this informal sector seems destined to grow even more important in the near future as it increasingly absorbs scientific data that were not released at publication as well as the data researchers continue to compile after publication. If Congress were to adopt a strong intellectual property right in noncopyrightable databases, this informal zone could expand further to include all the published data covered by an exclusive property right that had not otherwise been dedicated to the public domain.

As previously discussed, actual secrecy is taken for granted in this zone, and disclosure depends on individually brokered transactions often based on reciprocity or some quid pro quo. These fragile data streams, which have always been tenuous due to personal and strategic considerations, have increasingly broken down owing to denials of access and to a trading mentality steeped in commercial concerns that is displacing the sharing ethos.

Left to themselves, the legal and economic pressures operating in the informal zone are likely to further reduce disclosures over time and to make the informal data exchange process resemble that of the private sector. That trend, in turn, undermines the new opportunities to link even highly distributed data holdings in virtual archives or to experiment with new forms of collaborative research on a distributed, autonomous basis, as digital networks have recently made possible. The positive synergies expected from organized peer-to-peer file sharing on an open-access basis cannot be realized if researchers decline to make data available at all out of a fear of sacrificing newfound commercial opportunities or other strategic advantages. Nor will these new opportunities fully develop if those who are nominally willing to make data available impose onerous licensing terms and conditions—reinforced by intellectual property rights—that multiply transaction costs, unduly restrict the range of scientific uses permitted, or otherwise embroil those uses in anticommons effects.

Here, the immediate goal of science policy should be to reduce the technical, legal, and institutional obstacles that impede electronic peer-to-peer file exchanges and to generally facilitate exchanges of data on the most open terms possible across a horizontal or quasi-public space. At the same time, the measures adopted to implement this policy must avoid compromising or inhibiting the interests of individual participants who seek commercial applications of their research results in a private or vertical sphere of operations. This two-pronged approach could stabilize the status quo and reinvigorate the flagging cooperative ethos in the zone of informal data exchanges as more individual researchers and small communities experienced the benefits of electronically linked access to virtual archives and discovered the productive gains likely to flow from collaborative, interdisciplinary, and cross-sectoral uses.

From an institutional perspective, however, organizing and implementing such a two-pronged approach to data exchanges in the informal zone presents certain difficulties not encountered in the formal zone of inter-university relations. Here the playing field is much broader, the players are more autonomous and unruly, and the power of federal funders directly to impose top-down regulations has traditionally been weak or underutilized. The moral authority of these funders nonetheless remains strong, and peer pressures in support of the sharing ethos would become more effective if a consensus developed that the two-pronged approach we envision actually yielded tangible benefits at acceptable costs.

Much, therefore, depends on short-term, bottom-up initiatives that rely on individual decisions to opt for standardized, research-friendly licensing agreements in place of the defensive, ad hoc transactions that currently hinder the flow of data streams in this sector. Here, the solution is to provide individual researchers with a tool kit for constructing prefabricated, exchange transactions on community-approved terms and conditions. The tool kit would contain a menu of standard-form contractual templates that individual researchers could use to license data, and the templates adopted would be posted online to facilitate electronic access to networks of nodes. These templates would cover a variety of situations and offer a range of ad hoc choices, all aimed at maximizing disclosure in both digital and nondigital mediums for public research purposes.

For this endeavor to succeed, however, the templates in question would clearly need to allow participating researchers and their communities to make data available on conditions that expressly precluded licensees from unauthorized commercial uses or follow-on applications. Although this suggests the need to deviate from true public-domain principles once again, one should remember that, in the informal zone as it stands today and is likely to develop, secrecy and denial of access are already well-established, countervailing practices. One can hardly argue that permitting conditional availability would undermine the norms of science in this zone, given the inability of those norms to adequately defend the interests of public research in unrestricted flows of data at the present time.

The object, rather, is to invigorate those sharing norms by reconciling them with the commercial needs and opportunities of the researchers operating in the informal zone, to elicit more overall benefits for public science under a second-best arrangement than could be expected to emerge from brokered individual transactions in a high-protectionist legal environment. This strategy requires a judicious resort to conditionality that would make it possible to forge digitally networked links between individual data suppliers and that would let their data flow across those links into a quasi-public space relatively free of restrictions on access and use for commercial purposes.

Given the larger number of players and the disparity of interests at stake, a logical starting premise is that only a small number of standard contractual templates seems likely to win the support of the general scientific community, at least initially. A true public domain option, of course, should be available for all willing to use it. For the rest, a limited menu of conditional public-domain provisions, such as those offered by the Creative Commons, should be sufficient. Clauses that delay certain uses for a specified period, or that delay competing publications based on, or derived from, a particular database for a specified period of time should also pass muster, so long as they remain consistent with the practices of the relevant scientific subcommunity. In the absence of any underlying intellectual property right, an additional clause reserving all other rights and excluding unauthorized commercial uses and applications would complete the limited, "copyleft" concept. We believe that even a small number of standard contractual templates that facilitated access and use of scientific data for public research purposes could exert a disproportionately large impact on the increasingly open, collaborative work in the networked environment.

In the scientific milieu, however, difficult problems of leakage and enforcement could also arise. To address these problems, the scientific community, perhaps under the auspices of the American Association for the Advancement of Science, would need to consider developing institutional machinery capable of assisting individual researchers who feared that their data had been used in ways that violate the terms and conditions of the standard-form licensing agreements they had elected to employ.

More complex or refined contractual templates are also feasible, but their use should normally depend less on individual choice and more on the consensus approval of discipline-specific communities. Moreover, in the informal zone, efforts to influence the terms and conditions applicable to private-sector uses seem much less likely to succeed than are similar efforts in the inter-university context.

Attempts to overregulate the zone of informal data exchanges should generally be avoided at this stage, lest they stir up unwarranted controversy and deter the more ambitious efforts to regulate inter-university transactions described above. The success of those efforts in the zone of formal data exchanges should greatly reinforce the norms of science generally. It would also exert considerable indirect pressures on those operating in the informal zone to respect those norms and to emulate at least the spirit of any agreed contractual templates that had proved their merit in that context. The more that universities succeed in amalgamating their government-funded holdings into an effective, virtual archive or repository, the more that pressures would be brought to bear on individual researchers, research teams, and small communities to similarly make their data available in more formally constituted repositories. As a body of practice develops in both the formal and the informal zones, the most successful approaches and standards would become broadly adopted, and the desire to obtain the greater benefits likely to flow from more formalized arrangements should grow.

Meanwhile, efforts to regulate the zone of informal data exchanges should be viewed as an opportunity to strengthen the norms of science and to facilitate the creation of virtual networked archives electronically linking disparate and highly distributed data holders. The overall objective should be to generate more disclosure than would otherwise have been possible if all the players exercised their proprietary rights in total disregard of the need for a functioning research commons for nonprofit scientific pursuits. If successful, these modest efforts in the informal zone could alleviate some of the most disturbing erosions of the sharing ethos that have already occurred, and they could encourage federal funding agencies to take a more active role in regulating broader uses of research data. A successful application of "copyleft" techniques to the informal zone of academic research could also serve as a model for encouraging disclosure for public research purposes of more data generated in the private sector.

THE PRIVATE SECTOR

Scientific data produced by the private sector are logically subject to any and all of the proprietary rights that may become available. Here, the policy behind a contractually reconstructed research commons is not to defend the norms of science so much as to persuade the private sector of the benefits it stands to gain from sharing its own data with the scientific community for public research purposes. The goal is thus to promote voluntary contributions that might not otherwise be made to the true public domain or to the conditional domain for public research purposes on favorable terms and conditions.

From the perspective of public-interest research, of course, corporate contributions of otherwise proprietary data to a true public domain are the preferred option. Although the copyright paradigm reflected in the Supreme Court's *Feist* decision presumably made the factual contents of commercially valuable compilations published in hard copy formats available for such purposes, the federal appellate courts have lately rebelled against *Feist* and made it harder for second-comers to separate noncopyrightable facts and information from the elements of original selection and arrangement that still attract copyright protection. Online access to noncopyrightable facts and data is further restricted by the stronger regime that prohibits tampering with technological fences that was embodied in the Digital Millennium Copyright Act of 1998, although the full impact of these provisions on scientific pursuits remains to be seen. Meanwhile, many commercial database publishers may be expected to continue to lobby hard for a strong database protection law on the E.U. model that would limit unauthorized extraction or use of the noncopyrightable contents of factual compilations, and it appears likely that Congress will again seek to enact a database protection statute in 2004.

In contrast to the research-friendly legal rules under the print paradigm, all the factual data and noncopyrightable information collected in proprietary databases are increasingly unlikely to enter the public domain and will instead come freighted with the restricted licensing agreements, digital rights management technologies, and *sui generis* intellectual property rights that characterize a high-protectionist legal environment. Under such a regime, open access and unrestricted use become possible if private-sector database compilers donate their data to public repositories or contractually agree to waive proprietary restrictions on controls that would otherwise impede access and use for public research purposes.

Some examples of both donated and contractually stipulated public-domain data collections from the private sector already exist. An example in the first category is provided in the presentation by Shirley Dutton.[5] An example of the second type of arrangement is also provided by Michael Morgan in his presentation.[6]

---

[5]See Chapter 27 of these *Proceedings*, "Corporate Donations of Geophysical Data," by Shirley Dutton.
[6]See Chapter 28 of these *Proceedings*, "The Single Nucleotide Polymorphism Consortium," by Michael Morgan.

Although pure public-domain models initiated by industry will no doubt continue to be the exception rather than the rule, the availability of data on a conditional public-domain basis, or at least on preferential terms and conditions to the not-for-profit research community, should enjoy far broader acceptance and ought to be promoted. Certainly, the existence of contractual templates, along the lines being developed by the Creative Commons, could help to encourage private-sector entities to make conditional deposits of data for relatively unrestricted access and use by public-interest researchers.

Scientific publications by private-sector scientists provide another valuable source of research data. However, these scientists labor under increasing pressures either to limit such publications altogether or to insist that publishers allow supporting data to be made available only on conditions that aim to preserve their commercial value. Although many academics in the scientific community oppose this practice, it is exactly what would proliferate if private-sector scientists held exclusive property rights in the data that allowed them to retain control even after publication. This sobering observation might induce the scientific community to reconsider the need to allow private-sector scientists to modify the bright-line disclosure rules otherwise applied to public-sector scientists to encourage them to disclose more of their data for nonprofit research purposes.

Even when companies remain unwilling to make their data available to nonprofit researchers on a conditional public-domain basis, there is ample experience with price discrimination and product differentiation measures favorable to academics. To the extent that the public research community does not constitute the primary market segment of the commercial data producer, either of these approaches will help promote access and use by noncommercial researchers without undue risks to the data vendor's bottom line. The conditions under which such arrangements might be considered acceptable by commercial data producers will vary according to discipline area and type of data product, but it is in the interest of the public research community to identify such producers in each discipline and subdiscipline area and to negotiate favorable access and use agreements on a mutually acceptable basis.

The terms and conditions acceptable to private firms operating in the vertical dimension that opt into a public access commons arrangement might be fairly restrictive in their allowable uses, as compared with the conditions applicable under the standard-form templates implementing any of the other options discussed above. However, the goal of securing greater access to privately generated data with fewer restrictions justifies this approach because it makes data available to the research community that would otherwise be subject to commercial terms and conditions in a more research-unfriendly environment.

Finally, the importance of regulating the interface between university-generated data and private-sector applications was treated at length above, with a view to ensuring that the universities' eagerness to participate in commercial endeavors did not compromise access to, and use of, federally funded data for public research purposes. Here, in contrast, it is worth stressing the benefits that can accrue from data transfers to the private sector whenever a framework for reducing the social costs of such transfers has been worked out to the satisfaction of both the research universities and the public funding sources. These arrangements are especially important if the exploitation, or applications, of any given database by the private sector would not otherwise occur in a nonproprietary environment.

Price discrimination and product differentiation can also facilitate socially beneficial interactions between the private sector and universities. For example, companies might consider licensing certain data to commercial customers on an exclusive-use basis for a limited period of time, after which the data in question would be licensed on preferential terms to nonprofit users or even revert to an open-access status. This strategy may work successfully in the case of certain environmental data, where most commercially valuable applications are produced in real time or near-real time and can then be made available at lower cost and with fewer restrictions for retrospective research that is less time dependent. Such an approach might not work in other research areas, such as biotechnology, however, where a delay in access may not be an acceptable trade-off or that delay is too long to preserve competitive research values.

21

Strengthening Public-Domain Mechanisms in the Federal Government: A Perspective from Biological and Environmental Research

Ari Patrinos and Daniel Drell[1]

I want to begin by noting that these remarks are based on a commentary that we wrote for the June 6, 2002 issue of *Nature* and reflect our views, not those of the U.S. Department of Energy (DOE).[2]

The office in which we serve, the Office of Biological and Environmental Research at the DOE, is the steward of about half a billion dollars. Our responsibility is to oversee and manage its spending for the greatest good to the U.S. science effort in service to DOE missions and, by extension, our citizenry. Our office supports research in various scientific areas, among them environmental sciences, global climate research, medical technologies and imaging, and genomics including the Human Genome Project that we started in 1986. Among other things, we work very hard to adhere to practices and policies that promote openness of data access because we believe that experience has demonstrated that, paraphrasing Ivan Boesky, openness is good, openness works, and this maximizes the benefits to us all.

But we have to recognize and adapt to new realities. The U.S. government expenditures for fundamental research total in the neighborhood of $45 billion per year. Recent figures for fiscal year 2000 for genomics indicate that the U.S. government invests about $1.8 billion in this area, but the private sector invests close to $3 billion. Glossing over the imprecision of these numbers, it is clear that the expenditures on "postfundamental," or exploitation-focused, research by the private sector are roughly comparable, at least in its order of magnitude, to expenditures by the federal government. What the private sector chooses to do with research results and data is up to them. As more private-sector funding goes toward "upstream" (more fundamental) science, and the distinctions blur, the challenge we face is not to decry this situation any further but to try to work out accommodations that promote science. We are optimistic that it can be done, and we are particularly encouraged by Professor Berry's presentation.[3] We want to emphasize his point that, before we move toward more restrictive data policies, we should experiment, collect data, and see what happens.

In February 2001, two significant papers were published, one in *Nature*[4] and the other in *Science*,[5] reporting on the "draft" sequence of the 3.2 billion base pair human genome. The *Nature* paper derived from the multiyear

---

[1] The authors acknowledge with gratitude the helpful comments of colleagues, as well as Robert Cook-Deegan (Duke University). The views expressed herein are those of the authors and do not reflect policy of either the U.S. Department of Energy or the U.S. government.

[2] See A. Patrinos and D. Drell. 2002. "The times they are a-changin," *Nature*, 417: 589-590 (6 June).

[3] See Chapter 17 of these *Proceedings*, "Potential Effects of a Diminishing Public Domain on Fundamental Research and Education," by R. Stephen Berry.

[4] The Genome International Sequencing Consortium. 2001. "Initial sequencing and analysis of the human genome," *Nature* 409: 860-921.

[5] J. C. Venter et al. 2001. "The Sequence of the Human Genome," *Science* 291 (5507): 1304.

effort of the International Human Genome Sequencing Consortium, led by the U.S. National Human Genome Research Institute but also comprised of the DOE, the Wellcome Trust, and many other partners, whereas the *Science* paper resulted from the more recent effort of Celera Genomics Corporation, a private company based in Maryland. Although the consortium had practiced, since 1996, a policy of unrestricted rapid release (and deposition into GenBank) of their sequencing data as they proceeded, Celera put its data up on its Web site (www.celera.com) upon publication in *Science*. Celera and *Science* worked out a limited-access arrangement. On April 5, 2002, *Science* published two papers[6,7] reporting the draft genome sequences for two subspecies of rice, *Oryza sativa*; one is from Syngenta International Inc.'s Torrey Mesa Research Institute and was published with limitations on data access essentially identical to those associated with Celera's human genome sequence publication. Considerable controversy has resulted from these policies; however, this presents a challenge to make constructive suggestions for ways to move forward that might reduce the impasse and perhaps promote greater data sharing.

One possibility that has been proposed in various contexts[8] is to start a clock on the deposition of certain data whereby a journal or other depository agrees to restrict access to the source data underlying a paper for a specified duration; other data could be housed with a trustee who ensures that the data were indeed deposited at the agreed time. Careful provisions would be required both for how long the clock is set to run as well as precisely when it starts, but the idea is to permit a set duration for commercial exploitation (including the filing of patent applications) on inventions derived from the data. The U.S. Patent and Trademark Office allows up to one year before a provisional patent application is converted to a utility patent application, giving an applicant time to perform additional research toward developing an invention while retaining the early priority date; thus one year might be a reasonable time for such a clock to run, but this would be subject to negotiation. This is similar to past practices with databases such as the Protein Data Bank.

The responsibility for implementing this scheme could rest with the journal or with a respected nonprofit foundation (e.g., the Institute for Scientific Information or the Federation of American Societies for Experimental Biology). In consultation with GenBank (or a relevant public repository), the journal (or foundation) could provide access to the necessary files upon the expiration of the clock. It would be very useful to know the consequences of varying clock "periods," as well as just how much privately held data actually contribute to the commercial viability of a company; such studies would provide valuable insights.

As time goes by, data lose value both as new discoveries (and, in particular, new technologies for reacquisition of the same data) are made and as science, as is its wont, proceeds unpredictably into new areas. This clock mechanism would allow a company to publish valuable data that would otherwise remain private while offering some protection for a limited duration for the company to use the data exclusively. This role might be uncomfortable for journals and trustees, so it is important to explore fully a mechanism that all sides would have confidence in. An added concern could arise if implications for national or international security (for example, potential detection signatures in a pathogen's genomic sequence) emerged while the data were held on deposit before publication.

There is an urgent need to find ways of giving incentives to the private sector, which now controls vast amounts of valuable data that have no obvious short-term commercial value but could be of great potential research value. Most of the human genome sequence (about 98 percent of it) is noncoding; allowing greater access to this part of it would not seem to threaten Celera's stated goal of discovering candidate drugs based on those portions of the genome that encode expressed proteins. In addition, as is becoming ever clearer, the sheer volume of data from high-throughput sequencing centers (such as the Sanger Center in Great Britain or DOE's Joint Genome Institute) challenges even the most advanced and sophisticated labs to mine it for value within a reasonable time frame.

Can incentives be defined that would induce Celera, Syngenta, and other similar companies to relax the access restrictions for some of their data? That question deserves to be explored because the benefits move in both directions, with academic expertise becoming more available to private-sector companies and the science carried out in the

---

[6] J. Yu et al. 2002. "A Draft Sequence of the Rice Genome (*Oryza sativa* L. ssp *indica*)," *Science* 296: 79-92.
[7] S. A. Goff et al. 2002. A Draft Sequence of the Rice Genome (Oryza sativa L. ssp. Japonica)," *Science* 296: 92-100.
[8] Petsko, G. 2002. "Grain of Truth," *Genome Biology* 3(5): 1007.1-1007.2.

private sector becoming more accessible to academic scientists. Ideally, this becomes a "win-win" for both sectors. Echoing Professor Berry's comment earlier, this could be subjected to some exploration and evaluation.

There are several precedents for successful public-private collaboration. The Single Nucleotide Polymorphism (SNP) Consortium[9] and the IMAGE Consortium[10] both involved partnerships that placed into the public domain valuable genomic sequence information (the first on single base pair variants useful for trait mapping, the second of complementary DNA [cDNA] sequences representing expressed human genes); the commercial partners became valued contributors. They made the assessment that the value of restricting such data was not sufficient in contrast to the benefits of making these data and resources freely available tools for the intelligence and creativity of the widest possible base of researchers.

Another instructive example, one that is representative without being unique, comes from the Keck Graduate Institute in California. Industry-sponsored research carried out at the Keck Institute, involving Keck faculty, is built upon a carefully negotiated contract. Both parties work out who does what, when and where, and who will own what results. Of particular concern is the role of students whose educational needs must take precedence; this means carefully negotiating disclosure issues and rights to publish so that the attainment of their degrees is not restricted. Fundamentally, this requires the parties involved defining a work plan, benchmarks and milestones, and the terms of a mutually acceptable contract. Although this may not be easy, and the exact conditions need to be tailored to the specific parties and their needs, the Keck Institute has succeeded in using negotiated contracts with private-sector companies to attract research support, advance both the scientific work of their faculty and the education of their students, and contribute to the commercialization of scientific knowledge. In fact, practices such as this are not uncommon, although they do not garner the publicity that the disputes do.

Are some constraints on data access preferable to not seeing the data at all? We believe they are. Is the academic scientific community willing to forego the science being done outside the groves of academe? If this is to be the policy, then an increasing fraction of the 60 percent or so of genomics research conducted in the labs of private firms will remain unavailable to academic and government scientists. That is, in our view, too high a price to pay.

Our case is reinforced by actual practices among most genomics firms. They do not publish their data. The many firms sequencing cDNAs and identifying SNPs, for example, have information that would be immensely valuable to academic researchers if it were publicly available. The firms have, however, chosen not to publish those data, preferring instead to patent genes as they are characterized and to sell access to their databases under agreements that protect data as trade secrets. That is their right. We should, however, be creating incentives for companies to publish data when they choose to and to facilitate such publication within proprietary constraints, rather than clamoring for policies that will push firms toward nondisclosure.

Sir Isaac Newton is widely credited with the observation that "If I have seen farther than others, it is because I was standing on the shoulders of giants." The steady progress of science is founded on the traditional concept that individual scientists assemble knowledge "brick by brick." We believe that full and unrestricted access to fundamental research data should remain a guide star of science because centuries of experience suggest that it is the most efficient approach to promoting scientific progress and realizing its many benefits. However, we must also accept the current realities.

At no time has science ever been the exclusive province of those in academia; however, today the proportion of high-quality science taking place in the private sector (e.g., the invention of polymerase chain reaction technology and the development of cre-lox recombinase gene knockout technology) is impressive as never before. The potential in the private sector for productively collaborating with the academic or government scientist is greater than ever before. We should not bemoan this development but should welcome it. Private-sector science has its legitimate interests too. The burden of argument is on the academic sector to attract and justify greater openness on the part of private-sector science and to state clearly what the benefits to the private sector can be.

---

[9]See the SNP Consortium Web site at http://snp.cshl.org/, as well as Chapter 28 of these *Proceedings*, "The Single Nucleotide Polymorphism Consortium," by Michael Morgan for additional information.

[10]See the IMAGE Web Site at http://image.llnl.gov/ for additional information.

We offer the following conclusions for consideration by the National Academies as they explore the role of scientific and technical data and information in the public domain:

(1) Openness, by and large and as a guide for public funding of fundamental basic research, is a very successful policy because it generates data that in unpredictable ways lead to exciting insights into nature's workings.

(2) It is the appropriate role of the private sector to exploit open basic research to develop and commercialize valuable products. It is what the private sector is good at. The amount of their investment is large (and may in certain areas exceed that of the public sector) and the quality of the resulting discoveries is very high.

(3) We need to explore aggressively compromises and quid pro quos to attract private-sector companies to loosen their hold on that portion of their data that could benefit fundamental research, but in ways that do not threaten their intellectual property concerns. By working together, in creative ways, everybody can benefit.

(4) We suggest some mechanisms, none particularly novel, that could be used to increase private sector-public sector collaboration. Importantly, we think this is an area of potential opportunity.

(5) Different schemes (e.g., timers, impartial trustees, incentives, bilateral contracts, public-sector–private-sector consortia) can be put to experimental test to learn which work better, with what partners, and under what circumstances. We do not pretend to have all the answers, but we do assert that the exploration is worth undertaking.

(6) If we succeed, the scientific and financial benefits can be enormous; if we fail, so too could be the costs. Science will continue to advance regardless; but for self-evident reasons, we all would like to see it advance as rapidly as possible in the United States.

22

Academics as a Natural Haven for Open Science and Public-Domain Resources: How Far Can We Stray?

Tracy Lewis

I would like to start with the observation that has been made at this symposium, which is that, until recently, academic institutions have been regarded as a safe haven or safe harbor for open science. I would like to briefly review the rationale for that point of view. Then I would like to ask, to what extent is it possible to export the norms of openness away from the ivory tower to a corporate setting? Finally, I would like to address the issue of how the academic sector tries to accommodate both private and public sponsors of research.

The argument for why open science is such a good fit for academics hinges on two perspectives. The first is the idea that scientists derive great satisfaction from posing questions and solving problems, and to maximize that satisfaction, they should operate in an open environment in which they can share their ideas with colleagues and base their solutions on information they receive from colleagues and students. A second rationale for open science in academia, which is a bit less transparent but equally important, is that open science is really the glue that holds together the academic job market for scientists.

The argument that Paul David put forth in a very convincing fashion is that, by their very nature, scientists work in specialized areas that laypersons—in particular, university administrators and sponsors—cannot know much about.[1] As such, there is an information gap between the people who are paying the bills and employing the scientists and the scientists themselves. This presents special problems then for how to evaluate the scientists' output, what scientists to hire, what scientists to retain, and so forth.

The ingenious solution that openness allows is that other scientists working in the area, the peers, can be used as an information source to evaluate the operative science. Of course, this requires that there be full and open disclosure and that the information banks or the public domain upon which peer review takes place is as complete and current as possible. The scientists themselves derive benefit from peer review. Not only do they get feedback they also derive personal recognition from their peers and establish a professional reputation, and they signal their value to the marketplace. In addition, the information value of peer review is one that sponsors can use in making resource decisions about which scientific programs to promote and which scientists should get grants. Therefore, the open-science norm can be seen as a social equilibrium held together by a number of self-reinforcing factors.

One question we might ask is, does this paradigm work well outside of academics? When I sought to answer that, I came up with two examples. One of them is quite current—the open-software movement. The open-

---

[1] See Chapter 4 of these *Proceedings*, "'Open Science' Economics and the Logic of the Public Domain in Research: A Primer," by Paul David.

software movement, I would argue, has many attributes that are quite similar to the academic setting. It started from academic origins in the 1960s and 1970s as collaboration between university and private foundation scientists. It benefited from some visionary leadership on the part of Richard Stallman and Linus Trevault, who had a vision of promoting software in an open-source way; that is, the stock of knowledge regarding software should be allowed to pass unfettered or unrestricted to downstream users and developers. Not only was the spiritual leadership and the institutional infrastructure in place, but there was also a very important contractual relationship established, the so-called general public license, which provided the legal mechanism for having this knowledge pass downstream for use without any restrictions.

Aside from this, there are a number of supporting factors in the open-source movement, many of which have academic types of characteristics. For instance, the developers of software often benefit directly from extending and modifying the software. Often the developers are working with the users. With openness it is possible for the developers of software to delegate to the users decisions as to how to improve the software. After all, it is the users who ought to have a much better idea of which aspects of the software code should be developed and changed. As a result of this interaction between users and developers, there is an interesting and valuable cross fertilization of ideas.

We see the same needs for gratification among programmers. They enjoy sharing their latest challenges in solving software problems, just as academic scientists derive similar benefits. This is important because, like academic scientists, there are probably relatively few people who can appreciate the efforts that programmers put forth and the value of their products. Thus openness—the ability to share new ideas and to document new approaches—allows the workers in this industry to gain personal fulfillment. It also allows programmers, like academics, to signal their abilities in the marketplace so that potential employers recognize their capabilities and value.

One interesting question for economists is to ask, "Would there ever be any corporate support for open software?" They already exist in the marketplace. There are firms like Red Hat, who derive benefit from supporting open software because they can sell complementary products, they can manage the software in such a way to make it more accessible, and they can provide instruction manuals and information to move the software from the developer to the user. There are other firms such as IBM who find it in their best interest to support open software because they can use that open software with their own proprietary hardware and software products to offer a much improved product to the marketplace. So there is a combination of the open product and the proprietary product to produce an even better product. The existence of side-by-side open software and proprietary software brings into question whether open software will survive and, if so, in what segment of the market.

Open software has some inherent advantages over proprietary software. As I mentioned before, it gives rise to greater progress and cross fertilization because it is open and because users of the software give direct feedback to the developers as to the kind of products that they want. A second major advantage, which I have already mentioned, is that it is more pleasurable for a programmer to work in an open-software type of environment. He gets more feedback, he gets more direct gratification from sharing his exploits with his peers, and he is also more mobile in the marketplace because his strengths and capabilities can be signaled more directly to other employers.

There are, of course, advantages as well for proprietary software. One is that proprietary software companies can directly recoup revenues from the sale of their product. They do not need to develop complementary products, as is the case in open software. Proprietary software is more likely to appeal to a general audience because the processes of developing the software, explaining how it is used, and making it more user friendly are activities that a proprietary software manufacturer can afford to undertake as the costs can be recouped from doing so. Open software, however, tends to be more difficult, less accessible, and is a product that in reality is confined mostly to information technology professionals.

Taking all this together, what we would predict, and I think what we are seeing so far, is that there will be a continuing role for open software. Most likely, it will be in the information technology professional segment of the market, but proprietary software will probably exist side by side with open software in a different market segment—the more general user segment. This certainly is an interesting example of how some of the norms of openness we find in academics may survive in nonacademic or corporate types of settings.

Another example I will talk about briefly is the experience of the early days of Silicon Valley in the 1970s and 1980s. During that period, Silicon Valley was an exceptional region in the sense that there was an unprecedented era of technological progress and innovation. It was characterized by free exchange of information whereby scientists and researchers at different firms were allowed, and in some cases even encouraged, to share information with their colleagues at other firms, either informally over coffee or after work, or in formal settings at conferences and seminars. The information exchange among firms resulted in a considerable amount of cross licensing of technologies and ideas.

This unusual degree of openness was supported by several factors. One was an academiclike preference for disclosure. Scientists and engineers were an integral part of the commercial success of the firms in this fledgling industry. In addition, scientists and engineers had a very strong allegiance to their professional academic affiliations, probably more so than the allegiance that they had to these young upstart firms that were very small, sometimes with just 50-100 employees. As a result, given the importance of the scientists and engineers in this industry, they demanded and received as an employment condition the right to operate in an open environment where they were free to exchange ideas with colleagues at other firms. They were relatively free even to move from one firm to the next and take some of their intellectual ideas and property with them. Although the firms in this industry perhaps did not like that, there was little they could do if they wanted to attract and maintain the very best scientific and engineering talent available in Silicon Valley at that time.

There were additional factors that gave rise to openness. One was that Silicon Valley benefited from what economists would call agglomeration economies, meaning that there was a large concentration of companies in Silicon Valley working on similar or complementary products. Coincident with that was a very large, well-trained workforce of scientists and engineers concentrated in the valley at nearby schools in the San Francisco Bay area such as Berkeley and Stanford. This meant that information flows between firms and between colleagues at different firms were very easy. Also, it was easy for scientists and engineers to move from one firm to another in the valley. All of these factors conspired to produce, at least for some period of time, a very open environment in a corporate type of setting.

What conclusions can we draw from this? The first conclusion is somewhat reassuring. One can find examples where degrees of the open norm do exist outside of academia. The second is that in each of these cases in which openness did survive the corporate settings had some striking similarities with academic settings. These were all settings in which the primary players in the industry were the people doing the scientific and engineering work. These workers demanded—and, because of their importance received—special treatment, in being allowed to operate in an open environment. They were sometimes even encouraged to exchange ideas at professional meetings and to consult with colleagues at different firms. Given the special circumstances of these examples, there also is a negative message I think which comes from this. I would not expect that the norms of openness are likely to overcome the proprietary norms of most corporate settings, except in exceptional cases where the nature of the industry is such that it resembles an academic setting.

I now would like to address the issue of how universities can accommodate both public and private sources of research support. The description of the scientific open norm is somewhat more of an idea than a reality. In reality, academic institutions have for some time been facing increasing pressure to privatize. There are a number of explanations for this. The most fundamental one is that public funds are in short supply. Not only do we rely on public funds to support such admirable goals as research in public goods, but there are a whole host of other public goods such as education, welfare reform, and national defense that likewise are deserving of funding and are competing for scarce federal dollars. In addition, relying on public funding, while in principle seems like a good way to solve a lot of the proprietary concerns that one incurs when relying on private funding, nonetheless, is not a costless activity. Some analysts estimate the cost of raising one additional dollar of tax revenue ranges between 30 and 80 cents. This reflects the distortionary impact taxation has on individual employment and investment decisions. So again, the vehicle of relying on public funding in some cases is a fairly expensive and costly way to go. It should not surprise one that universities face a growing gap between their research desires and the quantity of public support for research. Universities need to fill this gap. Increasing reliance on private funding is the obvious solution.

Other factors, including the passage of the Bayh-Dole Act have created the infrastructure for universities to transfer intellectual property to the commercial sector. Coincidentally technology shocks in biomedical and computer science research have enhanced the commercial value of university research. This has induced universities to seek and obtain greater private research funding.

In addition, one can argue that the presence of private funding is likely to be disruptive to the fragile social equilibrium that I described earlier, which supports the norm of openness in academics. A number of speakers have described already how that process would unwind. However, I would like to point out that, although there are some obvious challenges to academic institutions to accommodate various sources of funding, there is also a silver lining. Private corporations, such as biotechnology firms who wish to establish a capacity to apply basic research for their own business, may find it worthwhile to strike alliances with universities and to establish their own research groups. To the extent that they wish to do so, they may be forced, if they want to get into that line of business, to accept some of the norms of openness that go along with academic research. They may also find that accepting these norms can be beneficial to them. It allows them to commit to do scientifically objective research. It gives them an advantage in attracting the best scientists and in establishing a reputation for being a leading technology firm. These opportunities aside, some very challenging steps remain for academic institutions to take to successfully accommodate different funding sources.

In concluding, I make three suggestions regarding strategies universities might undertake. First I suggest that universities carefully manage the portfolio of private and public research they undertake. Here, they could learn a lesson from accounting firms. Accounting firms have learned recently that packaging auditing and consulting services at the same time to the same clients is a bad idea. This situation positions the accounting firm in a huge conflict of interest. It is difficult for the auditing arm of an accounting firm to issue an honest statement about the financial health of a client, knowing that in doing so, it may risk losing the lucrative management consulting business that it has with that client. We have seen in recent months some of the abuses that can occur. Accounting firms have learned that it makes sense to sell off the management consulting activity to an independent firm thereby breaking up these two activities. Why? Because these two activities when grouped together just do not mesh. They present such perverse incentives that one could not possibly expect one firm to perform these contradictory activities in a satisfactory way.

This principle applies to the university as well. Universities should discipline themselves to reject private research that would enlist their advocacy or that would restrict their ability to disclose research findings. They should separate out research that hinders the universities from providing public education and research. One should apply the same principle to individual faculty. Faculty should not be asked to undertake multiple tasks, which inherently conflict and interfere with each other.

A second, and related, suggestion is universities should adopt job-related compensation. Universities undertaking research from different sponsors are going to ask their faculties to engage in various activities. It makes sense to compensate a faculty member based on his performance on the tasks he undertakes. Fine-tuning compensation to the particular job each faculty performs allows the university to target salary and resources to the most valuable areas. It also allows the university to compensate faculty according to whether their research is privately or publicly sponsored.

My final suggestion concerns the transfer of research and technology to the corporate sector. I recommend the university tailor revenue-sharing arrangements to suit the type of research transfer it undertakes. Given the different research output the university may transfer, it is unlikely that one arrangement, such as exclusive licensing, will fit all applications. Instead the university should develop a menu of transfer mechanisms conditioned on two important factors. One would be the corporate sponsor's requirements for cost recovery and exclusive access to research findings. The second factor would be the opportunity costs to other researchers of having incomplete or delayed access to the research findings. Transfer agreements should reflect these factors in computing compensation for transfer of research findings to corporate sponsors. Transfers permitting greater circulation of research to the public domain should be performed at lower cost to the sponsor.

Some of the symposium speakers have suggested one cannot expect universities acting alone to be faithful agents for the public good. If this is true, one might establish a standard for sharing arrangements. This would prevent a "race to the bottom" where universities offer overly attractive transfers to compete against others for private funding. Standards should, however, provide universities enough flexibility to tailor transfers to different types of research.

23

New Legal Approaches in the Private Sector

Jonathan Zittrain

In this presentation I hope to shed light on the current state of the public domain, especially with respect to technical and scientific data, and then describe several approaches that tend to eschew the public domain in its legal sense in favor of a rights regime that more subtly allocates power between author and downstream users.

When we talk about access and the public domain generally, what do we mean? It might be useful to discuss them in light of scientific and technical data that have occupied a large part of this symposium.

THE PROPRIETARY, THE PUBLIC DOMAIN, AND THE SPACE BETWEEN

For much scientific and technical data, typically one cannot assert a copyright—at least within the United States. These are the type of data for which there is not enough creative work in the expression to merit copyright protection. As a result, those who want to protect (or, depending on one's point of view, hoard) these data are left to other devices, such as secrecy—sharing the data with some, but not the rest of us—and contract. In this case, when the proprietor chooses to let others access the data, they impose extra "private" law, created in the transaction between publisher and consumer.

Contract, however, is always limited by "privity." I might have an agreement with you that you promise not to further share my data, but then once you do and the person with whom you have shared the data further shares them, it is no longer easy for me to limit consumption. That is, I might have an actionable disagreement with the person who violated the contract with me, but not usually with consumers downstream.

Furthermore, there is digital rights management, which Julie Cohen has addressed.[1] Even if a work is not protected legally, one may simply "privicate" it—that is, publish it far and wide, but publish it in a manner, thanks to technology, that makes it hard for people to do with it what they will, simply because their respective computers will not let them.

Finally, tied to technologies of digital rights management are laws concerning circumvention of those technologies. If a person attempts to figure out how to do something that their computer will not permit, and if they then seek to apply or share that knowledge, in many instances they could go to jail.

---

[1] See Chapter 15 of these *Proceedings*, "The Challenge of Digital Rights Management Technologies," by Julie Cohen.

In the case of creative work, one can have all the protections previously mentioned for the scientific and technical data, with the addition of copyright protection. Because copyright is the default rule for creative works, copyright holders avoid the privity problem that occurs with contract.

How does this work? If I create a work, I can assert copyright in it, provided that it is creative enough that I can. I "give it to you"—I license a copy to you or you pay me to own a copy of the work—but if you further make copies and the people downstream from you make unauthorized, unprivileged copies, I can go after all of you as a matter of law. In fact, since 1976, I do not even have to put the copyright mark in a symbol on my creative work to have the copyright attach.

So if you are surfing the Internet and encounter a wonderful haiku and there is no copyright symbol on it, it does not mean that the work is not copyrighted. There are plenty of people—lawyers usually—who get up at lecterns like this one and sow fear, uncertainty, and doubt, warning that works published online are in all probability copyrighted, so if in doubt, do not do anything with the work.

If this is the regime we have for these two types of data, what kinds of material do we have in the public domain today? We have work created by nonhoarders, those "crazy" people who give their work away for reasons that in the last session were explained to be actually quite rational. That sort of work can become part of the public domain either by choice or by patron encouragement or even requirement. For example, if a researcher receives a U.S. government grant, the terms of the grant might require that the researcher share the data and let others make derivative works from them. Yet often the patron in these types of arrangements is a university—and as others have discussed during this symposium, universities today are torn between whether they want to be dot edu or dot com. While they are trying to figure that out, universities may not be the ones to rely upon to encourage or require materials to be shared freely; indeed, they may have the opposite agenda.

So that is how the system, generally, is working. For someone wishing to release into the public domain scientific and technical data for which one cannot assert copyright in the first instance, that person need simply fail to take the previously discussed steps to protect that type of data—that is, fail to keep the data secret, fail to write a contract, and fail to create and apply a digital rights management system.

But if the work at issue is a copyrightable work, under the current system one has to take certain steps to disclaim it. A person can choose to distribute this type of material to others, either formally or informally; however, they may not know the legal steps necessary to enter it into the public domain. Instead, it is more often the case that the person chooses not to enforce the rights that are legally retained and that other people come to know that.

A final way that creative material enters the public domain is through copyright expiration. Copyright is for a limited time, which at this moment for corporate-created works in the United States is 95 years. Of course, these copyrights have been extended retroactively repeatedly, rendering such entry mostly theoretical—unless the Supreme Court holds for the plaintiffs (for whom I am co-counsel) in *Eldred v. Ashcroft*, a case currently pending before the court.[2]

We also have the informal, *de facto* public domain, brought about by the existence of photocopiers, personal computers, and the Internet. Even with a set of "background" rights reserved to the author in any creative work, the fact is that most published works—whether or not they are published online—are largely available for use. We have had a culture that permits a certain amount of copying for personal use, and many activities that would count as legalistic violations of copyright are neither frowned upon nor fought against by copyright holders, much less the general public.

The situation as it stands, then, is in flux. The large-scale publishers who usually benefit from some level of copying and sharing are now well aware that photocopiers, personal computers, and the Internet exist—and that in their current incarnations, they represent a threat to prevailing business models. These publishers are unhappy, they are litigious, and they are hiring good coders to write digital rights management systems.[3] They have a certain zealous righteousness to their position and freely use the language of theft to describe what is going on when, for

---

[2]See *Eldred v. Ashcroft* 537 U.S. ___ (2003). The Supreme Court decided in favor of the defendant. For additional information, see http://www.supremecourtus.gov/opinions/02pdf/01-618.pdf.

example, consumers use peer-to-peer networks to transfer copyrighted music files among themselves. It seems that they hope that their indignation—arguably borne out of a newly expansionist cultural view of copyright—will be adopted by future generations.

An acknowledgment of the ever-widening gulf between the attitudes of intellectual property producers and consumers is evident in a recent report on the creative industries published by the British Patent Office.[4] The report sought to consider ways of improving "the public's perception of the need for intellectual property and its relevance to so much of what we do in our jobs, at home, at leisure and in education."[5] The report states: "There is general acceptance that terminology used by IP practitioners as a result of legislative authority is cumbersome and not user friendly. An important step in achieving greater understanding and acceptance of licensing and the value of intellectual property is therefore finding an alternative way of referring to intellectual property." (The British Patent Office evidently has no favored replacement for the phrase, suggesting only that the matter be placed under further consideration.) Another recommendation in the report is that "school children should recognize their own creativity by including the copyright symbol on their course work."[6]

This reveals the cultural battle between society looking at the intellectual property regime as something that is merely holding them back from what they reasonably want and by all rights could do (before the sleeping giants awoke) and society viewing the intellectual property regime as a useful instrument with which to protect their intellectual fruits from misappropriation.

Attempts to manipulate cultural views aside, the current situation remains that there are two separate baselines for making use of others' work. For informal use—the kind of use that I have made by including others' clip art in my accompanying Powerpoint presentation—our consciences are basically the limit. For formal use, however, we turn to lawyers. If I wanted to take this presentation and publish it as part of a book, I would have to obtain copyright permissions. I am working on an Internet law casebook right now with four coauthors. Putting this book together, I am obliged to send out clearance letters for every fragment of others' work that I want to incorporate. It is a formal publishing enterprise; in a formal situation, with a company that represents a viable target for legal action, all of the defaults are reversed. Unless we literally get clearance, I cannot include the fragment in my book. Yet, again, if the book were simply a presentation for my students, all the defaults would flip back.

It is in light of these tensions between producers' and consumers' perceptions of IP and between the different standards for using other's work that we turn to approaches to promoting the public domain and open access to public domain or near-public domain materials.

NEW APPROACHES FOR PROMOTING THE PUBLIC DOMAIN

Freeing Code

You have heard in this symposium about approaches used in software licensing. As to that, I believe software approaches are blazing a trail, yet they are also legally untested. In other words, we have no idea whether some of the approaches I am about to describe actually work as a legal matter.

---

[3]While most recent cases have involved music and movie publishers, other major publishers have been waging this battle for a long time, and other speakers at this symposium noted the proprietary mindset of many scientific publishers, especially the elite ones. Cases from the mid-1980s involving Mead (Reed Elsevier's predecessor) and West arose in large part because of the new dimensions of infringement that were made possible by advances in technology. In addition, in *Williams & Wilkins Company v. U.S.* (487 F.2d 1345 (1973) (aff'd by equally divided U.S. Supreme Court at 420 U.S. 376 (1975)) claims by a publisher of scientific and medical journals against the National Institutes of Health and the National Library of Medicine for royalties on articles photocopied from copyrighted publications were denied. *Williams & Wilkins* was an early and influential case on the subject of the ramifications of the availability of photocopiers on infringement claims and fair use defenses. More recently, Reed Elsevier and Thomson have been among the most forceful lobbyists for strong database protection legislation in the United States and elsewhere.

[4]Report from the Intellectual Property Group of the Government's Creative Industries Task Force, available online at http://www.patent.gov.uk/copy/notices/pdf/ipgroup.pdf.

[5]Id. at pg. 3.

[6]Id. at Recommendation 7.11, pg. 23.

The first approach has been employed by the GNU organization to promote the free software movement. What is free software? There are four components to it: the freedom to run the program—even if you did not write it yourself—for any purpose; the freedom to study how the program works and adapt it to your own needs; the freedom to redistribute copies so you can help your neighbor; and the freedom to improve the program and release your improvements to the public so the community as a whole benefits.

Freedom here is of course described not in terms of author but rather consumer of the work, or someone who may want to do something with it. The ideology of GNU's founder, Richard Stallman, is that it would be best if all software worked this way—and copyright is not something, if all its rights are asserted, that allows those freedoms. So what is Stallman's approach, short of repealing copyright law for software (which, to be sure, he would like to do)? He does not urge software authors to release their code into the public domain simply by failing to copyright it. This approach is disfavored because in the absence of any information about copyright, the consumer may be confused, not knowing whether or not the program is truly free to be used as he or she wishes. Another concern that would arise if software were released into the public domain is the "proprietizing" of that software. If somebody takes a piece of software released into the public domain and makes something even better with it—one starts with Mosaic and ends up with Internet Explorer—she can copyright the result. It is entirely legal to take material from the public domain and use it as the basis for new, derivative, copyrighted work. So to prevent the privatization of what had originally been free, Stallman rejects the idea of releasing work into the public domain. Instead, he suggests asserting copyright with a carefully crafted "copyleft" license. One restriction of copyleft is that when software is based on an original work under copyleft, the new work must inherit the copyleft license. This makes it so that all derivative works are covered by a license whose substantive terms operate to keep the work free—free from downstream proprietization.

We have seen some other approaches of this sort in the private sector. Mozilla.org is a recent entry to the marketplace, providing the Netscape source code but putting certain restrictions in its corresponding public license so that downstream changes must themselves be free under the same kind of license. Sun Microsystems has taken a similar approach, using a self-described open-community process—and trademark as the instrument of control. With this approach, people can do what they want with certain implementations of Java, but if they stray too far in ways that might proprietize Java, Sun asserts trademark infringement. The derivative product can then no longer be called Java.

We also have a third example from the private, nongovernmental sector: the Internet Engineering Task Force, a group that has developed certain fundamental protocols, such as simple mail transfer protocol and transmission control protocol/Internet Protocol, that make the Internet work.[7] This is an example of what James Gleick called "the patent that never was."[8] The Internet Engineering Task Force does not release these standards into the public domain. Instead they have an organization, namely, the Internet Society, that holds the copyright for the purpose of replicating those standards and keeping them open a la copyleft. Internet standards are free for use, but they in fact are copyrighted by the Internet Society, and you would have to answer to them were you to try to proprietize derivative standards.

From Code to Content

Can the model formalized by Stallman for software be applied to other creative work? To address that question, I turn to a discussion of Creative Commons.

Creative Commons is a relatively new organization, founded to seed the lessons learned from the open-source (or "free software") movement.[9] Creative Commons starts with the conceptual understanding that copyright itself is a bundle of separable rights. Copyright need not mean one holds back all rights. It can mean that one holds back some rights while giving up others. Creative Commons aims to create a standardized—indeed, machine-readable—way for people to set elements of their works free categorically; that is, cleanly and clearly, but not necessarily wholly.

---

[7]See the Internet Engineering Task Force Web site at http://www.ietf.org for additional information.

[8]See James Gleick. 2000. *The Patent That Never Was*, available online at <http://www.around.com/patent.html>.

Creative Commons does have a human-readable as well as a machine-readable component; it is a basic "commons deed" that is attached to a work using Creative Commons templates. This deed explains in plain language what can and cannot be done with the work, either by consumers or by would-be authors who wish to create something new from it. There is also a lawyer-readable component, which is the legal code, the actual text of the license. The machine-readable component is the metadata, tags that make it so that computers can index, search, or display the work for others.

To keep the process simple, Creative Commons templates allow control by an author along five axes.[10] First, one can introduce the work entirely into the public domain using language provided from Creative Commons to make it clear that it is so released. If someone wants to put a work, or an excerpt of a work, in their textbook, they need not worry about sending a permission letter to anybody because the work is declared public-domain material.

Another axis deals with attribution; there is a license signifying that others can do as they please with the work, but that they must also give the original author credit for having created it.

The third axis is "noncommercial"—meaning that, in general, one can copy or otherwise exploit the work, but only for noncommercial purposes. I am not sure Creative Commons has worked through exactly what the boundaries of noncommercial are, so this is likely to be developed further as the organization matures. The endpoints, at least, are clear: one could not put a work so licensed into an anthology and sell it, but if somebody wants to take the work and talk about it or even repost it in full on a Web log, that would be fine.

Note, by the way, that the noncommercial axis is very different from the conceptual orientation of the free software movement. Richard Stallman does not care if software (or its derivatives) gets sold for money, so long as, once it is sold, the person who buys it can make as many copies as he or she wants and can see how the code works.

A fourth axis among the Creative Commons license attributes is "no derivative works," which means that a work can be copied, but it has to be copied exactly as it is found; one cannot incorporate it into any derivative work.

Finally, there is copyleft, which is to say—presuming that the author has not said "no derivative works"—that consumers are required to inherit the licensing terms, replicating them in any derivative work that they create.

Given this general plan, you can imagine Creative Commons as an organization functioning under a few different models:

(1) A central conservancy. The idea would be for creators not only to adopt the licenses, but also store the work with Creative Commons. If people went to <CreativeCommons.org> and searched for "owl," they could find pictures of owls that people have taken or rendered, with the respective license terms attached. Creative Commons has rejected this model for a number of reasons—such as the fact that they cannot necessarily build a functioning digital library given the kind of infrastructure that would be required to support it. Other reasons include the fact that Creative Commons could not necessarily verify or validate the works coming in—and perhaps the work would not really be authored by the person submitting it, thereby subjecting Creative Commons to possible liability for hosting it.

(2) Another option would be a distributed conservancy. In this conception, the material is hosted somewhere on the Internet, or perhaps in a library as a physical object, but Creative Commons would maintain a central index of works under Creative Commons licenses. One could go to <CreativeCommons.org> and look for owls, find several owls that meet the description of the clip art one wants, and follow the links to the work. That might attenuate the legal liability that Creative Commons could face for contributors misrepresenting or simply being mistaken about what rights they can convey or release with regard to the material in question.

(3) Finally, Creative Commons could build a completely distributed conservancy with no index at all. The main product of Creative Commons in this conception would be the licenses themselves. How would there then be an index? One would simply go to a search engine like Google, type in "owls" and add search terms matching metatags indicating the presence of Creative Commons licenses.

As far as I can tell, Creative Commons is somewhere between models (2) and (3) in this taxonomy.

---

[9]See the Creative Commons Web site at http://www.creativecommons.org for additional information.
[10]See http://www.creativecommons.org/learn/licenses/ for an explanation of the Creative Commons licensing options.

My own sense so far is that Creative Commons is a spectacular project. It is interesting to note that it is not necessary or even likely that a work associated with Creative Commons will literally enter the public domain; the project is more complex and more flexible. Should we then ask whether the glass is half empty or half full? Creative Commons gives people an easy means—a computer-readable, searchable means—to assert what they desire with respect to content they create. This may result in more content being more clearly omitted from the public domain than clearly placed there, as candidates for Creative Commons licenses could be drawn from works that would otherwise be released entirely into the public domain, rather than works that would otherwise be copyrighted in the traditional way. Digital rights management systems could easily be placed on top of the machine-readable code, making the licenses self-enforcing. So you might even think of Creative Commons as converging with what the British Patent Office report called for: a greater understanding by people of the rights they have and can assert in the works that they create; an easy, simple, nonlawyerly way of expressing that desire; and, to the extent possible through machine execution, having it be so. One should note that this is a goal different—perhaps laudable, but different—from the goal of having as much material as possible enter into the public domain.

Another aspect of this discussion is important to note: the problem of derivative works is less a problem in relation to creative works than it is in software. In creating software, the purpose of using another's source code is to create a new code that does something else. In this sense, one directly "builds on the shoulders of giants" with software—something one does more conceptually than literally with creative work. An author does not literally have to incorporate another's poem into her own to have a successful or meaningful new work.

Finally, in the area of the scientific and technical data not subject, at least in the United States, to copyright protection, Creative Commons offers an opportunity to affirmatively and publicly catalogue the data, allowing the author to make clear the fact of his or her original production of them for attributive purposes. Scientific researchers thus can encourage maximum dissemination of their discoveries and methods, without sacrificing the fact of their own contributions to those discoveries.

What, then, is the real value of Creative Commons? First, it helps us to identify works intended for the public domain. Second, it helps people join the cultural melee. This is a battle over the description of rights, an assertion of copyright as an instrument, and not just an instrument for control. That part we understand quite well. Creative Commons helps to underscore the fact that a legitimate use of copyright is not simply to stop others from copying, but also to give permission—to imagine that something other than "All Rights Reserved" could be the phrase that follows one's assertion of copyright ownership.

Will Creative Commons work? That will depend on the value of the work committed to the public domain, or at least to public use, under the Creative Commons system. Exactly what sorts of authors and work it will attract, we have no idea. This is one of those ideas so new that one really has to make a leap of faith to see if it is going to work. And then, if it does work, it will seem obvious that it was a great idea whose time had long since come.

24

Designing Public–Private Transactions that Foster Innovation

Stephen Maurer

Most university technology licenses are extremely conventional. The university selects one partner and gives it an exclusive patent license. The partner—who now has a monopoly in the university's technology—promises to pay royalties. However, this "exclusive licensing" model is only one possible transaction. Many alternative business models are possible and some have already been tried. We need to figure out which of these new ideas are wise and which would be utter disasters.

I will start by reminding you that patents have important drawbacks for society and that it is often preferable to leave discoveries unprotected—that is, putting them into "the public domain." I will then look at why universities focus so heavily on exclusive licenses. Finally, I will discuss 10 alternative licensing models that can often do a better job of spreading knowledge.

INTELLECTUAL PROPERTY VERSUS THE PUBLIC DOMAIN

It is easy to forget that intellectual property (IP) rights are, in fact, monopolies. They create incentives by letting the inventor stop others from using his invention. This creates an artificial scarcity in knowledge in exactly the same way that a baker's cartel creates an artificial scarcity in bread. Legislators and judges have always known this. In fact, the first patent statutes were at least as concerned with limiting the IP monopoly as creating incentives. That is why the English Parliament called its first patent statute "The Statute of Monopolies." Thomas Jefferson agreed. When he set up the American patent system, he said that his central task—which I think is our central task, too—was to draw "a line between the things that are worth to the public the embarrassment of an exclusive patent, and those which are not."[1]

In addition to monopoly, patents have two other potential drawbacks. If I am allowed to patent a particular idea, how do I make money from it? The most obvious way is to hire someone to develop it, i.e., to turn it into products. But if I want to make a profit, I must keep my costs down. So I am not going to hire everyone in the world. Instead, I will I hire one person or maybe two, and see what they develop. This approach works well when development is obvious and straightforward, but that is not always the case. In fact, there are two reasons why it may be better to leave the idea in the public domain. First, suppose that the product made from the idea is patentable. In that case, the whole world ends up racing to develop the idea. This means that society gets the product faster. This is a very valuable benefit when the underlying idea represents a fundamental advance like, for

---

[1] Quoted in *Graham v. John Deere Co.,* 383 US 1 at 11 (1966).

example, the laser. If we believe that universities produce more than their fair share of fundamental advances, we should put their discoveries into the public domain.

Second, some ideas are "embryonic," i.e., the strategy for developing them is not immediately obvious. Again, many university inventions fit this profile. Suppose that a biologist discovers that a certain protein binds to the outside of a cell but has absolutely no idea what it does. Turning that kind of idea into a workable product is fraught with peril. If the idea is patented, the patent holder will hire one or two people to attempt development. They could easily fail. But if the idea is not patented, 500 people may decide to take a shot at the problem. If just one of them succeeds, society will receive a huge payoff.

WHY UNIVERSITIES LIKE EXCLUSIVE LICENSES

If patents and exclusive licenses have such drawbacks, why do universities favor them? There are at least four reasons. The first reason is ideological. Congress passed the Bayh–Dole Act because it believed that industry would not develop university inventions unless it received IP rights. There is some evidence for this position. If you go out and ask companies if they would have done a particular investment without licensing rights, they sometimes tell you—and it may be an honest answer—that they would have said "no." But there are also many cases where companies were already using the invention before the university got wind of it and demanded royalties. Now the university has gone from disseminating knowledge to taxing it. At least potentially, Bayh-Dole has become a drag on innovation.

The second reason that exclusive licenses are popular is that they have a powerful constituency. There is the licensing officer, who is praised for bringing in cash. There is the faculty entrepreneur, who would like to get rich. And there is the university administrator or state legislator, who hopes to make the university at least partially self-supporting. All of these groups want to maximize revenues. But the best way to maximize revenues is to act like a monopolist. In other words, stick with the exclusive licensing model.

The third reason that exclusive licenses are popular is that universities tend to produce embryonic ideas. University technology officers and faculty members frequently say that they have no way of knowing whether an individual idea will be valuable or not. So they obtain as many IP rights as they can. Now it is one thing to patent an idea because you plan to develop it. That is an investment. But it is quite another to patent an idea *in case* it later turns out to be valuable. That is a lottery. The result is overpatenting and a shrunken public domain.

The final reason is that many people think that "exclusive licenses" and "start-up companies" are evidence of economic development. This is misguided in my view and comes from the fact that it is much easier to count licenses than to track the number of people who use the public domain. Nevertheless, the fact remains that many universities subsidize patents and exclusive rights transactions. At the same time, hardly anyone subsidizes the Creative Commons or other efforts to expand the public domain. So the net effect is that society is paying people to shrink the public domain.

LOOKING AT THE ALTERNATIVES

I want to describe three categories of alternative transactions. I will argue that the first category consists of transactions that are unambiguously good for society. The second category consists of transactions where the "core" value of scientific data—the right to use, modify, and republish information—is unimpaired. I argue that this situation is always preferable to a conventional exclusive license. Finally, I will describe a variety of transactions that are *usually* preferable to an exclusive license. Even though these need to be judged on a case-by-case basis, we should probably encourage them. One way to do this is to put such transactions into some type of "favored" or "safe harbor" category.

Situation 1: No Public Funding

The first category involves situations in which the experiment cannot be done with public money or charity. In other words, nothing will happen without the private sector. Of course, you should still try to negotiate the least restrictive license you can. But suppose that your negotiating partner insists on an exclusive license. Should you agree to give the private sector a monopoly right? Many people are surprised to learn that the answer is "absolutely, yes." A monopoly price is bad, but it is still preferable to not getting the information at all. You should do the exclusive license.

The clearest historical example I know about is cyclotrons in the 1930s. It was the first big physics research program. It absolutely could not have been done with government money. E.O. Lawrence made a deal with the Research Corporation. The Research Corporation figured that if it got enough cyclotron patents it could make cheap versions of radium and corner the medical isotopes market. Now, as most of you know, World War II and the Manhattan Project intervened, so that medical isotopes became plentiful. But suppose they had not. Suppose the Research Corporation had cornered the market. Even if the new isotopes were only a few pennies cheaper than radium, society would have still been better off. Would they have been as well off as if the government had funded the whole thing? No. But at least we would have had the information.

There is a point here about the federal science agencies. We have to reinvent the way that agencies do things. In this example, I assumed that the government was never going to fund cyclotrons. In many cases, however, scientists do not really know whether the government is willing to fund their projects or not. Only the agency knows. So my suggestion requires candor: The agency must be willing to say "We're never going to fund this—go out and make the best deal you can."

Situation 2: Tangential Applications

It turns out that scientific data have uses besides "doing science." Can we sell those uses while leaving the core ability to consult, modify, and republish data in the public domain? And, if we do, can we earn enough money to support the database? The dot coms have produced several business models that are worth considering.

The first model is to sell companies the right to use the database as content or a traffic builder. The basic idea is that posting data on a website "attracts eyeballs" to the host's other products. I will discuss an example of such a deal below.

The second model is to run an alert service. Suppose you have a database and you keep putting new data into it. If you know that somebody is interested in a particular category of data, you might send them an e-mail alert whenever relevant information is added. That service is worth something.

The third model is to sell advertisements. In the general dot com world ads are incredibly lucrative. Although ads are less lucrative in science, some journals have said that they can generate about $250 per published paper.

Finally, there is data delivery. Companies sometimes pay hundreds and even thousands of dollars to receive data in special formats. For example, you can sell them a CD-ROM version that they can archive or make special arrangements to deliver the data behind their firewall.

I can see the emergence of a principle where we should always choose transactions that leave people free to do science. Such deals are always preferable to traditional exclusive licenses. They are also preferable to some recent experiments designed to make academic databases self-supporting. The most famous example is probably the SWIS/PROT database in biology. It imposes substantial restrictions on academic users' ability to modify and republish entries.

Situation 3: Safe Harbors

The third category consists of business models that are usually preferable to exclusive licenses. Unlike the first two categories, you cannot say that such terms are *always* preferable. However, it still makes sense to create some type of favorable presumption or "safe harbor" that such deals ought to be approved.

The first type of business model involves selling updates. Dr. Bretherton talked about the SeaWiFS (Sea-Viewing Wide Field-of-View Sensor) experiment, where commercial fisheries receive data 2 weeks before the scientists do. NASA paid approximately 40 percent of what this mission normally would have cost. Frankly, it sounds like they got a good deal. Updates have also been tried in biology. For example, Cardiff University's Human Gene Mutation Database has been very controversial. Celera's commercial subscribers receive Human

Gene Mutation Database updates 1 full year before the general public does. Biologists are up in arms about that, and maybe they should be. But now we are arguing about how long the embargo should be. The answer will almost certainly vary from case to case.

The second model is to provide an enhanced version of a basic (no-charge) database. Companies will often pay a lot of money for extra bells and whistles. In principle, the money can help fund your basic service. My background paper for this conference describes a deal between a corporation called Synx and a biology database called the Human Genome Database.[2] Synx would have received the right to produce a premium version of the Human Genome Database; in return, it would have paid royalties and delivered certain software to make the public-domain version stronger.

The third model involves offering academic users a different, cheaper subscription rate. This strategy is ambiguous because it can be used to maximize revenue. For example, Celera's academic licenses generate a big percentage of the company's total income stream. On the other hand, you can also imagine setting prices to *minimize* the cost of academic subscriptions subject to the database breaking even. Should you do that? There is a large group of people who say that data should always cost exactly nothing. On the other hand, no such rule exists for other inputs you need to do science. For example, hardly anyone complains that *Nature* charges for a subscription. So it is a plausible question: Do data prices become reasonable when they are comparable to journal subscriptions?

The fourth model involves disappearing IP rights. Just because patents last 20 years does not mean that exclusive rights have to. During the 1980s, Harvard did a deal with Monsanto where the exclusive right disappeared long before the patent did.

The fifth model involves fixed-fee contracts. For reasons I will not get into, many of the economic distortions caused by monopolies go away if royalties are fixed in advance. What is really pernicious is the so-called running royalties, where I get a slice of your sales. So to the extent that universities negotiate fixed fees, that is a good improvement.

The last model is nonexclusive licenses. These are tricky. If a university charges high enough royalties, it can end up licensing everyone in the world and still enforce a monopoly price. So nonexclusive licensing is only a solution if the university also charges royalties that are something less than the full-blown monopoly price. One interesting feature of the Novartis and Monsanto deals is that they both stipulated that the corporate sponsor would receive a *nonexclusive* right to discoveries. A cynic might say that this did not matter, because if I already have a nonexclusive right I am in an excellent position to demand exclusivity. After all, the university knows that this is the best way to maximize revenue. So these deals could be a kind of cat's-paw arrangement. But what if the university committed itself to give out at least one nonexclusive license? Variations on this theme are worth exploring.

THE POLITICS OF VOTING "YES"

At this point, I want to make a confession. I once negotiated a traffic builder agreement between a worldwide community of 600 mutations biologists and a company called Incyte. The deal we made was that Incyte would pay the community $2.3 million to build a worldwide database of human mutations information, which currently does not exist. In return, Incyte would have received one—and only one—exclusive right. It would be the only commercial company allowed to post the database on its Web site. There may be a more minimal version of a right that you could give to a company in exchange for real money, but it is very hard to imagine. Politically, I think that this is a bit of a test case. If you cannot get a community to agree to this deal, you probably cannot get it to agree to anything.

So what happened? The community held a meeting to discuss Incyte's proposal—and, after protracted wrangling, decided not to hold any vote at all. As far as I can tell, there were two reasons for this. First, most members were hesitant: "Incyte's proposal looks fine to me, but I just saw this 20 minutes ago. Maybe there's a catch here.

---

[2]See S.M. Maurer, "Promoting and Disseminating Knowledge: The Public/Private Interface" at http://www7.nationalacademies.org/biso/Maurer_background_paper.html.

Give me six months to think about this and I'll get back to you." Of course, six months later the deal was no longer on the table.

The other problem was that members were afraid that they would be punished for voting "yes." "Even if it is a good idea, this deal is so novel that somebody's bound to criticize me in the pages of *Nature*." This problem only got worse after the representative from the National Institutes of Health (NIH) gave her opinion. She completely ignored the Incyte proposal and suggested that her agency might consider a grant application instead. People asked themselves, "If NIH won't venture an opinion, why should I?"

WHAT AGENCIES CAN DO

Both of the foregoing problems involved a failure of leadership. Society needs to decide—in advance—what types of transactions it wants. And it needs to give people the confidence to say "yes" if somebody goes out and obtains a suitable offer. Neither of these things is likely to happen as long as saying "yes" requires a personal decision by 600 individual members.

I believe that funding agencies have a role here. However, it is not a traditional one. Deciding whether a particular transaction is in society's interest is not like peer review. You cannot answer it by appealing to scientific merit. Instead, the agency has to decide whether it has enough money to fund a particular experiment and, if not, how many rights it is willing to give the private sector to get the job done. You can even imagine a day when NIH shows up at negotiations between the private sector and the academics and says, "I'll chip in some money if you loosen these restrictions." That will require some heroic changes, but it is not fundamentally unreasonable. In fact, NIH has already become much more willing to write regulations that tell people to make sure that any private-sector deals include particular provisions. So they are getting into a hortatory mode.

Finally, the idea of doing new types of private–public deals is not a theoretical subject. My own experience is that industry will stand in line to talk to you. As soon as we talked to Incyte, Celera got jealous and asked to get involved. In the end, we had three or four firms offering to help. So these deals are feasible. It can be done. The only question is whether individual scientists feel that they have a green light to do them. If you look at the statistics, individual scientists originate most grant proposals, put together the most exclusive license deals, and create most start-up companies. So my final advice is that the funding agencies need to decide which deals are desirable, announce some clear guidelines, and then stand back. The scientists will do the rest.

25

Emerging Models for Maintaining Scientific Data in the Public Domain

Harlan Onsrud

After sitting through the presentations yesterday, I went back to my room and wrote a new talk. And after hearing further presentations today I think we need a reality check, or perhaps more accurately a view from the trenches. As we all know, the price that private scholarly publishers set for maximizing profits is not the same price that one would set for maximizing the scientific uses or distribution of the scholarly literature or databases. If a publisher can double the price for access to their journals or databases and lose fewer than half of their subscribers, their overall profits will increase. So this is of course what they have been doing.

To get my biases on the table, I teach at one of the universities that has been marginalized by this process. In this new information age, at my university, our ability to access the scholarly literature actually is decreasing each year as our library is forced to subscribe to fewer and fewer electronic journals. In addition, unlike paper journals where we still had a copy on the shelf, what we had access to last year is now gone.

So how should we as scientists react to this situation? One of my first reactions was tell our dean of libraries to simply cancel wholesale all the publications of those publishers that are the worst offenders, to take a political stance. She says that it is alright for the Massachusetts Institute of Technology (MIT) to do so but if she does it she will look like a crackpot. She has professors literally begging her to continue to subscribe to journals that now annually are approaching the cost of cars. Of course, no matter how high the price goes, the MITs and Caltechs will be in that upper portion of the academic market best able to afford access to such journals. It is those of us in the lower third of research university wealth that are particularly hard-pressed.

Should we simply abandon scholarly work and research in our poorer states? That is, should we allow the marketplace to determine which universities will have access to the core research literature and which will not? When you look at a state like Maine we have one of the highest high school graduation rates in the nation, and those graduates rank very high in the National Assessment of Educational Progress in Science. Should we admit that these highly qualified students should not have the opportunity to contribute to the advancement of science at the university level? With a statewide average per capita income well below the national average, few of our Maine high school students can choose to attend universities at out-of-state or Ivy League tuition rates.

So if you believe in the democracy of education, what can we as individual scientists and professors do? I argue to my peers that support of the public domain begins at home. I webcast my graduate course class sessions openly on the Web. I publish my syllabi, class notes, course materials, and most of my peer-reviewed journal articles openly on the Web. If we can do it at poor universities why is it that our leading universities do not already have hundreds of their courses openly webcast and openly archived? It is not a big technical burden. Just do it.

The faculty senate at my university proposed, and the university system administration adopted, a new formal intellectual property policy with a strong presumption of ownership by professors in the copyrightable teaching, scholarly, and multimedia works they produce. Furthermore, all professors are highly encouraged by this policy to make their works available through open-access licenses or to place them in the public domain. Thus, at my university, it is now clear that faculty members do have the power to place works into the public commons.

Let us assume that the Creative Commons project proves to be a great success and hundreds of scientists start attaching open-access licenses to their articles and datasets before submitting them for peer review. I already attach such licenses to my submitted journal articles, and those articles are rejected summarily by most publishers without even being subjected to the peer review process.[1] I receive letters from corporate attorneys telling me why they cannot possibly accept the open commons license. Will increased submissions by other scientists help place pressure on the publishers? Possibly, but my guess is that most scientists will simply buckle.

Therefore, perhaps my university should pass a formal policy stating that any peer-reviewed publication that is not allowed to be within an openly accessible archive six months after publication shall constitute a low-grade publication comparable to a nonrefereed publication. After all, such publications inherently are of less value to the global scientific community due to their limited accessibility. Thus, on any applications or nominations for promotion, tenure, or honors, such a publication could not be listed as a qualifying prestigious publication. It does not qualify as being in the peer-reviewed, open-archival literature. If we impose this requirement at my university alone, we will indeed probably look like a bunch of crackpots. If, however, the Caltechs, MITs, Harvards, Columbias, Dukes, and other elite universities represented in this room start pursuing this approach, it very well may have an effect. However, those are the same elite universities that actually benefit from the current research and publishing paradigms.

Administrators at these first-tier universities have very little incentive to make waves. Ten years ago the Office of Technology Assessment reported that 50 percent of all federal research funding went to 33 universities, and my impression is that those numbers have not changed much.[2] These are the same universities that receive the major share of corporate funding and are the primary beneficiaries of Bayh–Dole. It would take very enlightened administrators believing in a broader sense of scientific peerage and long-term preservation of science to actually risk altering an incentive system of which their own faculty are the primary beneficiaries.[3] Will it happen? Who knows? Hope springs eternal.

Perhaps in reporting the progress of past work when submitting research proposals to the National Science Foundation and the National Institutes of Health, scientists should be requested to list only those published articles that are available in openly accessible electronic archives. Right now when scientists at universities in the medium to lower tiers of wealth are requested to peer review research proposals, we have no way of accessing much of the referenced work because our libraries do not have access to those works. This degrades the peer review process and the overall quality of science. In my department, we serve as editors for and regularly publish in journals that our library says it cannot afford to subscribe to. The National Science Foundation and the National Institutes of Health have the power to fix this situation. However, it is not even on their radar screens. Why not? Because it is not a high priority for the top 50 research universities because those universities will never be cut out of the literature access loop by marketplace dynamics. The current situation, however, is perpetuating a caste system in the ability to do high-quality research in our nation. We can talk all we want to about economic and legal theories. However, to arrive at a sustainable intellectual commons in scientific and technical information, we will need to

---

[1] Self-archiving of an electronic prerefereed version can help circumvent some legal issues. See, by example, frequently asked questions at http://www.eprints.org. However, this approach currently when applied generally results in cumbersome metadata and corrigenda maintenance issues.

[2] Fuller, Steve. 2002. "The Road Not Taken: Revisiting the Original New Deal," Chap. 16 in Mirowski and Sent, eds., *Science Bought and Sold*, University of Chicago Press, Chicago, IL, p. 447 referring to Office of Technology Assesment, 1991. *Federally Funded Research: Decisions for a Decade*, U.S. Government Printing Office, Washington, D.C., pp. 263-265.

[3] We obviously have some enlightened administrators. By example, Caltech has been a leader in research self-archiving (http://caltechcstr.library.caltech.edu) whereas MIT has been a leader in making teaching materials accessible (http:// http://ocw.mit.edu/index.html). Yet motivations and constraints vary among universities. Thus, partial solutions by one university in addressing academic literature access problems may be impractical or not as useful for use or emulation by many others.

provide incentive systems whereby it becomes very obvious to individual researchers that they will be far better off in terms of prestige and other rewards if they publish in forums where their works will be openly and legally accessible in long-term archives. Again, I advise the funding agencies to just do it.

Now, please do not misunderstand my comments. There are actually some great benefits in pursuing research in the hinterlands of science. I work with a small group of scientists that have managed on average a couple million dollars of research funding over the past few years. All of these professors regularly receive offers to move elsewhere. But, as Paul David said yesterday,[4] to do truly innovative work that is on the fringes of established research fields you are sometimes far better off to actually break away from those fields. That perfectly describes the researchers I work with. They are doing very high-quality research. Access to scientific data and information is just as critical to our faculty and students as it is to those at the top research universities, but we currently work under a very different access environment.

Go ahead and point at the publishers but, as Pogo said, "We have met the enemy and he is us." We hold much of the solution to our own access problems in our own hands. The solution rests in the hands of scientists, funding agencies, and university administrators. Our goal should be to provide all university students with the same access to scientific and technical data and literature that the leading research universities have. We would all be far better off.

I did not come here to talk about any of what I have just talked about, however. I came to talk about some success stories, some emerging approaches for the widespread sharing of data and information that are actually working and hold out great promise for scientists globally. In particular I wanted to talk about CiteSeer (formerly Research Index), which purportedly provides access to the largest collection of openly accessible full-text scientific literature on the Web. Its legal and technological approaches are very different from most other archiving efforts on the Web, and the social dynamics it has created in the scientific community also are very different. This system currently contains over five million citations and a half-million full-text articles. How can it be legal to have a half-million full-text articles openly accessible through this system when no one gave Research Index permission to copy those articles?

CITESEER[5]

Most of you are familiar with at least some of the major specialty collections of full-text journal articles that are freely accessible on the Web. For instance, the NASA Astrophysics Data System has 300,000 full-text articles online; Highwire Press has about the same number but focused in the biomedicine and life science fields. There is also the ArXiv at Los Alamos, PubMed Central, etc. Most of these online archives deal with intellectual property issues on a journal-by-journal negotiation basis or have scientists submit original work directly to their archive.

Scientists and graduate students in my research field typically need to access articles and datasets across a broad range of disciplines, including various branches of engineering, computer science, the social sciences, and even the legal literature. Many of us would prefer the ability to cite across any and all scholarly domains and link from any citation we find on the Web to the full-text copy of that article on the open Web. One approach that is being used to index and access the computer science literature is to search and crawl the entire Web. They do this using an algorithmic approach to find citations that are germane to the computer science literature, and then the system allows you to directly link to any full-text article that is found. It works on a citation-to-citation basis. From my perspective this is far preferred for indexing and accessing literature across and among scholarly domains. If you go to the CiteSeer Web site[6] today you will find about five million distinct citations within the computer science literature that have been drawn from about 400,000 full-text online articles. The system also has some

---

[4] See Chapter 5 of the *Proceedings*, "The Economic Logic of 'Open Science' and the Balance between Private Property Rights and the Public Domain in Scientific Data and Information: A Primer," by Paul David.

[5] The following description of CiteSeer and the legal foundations of the approach are based on a presentation by the author at the Duke University School of Law Conference on the Public Domain on November 10, 2001. For more information, see http://www.law.duke.edu/pd/realcast.htm.

[6] See http://citeseer.com for more information.

useful automated tools for sifting the wheat from the chaff—in other words, for getting at the most cited and respected articles within a specific subdomain of interest.

The legal problem with this approach is in obtaining permissions to copy the half-million articles. You need to automatically copy the journal articles to test the article against your profile conditions, extract and index the citations, as well as then host copies of the full-text PDF or postscript files. The developers have not bothered asking publishers for copyright permissions, and no publisher in the computer science community has yet to complain. The system developers appear to have taken the position that (1) they gain at least some legal protection granted to Web crawlers by the Digital Millennium Copyright Act (DMCA); (2) if publishers or authors do not protect their Web sites from Web crawlers, that is their fault; and (3) if you object to the Web crawler copying any of your articles, the system managers will be happy to remove those articles but please protect your site in the future or the crawler is likely to pick them up again.[7]

Many of the full-text articles that the crawlers have copied from Web sites were of course placed there by the professors and scientists who wrote them. Can one assume that these professors retained the copyrights in their published works? Or should one assume that scientists transferred all or a portion of their copyrights to the publisher? If authors did transfer their rights to publishers, which is certainly very common in my research field, does that mean that CiteSeer is acting similar to a Napster for the computer science literature? After all, it is a facility that contributes to the illegal sharing of copyrighted articles among scientists. However, unlike Napster, the original authors or talent are not complaining or losing any money because scientists typically are not into publishing articles for compensation. Furthermore, unlike Napster, many of the lead publishers in the computer science community are member organizations whose members would rise up in revolt if their professional organization objected to the system.

This is an extremely valuable resource and it is used by thousands and thousands of scientists every day. Although you and I might use Google, our faculty and graduate students run to check CiteSeer. In fact, they will always check CiteSeer before resorting to the commercial online databases that the university subscribes to. Lee Giles at Penn State is one of the people who set this crawler running. The project started as a side assignment to one of his graduate students to find some articles on the Web that they figured must exist but that they could not find through normal channels. The code found what they wanted, and then they started to use it for broader and broader searches. So there was no initial grand scheme to create this capability. It has evolved over time as various people found it useful, improved the code, and let it run.

With 5 million citations and growing, you can now ask questions like who is the most cited person across all of the modern computer science literature? You can come to a conference and know the general citation ranking of every person in the room, who has the most cited article addressing a topic like the public domain, which journal is the most influential on the topic, and who has the most cited article in the most respected journal. You can compare the citation records of those articles that are available online with those that are not and discover that you are 4.5 times more likely to be cited if your articles are openly available online. Professors are now actively shipping in the URLs where their articles may be found so the crawler can pick up their missing full-text articles. If you are an academic, your goal in life is to be cited, to be a recognized authority, to know that your work matters. CiteSeer has created a dynamic of professors making certain that all their articles are available and readily accessible on the Web.

So far, the private scientific publishers have not been complaining about this situation. My guess is they are not likely to unless they want to be boycotted by the general computer science community. I also ponder whether the scholarly community reactions would be the same if a crawler was currently indexing and copying all openly available law review articles or biology articles.

Let us assume that you are in one of these other scholarly domains and you want to solve the legal dilemma for your own discipline. How would you do it? In my case, I would set up a Web site and call it the Public Commons of Geographic Information Science. We actually have a mock site up, but eventually I would want this site hosted by the University Consortium for Geographic Information Science, the group of 70 universities and research institutions housing the leading GIScience research programs in the United States, to give it credibility.

---

[7] Terms of Service at http://citeseer.nj.nec.com/terms.html.

The commons or open library has four components: (1) open access to GIScience literature, (2) open access to GIScience course materials, (3) the public commons of geographic data, and (4) open-source GIS software. Our basic rule in designing this online material is to keep it very simple for the scientist. In a single paragraph we tell them what is wrong with the current publishing paradigm. In the second, we present a solution. Third, we walk them through four steps that solve their journal copyright and access problems.

Step 1 focuses on submitting articles to GIScience journals. In submitting your work to a scientific journal for peer review, we recommend that you place the following notice on your work prior to submission: "This work, entitled [xxx], is distributed under the terms of the Public Library of Science Open Access License...." That license essentially says anyone can copy the article for free as long as attribution is given. Then we provide an optional statement: "Although this license is in effect immediately and irrevocably, the authors agree to *not* make the article available to any publicly accessible archive until it is first published, is withdrawn from publication, or is rejected for publication." Note that we give scientists one recommendation and no options. We do not give them a suite of open-access licenses to choose from. Most research scientists and engineers could care less about analyzing the law and social policy. They just want your best shot at supplying a legal solution for the discipline. The Creative Commons project is taking a similar approach by offering only a very limited set of license provisions for users to chose from.

Step 2 concerns which GIScience journals will publish articles subject to prior rights to the public existing in the article. Ultimately, I think most journals in my field will have to come around. However, the solution is not to beg journals in your discipline to accept an open-access license. One potential solution is to list the primary journals in the field and then have scientists report back whether use of the license was accepted. Those journals that allow open-access licenses will have a substantial competitive advantage in attracting submissions, particularly when most scientists discover that they are four times as likely to be cited by other scientists if their articles are openly available.

Step 3 focuses on how the researcher can ensure that others will find their published articles through a widely accessible citation indexing system. This step includes a description of CiteSeer. The problem in the GIScience discipline is that our scientists publish across a broad range of literature, not just the computer science literature. Therefore, we either need to set up a CiteSeer capability with algorithms and keywords developed for our specific domain or hope that someone comes along to scale up CiteSeer to cover all science literature on the web.

Step 4 involves ensuring that a researcher's article, once published, is maintained in a long-term public electronic archive in addition to sitting on a server at his university or on the server of his publisher. Of course there is no longer a default right to archive in a world of electronic licensing. Scientists can overcome the legal impediments to archiving by following the open-access licensing approach recommended in step 1. However, many long-term technical archiving issues still remain.

My primary interest is in developing a public commons of geographic data. The challenges in that instance are far greater than for open-access sharing of journal articles, particularly if the vision is one of hundreds of thousands of people creating spatial datasets, generating the metadata for those works, and then freely sharing the files.

Libraries in our local communities do not exist as a result of operation of the marketplace. Public libraries exist in our communities because a majority of citizens agree that the tax money we spend on them results in substantial benefits for our communities. Similarly, do not expect digital libraries that provide substantial public-goods benefits to be developed or maintained by the marketplace.

FURTHER LEGAL DISCUSSION

The computer science literature implementation of CiteSeer links to and maintains copies of over a half-million full-text online journal articles that may be freely accessed. Yet CiteSeer is in a different legal position to that of most other online archives. No authors or publishers are asked permission regarding whether the CiteSeer Web crawler may copy and retain articles. How can the copying and provision of access to over a half-million articles without gaining explicit permission of the copyright holders be legal?

CiteSeer finds the articles that it indexes by crawling the Web. To index found articles, the software copies each PDF or postscript article it finds, converts it automatically to ASCII, searches for keywords, and then extracts and processes appropriate indexing information. A link is automatically provided in the database created by the software to the URL where the article was found. Typically this link is to the Web site of the author. As a backup, in case the link to the author's Web site is down, and as a means to provide more efficient access to the author-hosted article, the system caches a copy of the article and provides it in various formats that may be directly linked by users.

Articles are seldom copied by the crawler from, for instance, commercial publishers of scientific articles because those sites are typically protected by passwords or other technological protections. Unless permission is granted, CiteSeer indexes only articles found publicly available on the Web without charge.[8] Furthermore, CiteSeer purports to adhere to all known standards for limiting robots and caching. As the system has evolved, most articles now being indexed are those submitted by authors.[9]

CiteSeer gains its legal basis for existence primarily through the DMCA of 1998. Title II of the DMCA added a new Section 512 to the Copyright Act to create four new limitations on liability for copyright infringement by online service providers. The limitations are based on conduct by a service provider in the areas of (1) transitory communications, (2) system caching, (3) storage of information on systems or networks at direction of users, and (4) information location tools.

The legal issues are varied and complex but, by example, one issue involves whether authors have authority to post their scientific articles on the Web. If CiteSeer picks up an article for which exclusive rights were given up by an author to a publisher, is CiteSeer liable or is the author liable for the violation? Section (d) of the DMCA on Information Location Tools states explicitly that "[a] service provider shall not be liable . . . for infringement of copyright by reason of the provider . . . linking users to an online location containing infringing material . . . by using information location tools . . . ." This exclusion from liability is followed by a list of three conditions that the operation of CiteSeer appears to meet. Similarly, in those instances in which an author specifically submits a URL to the system so that material can be picked up by the automated system, Section (c) of the statute on Information Residing on Systems or Networks at Direction of Users states "[a] service provider shall not be liable . . . for infringement of copyright by reason of the storage at the direction of a user of material . . . ." This exclusion from liability is followed by a list of three conditions that the operation of CiteSeer again appears to meet.

The most tenuous part of the legal position of CiteSeer relates to caching of the full text of articles. It is not clear that the exclusion from liability under Section (b) on System Caching applies to CiteSeer. Nor does it appear to apply to most other nonprofit or academic service providers because the requirement of passwords or other controls are often not applied in these exchange environments. However, a similar exclusion for caching by the typical nonprofit service provider can be argued under other sections of the Copyright Act. Furthermore, if such a challenge were ever raised under the DMCA, CiteSeer operations could convert immediately to a free subscription and registration operational environment to subvert such a challenge. Another problematic legal issue is defining the point at which allowable temporary storage of material (i.e., caching) crosses over to become longer-term storage or archiving. Even though making backups of materials on the Web sites of others is widespread and commonplace across the Web, "archiving" that exceeds "caching" arguably requires explicit permission from copyright holders.

The World Wide Web, when initiated, was clearly illegal from a plain language reading of the law by most attorneys because the Web allowed and in fact required the copying of documents without explicit permission. However, the Web was found by society to be so useful that ways were found to reinterpret and clarify the law to allow the innovation to spread. In a similar manner, the consensus appears to be that even if some lack of clarity exists in the law today with regard to the operation of systems such as CiteSeer and Google, highly useful Web-wide indexing systems are likely to be looked upon with favor by interpreters of the law. As long as the provisos in Section 512 of the DMCA are met, it appears that the approach is on relatively sound legal footing.

Fortunately, most scientific publishers have revised their copyright policies in recent years to allow authors to post their authored articles on their own Web sites. This largely negates the potential argument that CiteSeer

---

[8] See CiteSeer's Terms of Service, Paragraph 2, at http://citeseer.nj.nec.com/terms.html.

[9] Correspondence from feedback@researchindex.org dated January 11, 2003. See also paragraph 4 of Terms of Service at http://citeseer.nj.nec.com/terms.html.

operates much like the former Napster in facilitating the illegal sharing of articles among scientists. Furthermore, as required by the DMCA, systems such as CiteSeer must remove articles in a responsible manner when requested to do so by a copyright holder [Section 512(b)(c)(d)].

One of the most promising options for addressing the legal clarity issue in the long run is to encourage and facilitate the ability of scientists to grant rights to the public to use their works prior to submission of those works to publishers that do not already allow open-access archiving of scientific works. By example, the Creative Commons is working to facilitate the free availability of information, scientific data, art, and cultural materials by developing innovative, machine-readable licenses that individuals and institutions can attach to their work.[10]

---

[10]For additional information, see http://www.creativecommons.org and Chapter 23 of these *Proceedings*, "New Legal Approaches in the Private Sector," by Jonathan Zittrain.

26

The Role of the Research University in Strengthening the Intellectual Commons: the OpenCourseWare and DSpace Initiatives at MIT

Ann Wolpert

The challenging environment that has been described during the course of this symposium has the potential to profoundly affect research universities. The economists among us understand that markets are not normally passive in the face of changing economic forces, and research universities do not have the luxury of standing passively by as intellectual property laws and norms change. Research universities are mission-driven, not-for-profit enterprises. They may host technology licensing offices, but their primary mission is education and non-profit research.

Those outside the academy sometimes talk about the research university community as though it were some kind of monolithic industry. In truth, not only is there considerable variation among research universities, there is also a distinct lack of uniformity within institutions. A personal story illustrates this point. Shortly after I joined the Massachusetts Institute of Technology (MIT), I was worried about a piece of legislation that was coming before the U.S. Congress. I approached the vice president of research at the institute to express my concern and urged MIT to take a position on the issue. His response was, "if you can figure out who MIT is, then maybe you can persuade them to take a position."

We need to consider this heterogeneity, as well as traditions of intellectual freedom, when we talk about the research university. It is difficult to generalize, because healthy research universities have many diverse activities going on simultaneously under one roof, which is entirely consistent with the mission of such organizations. The mission of MIT, for example, is "to advance knowledge and educate students in science, technology and other areas of scholarship that will best serve the nation and the world in the 21st century." The mission statement goes on to say that the institute "is committed to generating, disseminating and preserving knowledge and to working with others to bring this knowledge to bear on the world's great challenges." MIT's Technology Licensing Office, like other university licensing offices, operated with the institute's mission, policies, and procedures.

Research universities play a significant role in the value chain of new knowledge creation in science and technology. At the risk of stating the obvious, research universities recruit and retain faculty and research scientists. They admit students and support them with financial aid. They cover the growing portion of the nonrecovered expenses associated with research. They invest in education and research technology, and they pay for the network infrastructure on our campuses, as well as the libraries. Libraries, by the way, are investing an increasingly significant percentage of their material and resource budgets in support of databases and database resources.

We have heard a fair amount at this symposium about the role of information technology (IT) and the growth in complexity of the scientific and technical data environment. The fact is that IT has affected other aspects of the university mission as well. Students expect to have ready access to the full panoply of digital content, data included. Faculty need an increasingly sophisticated work environment and a fair amount of IT support. Faculty also need bandwidth, systems support, and library resources delivered to the desktop; all of which require institutional investment—as do new labs, and redesigned, updated existing labs, and new buildings in which to put big science.

Research universities also need academic disciplines to consider how new research methods should be calibrated in the context of promotion and tenure decisions. There was an interesting conversation at MIT not too long ago about how faculty members can subject their work to the scrutiny of peer review when the scholar is working in an environment that is entirely electronic and on the Web.

Finally, IT in the legal, regulatory, and compliance context has caused a distinct rise in the cost of doing business for research universities. For example, libraries now license databases. The MIT libraries are probably not unusual in dedicating two full-time people to license agreement negotiation. Likewise, MIT's information systems department has a staff of lawyers who negotiate software licenses for the university. And there is now a staff member whose job it is to respond to calls relating to the provisions of the Digital Millennium Copyright Act (DMCA). Bear in mind that people can call up and demand that you take content down, and the bias is in their favor. Research activities also incur new expenses. Vice presidents of research work hard at protecting the scientists and students in their institutions.

These new costs of doing business are all total overhead. They are defensive, and they are dead-weight costs to the university. There has been some question earlier in this symposium about whether the DMCA costs money. The answer is yes, the new legal environment costs a great deal of money in dead-weight overhead.

As research universities have confronted the reality of rising dead-weight costs and diminishing flexibility, in an increasing Draconian intellectual property (IP) environment, they have become increasingly aware of the importance of advocating for and supporting openness. The best students will be diverse in their nationalities and religions. These students need access, as we have heard from Harlan Onsrud,[1] to high-quality information, and they should not have to choose between buying lunch, buying a dataset, or buying an article. Faculty should be able to teach the best way they know how, without the requirement to plow through endless permissions and approval processes to obtain the ability to use in their courses information that is now (by default) protected by IP regimes.

You have already heard a great deal about the importance of openness in research and about the need for work to have visibility and impact. I want to second Harlan Onsrud's comments on subscriptions.[2] Because as a practical matter, if the size of a subscription base is reduced to 50 institutions, and the license terms of digital access to that database prohibit interlibrary loan, then one really has to ask the question as to whether publishing in that particular journal does indeed provide appropriate visibility and impact.

One might reasonably conclude from these remarks that research universities, and the home that they provide for many scientists and engineers, are in deep trouble. Imagine for a moment that universities are not focused on teaching students, are not conducting not-for-profit research, but rather are engaged in some other enterprise. Imagine a business trying to operate under the various constraints and uncertainties that apply to research universities. Imagine that MIT was not MIT, but rather a major metropolitan newspaper publisher. The business would be in a situation where its staff authors were writing material on the premises and then sending that material to an outside third party. This newspaper would then have to buy back the work of their staff authors at an arbitrary price so that it could be used in the publication of the newspaper. Imagine, worse yet, that the third parties were near monopolies. Imagine, moreover, that the newswire services it dealt with chose to license content to it under arbitrary and unilateral terms, so that it had no control over the data stream. Then imagine that compliance with all the IP requirements that surround the content that goes into the newspaper had become unrecoverable and unsustainable. This is the situation that universities find themselves in today.

We believe, at least at MIT, that new educational and information data management strategies are required. There are two initiatives under way at MIT, which reflect a market response based on a commitment to openness.

---

[1] See Chapter 25 of these *Proceedings*, "Emerging Models for Maintaining Scientific Data in the Public Domain," by Harlan Onsrud.
[2] Ibid.

Both initiatives emerged from faculty desires and needs as they were articulated and are intended to give faculty a new set of tools that will enable them to create new approaches to and methods for managing their intellectual work. Both initiatives illuminate how profoundly the post-DMCA environment has already distorted work in the academy, and both point to the importance of initiatives such as the Creative Commons.

The first of these initiatives is the OpenCourseWare initiative.[3] OpenCourseWare intends, eventually, to put all of MIT's courses on the Web, free of charge. It illuminates the intellectual framework for how MIT approaches the challenge of teaching MIT students. In a sense, we are publishing MIT courses on the Web so that the educational strategy that MIT uses can be shared openly across the world. This is an effort to create a public good, to put into the public domain what MIT knows about teaching the kinds of students who come to MIT.

In developing this initiative, we encounter the post-DMCA problem. For example, who knows what agreements or license terms apply to the material that is embedded in the faculty member's course notes? Courses are littered with IP that may have no bonafide reference, where there is no way of tracking the ownership, and for which there is no way of understanding whether the person who contributed that information to a colleague's teaching activities also intended that it should be put up on the Internet free of charge for the world to see. These issues are very complicated, given that everything is in all probability owned by somebody.

We are working through a variety of astonishingly complex issues as a result of our attempt to make a public good out of a traditional way of teaching. Clearly there are some things that are more problematic than others, just by virtue of the way the law works. Recommended and required reading is going to be difficult to post on the Web without permission from publishers. In our first efforts to obtain permission from publishers, 80 percent of publishers denied permission to post materials free of charge on the Internet in this context, even though what was being posted was a minimal part of any one publication, and even though you could imagine it operating as advertising for that publication.

There are also complexities around software. If a faculty member uses a piece of software in his or her course, and that software comes with a particular licensing agreement to MIT, what were the terms of that licensing agreement? Was it negotiated by the department, or by the faculty member, or by the institute? Is the faculty member using a site license or an individual license? What are the terms and conditions of use of the software that faculty use to manipulate and create content that they would ordinarily consider as essential to the course?

Data are an equally big issue in the OpenCourseWare context. Faculty would like to be able to provide actual access to raw data, particularly in the case of social sciences and hard sciences, so that a student visiting the site could understand the pedagogical intention of the faculty member. Given what we have heard during this symposium in terms of reach-through claims in the patent environment, there is a similar concern about the capacity of original publishers to reach through the teaching environment and constrain what can be put on an open Web site.

In the legal environment in which we currently operate, OpenCourseWare is a publication mechanism of MIT. As such it becomes a highly visible target for those who would object to the use of anyone's IP in an environment that would otherwise pass the four tests of fair use in the institute's internal teaching activities. OpenCourseWare is not intended for profit, it does not use a significant percentage of an individual work. It is intended to be factual rather than creative. The market impact of any item is minimal at best. Yet commercial publishers are concerned about putting content up in the OpenCourseWare environment. As a consequence, we are inspired at MIT to think about enabling new tools so that faculty can behave differently. Perhaps the rules of the game can change as well.

The second initiative at MIT is an initiative called DSpace.[4] This is not a publishing enterprise. It is an institutional response to faculty having called the libraries and asked "Can I put my stuff with you? I have all this digital content, and no place to keep it. Will you take it for me?" DSpace is a way for faculty to put the good material that they have prepared and are ready to share with the world in a secure stable, preservable, dependable repository with distribution capabilities. DSpace was built in partnership with Hewlett Packard and with additional support from the Mellon Foundation, and it is being written in open source.

---

[3] For additional information see the MIT OpenCourseWare Web site at http://ocw.mit.edu/index.html.
[4] See the DSpace Web site at http://www.dspace.org for additional information.

As Professor Bretherton was describing a structure of trees, roots, and branches in his talk,[5] it seemed to me that he was characterizing a functionality such as DSpace. Institutional repositories of this kind present an opportunity for robust roots in that tree structure that would enable faculty to build repositories of work that they would like to share through just such a model of distribution and management.

There are a number of interesting issues that arise from the design of DSpace. For example, we are interested in the prospect of using Creative Commons licenses as a way of helping those who deposit material in DSpace signal the way they would like to have their material used. There are no conventions such as Creative Commons licenses available for submitters right now. So if a faculty member wants to deposit his or her work in a digital repository that will serve it to the world, and maintain it over time, there is no existing set of licenses that can be built into the metadata that will tag to identify how the work can be used going forward.

We believe that a federation of interested institutions will be needed to establish and maintain sustainability for a digital repository. So our great hope, and our reason for writing the code in open source, is that there will be sufficient interest, first across the United States and perhaps internationally, in the idea of building digital repositories at the institutional level. Despite what one hears about how easy it is to create digital content, preserving, maintaining, and keeping digital content persistently available is a research challenge. Our hope in federation is that we will be able to share that challenge across multiple institutions.

There is also no clearly established model for a relationship of this kind. Libraries themselves have quite a fine and interesting experiment that is now well over 30 years old called the Online Computer Library Center in which libraries have banded together to share cataloging data in a not-for-profit library-managed enterprise.[6] That enterprise has been an interesting model for us as we think about how one would federate digital repositories across institutions. So we think that the library community can figure out how to do this.

A final challenge to DSpace is that disciplines vary greatly in terms of what their expectations are for a repository. Some of the early adopters of the DSpace repository are faculty in ocean engineering, and they deal in datasets that are terabytes in size. Some faculty have large collections of images. Other faculty have much smaller, more text-oriented expectations, which illustrates the fact that scientists like science, not database administration. They have always expected that libraries would be there for them, and so we are. On the other hand, the challenge to us is to help scientists take advantage of new tools.

At the end of the day, research universities, scientists, and funding agencies need a new alliance. We need strategies to advance and expand research-based education. We need to be able to educate and conduct research without Draconian external rules. This probably means developing our own systems for the exchange of data and information on a direct institution-to-institution basis. We need to assure persistent availability and accessibility of research data. This probably means keeping it as close to scientists as possible, and it means new IP options like the Creative Commons need to be deployed.

We need to solve the challenge of the born-digital world. Researchers and educators now routinely produce work that has no paper analog. We know that a great deal of work already has been lost, and we are deeply concerned that there are no easy ways to approach the long-term archiving of work that is digital only. As such, we are faced with the prospect of sentencing work to a five-year shelf life—or only for as long as the proprietary software is interested in addressing the problem.

Last, we need to solve the archiving problems. Bruce Perens and I were talking about some work that he has done in restoring works that Disney owns that were damaged. The cost and effort were phenomenal. Clearly, the losses are mounting similarly in the higher education and scientific communities. Yet we do not have Disney's money. We need a long-term solution to archiving.

Through OpenCourseWare and DSpace, MIT is working hard to develop some prototypes, to share the ideas and the software behind those prototypes, and to interest others in joining us in meeting the challenge.

---

[5]See Chapter 8 of these *Proceedings*, "The Role, Value, and Limits of S&T Data and Information in the Public Domain for Research: Earth and Environmental Sciences," by Francis Bretherton.

[6]See the Online Computer Library Center Web site at http://www.oclc.org for additional information.

27

New Paradigms in Industry: Corporate Donations of Geophysical Data

Shirley Dutton

I would like to discuss some recent donations from private industry of geophysical data that had been completely proprietary and now are in the public domain. Three energy companies recently donated large collections of rock samples to The University of Texas at Austin. I am a geologist at the Bureau of Economic Geology, which is a research division of The University of Texas. As a geologist, I find these donations of unique rock collections to be very exciting.

These data had been held completely proprietary from the time they were collected, but they are now in the public domain. I will stress in my presentation that these donations were made possible by the accompanying financial gifts that have been given along with the data, and also by government grants that have made it possible for The University of Texas to accept these data.

Before I tell you more about these donations, I would like to tell you a bit about what these data are and why we consider them so important. Geologists need rocks. They are the key for conducting academic research in geology, and they are also important in developing natural resources. Many of these rocks are very expensive to acquire and would be very difficult to replace. Obviously, this is not true about all rocks. A geologist can go out with a rock hammer and collect samples from an outcrop; these samples are not expensive to acquire and would be easy to replace. But rocks collected from the earth's surface only sample the thin skin of the earth. Geologists need three-dimensional data. The rock samples in the industry donations were taken in wells that were drilled two or three kilometers, or even deeper, into the earth. These are very expensive to obtain and would be very difficult to duplicate.

There is an important constraint that makes geologic data perhaps a little different than other data we have been discussing in this symposium. We are not talking about digital data; these are actual physical collections—rocks that are heavy and take up a lot of room. It takes strong people to get them out of storage and to lay them out for the researchers. These are real constraints to the ability of The University of Texas to accept a data collection. It is not simply a matter of acquiring computer data disks.

When oil companies drill wells, they can be very expensive; for example, a shallow well onshore in the United States can cost several hundred thousand dollars. Deeper wells and wells drilled offshore can cost more than a million dollars to drill. Many companies will use special drill bits to cut rock samples (cores) as a well is being drilled. In the United States, these cores are owned by the company that acquires them. They are able to keep the cores completely proprietary and use them for their own purposes. For the last 50 years, most large oil companies in the United States had their own private core collections that they would use internally within that company. In most cases, these cores were not made available to anyone outside the company.

Currently, in the oil industry, many of the research labs and private core facilities are closing. The companies are finding that they have several choices of what to do with their core collections. They can keep them, but there is an expense involved in doing this. It is expensive to maintain a facility or rent warehouse space to store rock. Another choice is to discard the cores. This may be the least expensive choice, but it is not without cost. Core collections represent a large volume of material, and discarding the rocks would involve some expense. Or they can choose to donate the data, and several companies have recently done this. As a matter of fact, I am not aware of an instance of any large oil company that has closed a repository and thrown away the rocks. Oil company executives realize that this is valuable material that would be a shame to throw away. So they are exploring the options and developing business models for donating it instead.

The decision to donate material to a public facility must make business sense to the company considering the donation. Although cost avoidance, such as staff salaries for maintaining the collection, rental of warehouse space, and other operational costs, is a big factor in the decision to transfer data from the private to the public domain, it is rarely the only reason. These decisions are usually based on several issues that include cost savings, as well as other factors such as continuing access to the data without the overhead, preservation of the data in case there is a return to it for unforeseen reasons (oil and gas data might someday be used to explore for water, for example), goodwill with the community (data may be used for education and outreach), future disposal costs, and, when endowments to fund the facility are included, tax incentives.

The first of these donations was made in 1994 by Shell Oil Company to The University of Texas at Austin. They developed a model that allowed the university to accept this donation. Shell had a huge amount of rock material to donate, and with no additional resources, the university would not have been able to accept it. However, in addition to giving the rock material, Shell also donated the Midland, Texas warehouse where they had stored the cores, so the university would have a place to keep the rocks. The other important factor was Shell's donation of $1.3 million toward an endowment; the money in the endowment could be used by The University of Texas to operate the facility and make the data available to anyone who wanted to use them. Thus, we have an example of a physical collection that had always been proprietary, which has now entered the public domain. It can be used by academic researchers, and it can even be used by geologists from other oil companies now.

The amount of money that was provided by Shell was very generous, but it was not enough to completely endow the facility. The University of Texas received a bridging grant from the Department of Energy (DOE) to operate the facility for five years. This was very important, and enabled the donation to proceed. With this government grant to pay operating expenses, the university could reserve the money that Shell had donated. Additional money was raised to increase the endowment, and now the amount of money in the endowment covers the operating expenses of this facility.

A few years after Shell made this donation, another energy company, Altura, also wanted to donate their cores. The Altura cores were also stored in Midland. It did not make sense to have two different warehouse facilities in different parts of Midland, so Altura donated money to build another warehouse adjacent to the former Shell warehouse. This way the cores could physically be in the same place, and the same staff would be able to curate both collections. This was another important model, a little different than Shell's donation, which allowed the Altura donation to happen.

In August 2002, British Petroleum (BP) made a large donation of cores and cuttings from oil and gas exploration wells. They also donated the warehouse in Houston where they had stored this material, along with a research building and 12 acres of land. Again, following the Shell model, they made a $1.5 million donation toward an endowment. This monetary donation makes a very important contribution toward an endowment to run the facility. The university will be attempting over the next several years to build up that endowment so it is enough to operate the facility. DOE has provided another critically important grant for the initial operating expenses of this facility.

One exciting aspect of the BP donation is the office and lab facility that they gave along with the warehouse. The building includes modern core examination rooms with roller tables and excellent lighting for viewing cores.

There are also lab rooms that can be used for various analyses. Finally, there are other facilities, including conference and lecture rooms.

In addition to giving the rock material, all three of these companies have provided data to go along with it. These rocks would be worthless if we did not have any information about them, such as where they came from. So a key part of each donation is the information that comes with it. The companies have provided the name of the well, the geographic location where the well was drilled, and the depth of the cored interval. That information is key for researchers to figure out what the samples are and whether they would be useful for a particular research project. In most cases the companies have also provided a unique well number, which allows a researcher to get information from some of the commercial databases. For example, if a researcher knew that there was a core available from a particular well, she could then purchase a geophysical log from that well from a commercial database that owns the rights to the geophysical data. The information that has been supplied in addition to the rock is key to its use in the academic community. All these data are now available on the Internet, so anyone anywhere can search the holdings and decide if there is material that they would like to use.

Why, you may wonder, would a company donate proprietary geologic data to a university? The answer is that the company had completed its use of the core for exploratory purposes and therefore felt the best use of the material would be to donate it to an academic research facility. As Lord John Browne of Madingley, BP Chairman and CEO, stated in announcing the BP donation, "these are valuable books, but we've read them." At the time these cores were taken, they were acquired for a specific purpose. The company may have been operating a field and needed rock information about that field to produce it more efficiently. Or the company may have been exploring a particular geographic part of the country and needed background information about what the reservoirs in general are like in that area, and these cores provided key information. Most of these companies no longer are exploring in the United States, and in many cases they have sold their fields to smaller companies. From their point of view, they have obtained the information they needed. However, they realize that the cores retain tremendous research value, and so they are willing to make these data publicly available.

As a result, these proprietary geologic data are entering the public domain, and they form important material for current areas of research, such as stratigraphy, sedimentology, and diagenesis, as well as areas of research that we have not even considered yet. This material is now preserved, so someday in the future, when somebody realizes how to use it in paleoclimatology research, for example, or some other field that we currently do not know how we can use the information, it has been preserved and future scientists can use it. If there is a new analytical technique that is developed in the future, the material will still be there, and new uses can be made of these rock samples. Of course, it is also a very important source of material for student theses and dissertations, and it is still of value in oil and gas exploration and production. Even though there are different companies operating in the United States now, usually smaller companies, these data are still of importance to companies producing from mature oil fields.

In conclusion, these companies have developed a business model for donating proprietary rock data to the academic community. These donations allow valuable, irreplaceable physical collections to enter the public domain. These donations were made possible by the accompanying donations of the physical space in which to store the rocks, and financial resources toward the cost of curating the cores and making them available to others. Finally, government grants have made a very important contribution to enabling The University of Texas to accept these donations by providing critical funds for the initial operating expenses while an endowment builds.

28

New Paradigms in Industry: The Single Nucleotide Polymorphism Consortium

Michael Morgan

I was asked to comment on my understanding of the European Union legislation protecting databases and what effect it has had. From a personal perspective in the biomedical sciences and as an academic user, my answer is none. I think it is fair to say that until it has some effect, we are not going to take much notice of it. European legislation is often a mystery, and I am not sure how much the United Kingdom had to do with that draft legislation. I believe the database directive has now been incorporated into U.K. law.

I want to discuss both the Single Nucleotide Polymorphism (SNP) Consortium and the Human Genome Project. I am afraid most of my presentation will be thin on law and possibly too high on rhetoric. Having been engaged in a personal and direct way with these issues as a trained scientist, I find it quite difficult to be always as objective as I ought to be. To paraphrase Winston Churchill, I have always thought that lawyers should be on tap and not on top.

The Human Genome Project is a consortium involving laboratories and funding agencies around the world. The major funding organizations in the United States are the National Institutes of Health and the Department of Energy, and in the United Kingdom a private organization, the Wellcome Trust. The U.K. government did not put very much resource into this initiative. One of the things that we first discussed as we began to think how to organize the Human Genome Project was what to do about the DNA, who does this DNA belong to, and was there something special about the fact that we were dealing with human DNA rather than mouse or rat DNA.

As you have already heard, we developed a series of principles that have been called the Bermuda Rules or Bermuda Principles. In essence, in return for the enormous largesse given to very few, selected sequencing groups, the sequence data would be deposited into the public databases every 24 hours.[1] The raw information would be provided for people to use as best they could. That was a grassroots movement. That was not imposed by the funding agencies. It is remarkable how in the scientific business we are so dependent on these agencies being champions and leaders. However, this policy was very much due to two scientists, John Sulston in the United Kingdom and Bob Waterston at the University of St. Louis. They had to persuade the scientific community, the leaders of all the major research groups capable of taking on this task, and there was eventually agreement on these principles, which were ratified at subsequent meetings held in Bermuda.

The Human Genome Project was progressing rather well, until the announcement in May 1998 of the establishment of Celera. In principle, of course, there is no reason why a private entity should not go about sequencing the human or any other genome and using that information as it chose to do. The real issue that set the scientific

---

[1] For additional information on the Bermuda Principles, see www.wellcome.ac.uk.

world alight was the fact that efforts were made to close down the public-domain activity. There was a meeting at Cold Spring Harbor, during which Dr. Craig Venter told the National Institutes of Health to give up their human sequencing program and focus on the mouse.

By pure chance, the Wellcome Trust was considering a proposal to double the financing available to the Sanger Center, now the Sanger Institute, to sequence the human genome. The governors approved the award, and John Sulston and I flew to Cold Spring Harbor to announce the news to the assembled scientists and to talk to the funding agencies. As we all know, the Human Genome Project survived and there was one project in the public domain and one in the private domain. But you need to ask yourselves, what would have happened if the Wellcome Trust had not by that time become reasonably wealthy and if there had not been a public-domain Human Genome Project. Would those data now be available to scientists around the world?

There were negotiations to try to bring the private and public programs together. There were exchanges of letters. Finally, because time was pressing and we wanted to move forward, the Wellcome Trust released a letter that in essence stopped the negotiations. Celera did not react very well to this, and a cartoon was published in *Nature* purporting to show that the gene of human aggression had at last been isolated. The following week, there was a joint statement issued by the White House and 10 Downing Street, which included a declaration supporting the release of data. Unfortunately, it was misinterpreted, particularly by the financial press. It had an effect that had not been foreseen, which was the suggestion that it was inappropriate in any way to use the patent system to capture intellectual property (IP) coming out of genome research, which was not the intent. There was also a simultaneous announcement by both camps of completion of a working draft. I want to make it clear that the Wellcome Trust is not opposed to the appropriate patenting of IP, I just wanted to mention that, whereas we regard the human sequence as something that itself should not be patented, the proteins derived from that and therapies to interfere with that are entirely appropriately protected by the various IP rules.

Each of us contains three billion nucleotides in our DNA. Each of us is extremely different, yet our DNA is 99.9 percent the same. But that means there are three million differences, and those differences will be distributed throughout the three billion nucleotides. If you want to do a bit of mathematics as to what the various combinations can be, you can see that it is truly astronomical and explains why we are all so different. As a species we are more closely related through our DNA than any other species on the planet, and we understand, or at least we have some reasonable understanding, of why that is so.

These are some of the reasons why SNPs were seen to be important by the pharmaceutical industry. They in some way will reflect our different susceptibility to disease, our different susceptibility to the action of drugs. If you can understand that information with respect to whom is susceptible, for example, to diabetes or hypertension, you will be able to segment the population, advise on different changes in lifestyle, and develop appropriate drug therapies. There are many other uses. As such, industry recognized the need to understand the variation in the human genome, the SNPs. They could see that if these could be mapped out and we could understand where they were and correlate them with disease, that would be very powerful. GlaxoSmithKline estimated that it would cost approximately $250 million to get a reasonable first map of SNPs. Even for a large company like Glaxo, $250 million is a large portion of the research budget. Multiply that by the fact that each pharmaceutical company would need its own SNP map and you can see why, from a purely economic reason, industry was interested in somehow mitigating those expenses. So they all got together to explore whether a combined effort would enable them to get a SNP map cheaper, or share the cost. It was only later in the process that the possibility of public funds via the Wellcome Trust became a possibility.

One of the reasons that putting the data into the public domain was appealing was, if you get a cabal of companies together to produce information that they will share only among themselves, you can run afoul, particularly in the United States, of antitrust legislation. So we developed a model for a not-for-profit company, a 501(c)(3) in the United States, where the partners would join, agree to the workings of the organization, work out a work plan, and fund it.

The companies involved were not only the major pharmaceutical companies in the United States and Europe, but also IBM and Motorola. Motorola wanted to put SNPs on chips, exactly what is now happening. The mission of the consortium was to gather these data to serve the medical community, the life sciences community, and the membership. Industry needs proper quality control. If this resource is to in any way be used in the drug industry, it will no doubt come under the rules of the Food and Drug Administration in the United States and the equivalent agencies around the world, so this needed to be an industry standard. As such, quality issues that do not normally worry academic groups had to be built into the business plan.

Very early on, it was agreed that we needed an IP legal task force to work out how these data should be made available and in what form. It became clear that there was a risk that simply releasing the information might enable other entities to download the data, subject them to some form of IP protection, and then sell the data back to the member companies. The senior executives did not want to go back and explain to their boards that they spent a lot of money and then now had to buy the information. So a series of data policies were initiated. By January, six months after the formation of the SNP Consortium, the information was released via a Web-based site.

By the end of the project, we had mapped 1.25 million of these SNPs, and all were released into the public domain. I want you to remember that number, 1.25 million. Glaxo's original program was going to cost $250 million and was going to end up with 150,000 mapped SNPs.

A number of scientific groups around the world were commissioned under contractual conditions to determine how the data were to be put together and to determine the quality assurances, and a data center that would capture all of the information was set up. None of the data would be released to the public or to the companies until they had been validated and mapped, and then they would be released to the companies and the public at exactly the same time. Nobody saw the data ahead of anybody else. The release policy made the data available at approximately quarterly intervals, which enables the SNPs to be mapped. This enables the patent to be applied for in bundles of SNPs to establish prior art, which is the only purpose of it—a protective patenting policy.

The IP policy was to maximize the number of SNPs into the public domain at the earliest possible date and ensure that they remained free of third-party encumbrances, so that the map could be used by all without financial or other IP obligations—no charge to access the data, no licensing fees, nothing. This plan was simply to ensure that there was a priority date for the SNPs, which would then prevent anybody else from capturing them. That has worked, but the legal task force, using lawyers from each of the companies and the Wellcome Trust and using very expensive external lawyers, took a long time to come up with a deal that we all thought would work. Again, it was extremely important that we got an agreement that nobody would have prior access to the information, and that remains the case.

As I mentioned earlier, this program from scientific and economic perspectives was spectacularly successful, in that with a budget of $44 million, not $250 million, we mapped close to 1.5 million SNPs rather than the target of 150,000. As such, industry was very pleased. And, of course, we thought maybe this model could be applied elsewhere.

We are in the process of setting up a structural genomics consortium that will provide a high-throughput resource for determining the structures of human proteins. The data release principles have been agreed on, although the details have yet to be fully worked out as part of the contract between the Wellcome Trust and the companies. However, our IP experience informs us that the raw structural data should not be patented and should be made freely available to researchers everywhere. Again, all these coordinates, once they are validated, will be released into the public domain on the Web. There will be no restrictions, and they will be provided to the public and the consortium members at the same time.

At the moment, we are working with the U.K. Department of Health and the U.K. Medical Research Council on a national population collection. We will try to recruit 500,000 volunteers, aged between 40 and 65, to ask them to donate blood, and then a DNA database will be established. Access rules for the database are still under development, but it will be freely accessible. There are many things that still need to be done, including further public consultation, because there are significant privacy issues that need to be resolved. But the principles of access have been agreed. The aim is to make the information available to anybody who wants to use it for health care and public health benefits.

One unique thing about this database is that people who use the data will generate new data. As a quid pro quo, those new data will have to be redeposited in the database, so that the database will continue to grow. And, of course, the very important thing for this database is to ensure that the way in which the samples and the data are used is consonant with the consent that has been obtained from each of the volunteers.

The Wellcome Trust is committed to advancing health care through support of biomedical research. We fund the research on the basis of scientific merit, not for direct commercial benefit, although we need funds to fund research. By that same token, we are supportive of the protection of IP in an appropriate way and an appropriate time in the value chain, and with appropriate licensing terms. We are under an obligation under the Charity Commission in the United Kingdom to ensure that useful results of research are applied to public good. You cannot have exploitation of research results without protection of the IP. As such, it is a challenge.

The release of this basic information is, in my opinion, definitely in the public interest, especially for genome sequences and protein structure. Access to this information furthers research progress rather than hindering it.

29

Closing Remarks

R. Stephen Berry

I will try to present a brief summary of the symposium from a very personal viewpoint. The first session of the symposium focused on the challenge that we are confronting in terms of scientific and technical data and information (STI) in the public domain, then Session 2 looked at how these are evolving and what are the new pressures. Session 3 examined potential consequences of changes and particularly the diminishing of the public domain, and the symposium ended with a final session on potential responses from the education and research communities in preserving the public domain and promoting open access. This structure gave a sense first of realism, then of foreboding, and finally, the last set of talks provided a sense of optimism and turned the perspective in a very positive direction. There are some points that kept being emphasized that we might keep in mind. First, openness, we now can argue, is a demonstrable good. The meeting began with the sense that there were many reasons to question the value of openness relative to the value of protection. The case for economic health being fostered by openness is a case that we can make, and one that needs to be made much more in the circles where decisions relative to the public domain or especially laws are being considered.

We also heard about the importance of experimentation, the importance of trying different ways of doing things, and particularly of the likely optimum course being one that has many different pathways. We heard over and over the dangers that a Digital Millennium Copyright Act type of approach produces by essentially forcing almost necessarily a single track. In addition, we heard from a number of speakers about the counterproductive character of inhibitory legislation versus the very desirable character of permissive legislation. This is a case that this meeting has argued needs to be made much more forcefully in the legislative community.

I think the point of despair during the meeting was when we heard about worst cases that restrict or could potentially inhibit open access to and use of public domain STI and the potential consequences. These cases illustrate the dangerous course with regard to the continuance of the scientific enterprise as we know it now.

The best cases, even the pretty good cases, by contrast, look quite positive. The final set of presentations from the education and research communities leaves us with a sense that the chances are pretty good that the world is going to turn out all right, provided that the points of view expressed and the evidence presented here is relayed to the people who are making these decisions. The rational decisions seem to be very clearly defined in favor of openness and public domain.

We need to continue to reexamine the institutional relationships that involve scientists, their universities or the government or their laboratories, and the publishers of books and journals. It was clear from some of the talks that, in many ways, universities and scientists can afford to get tough, for example, to decide not to publish in the journals or with the book publishers that are forcing universities like the University of Maine to drop subscriptions. If a significant fraction of the scientific world is not going to cite my papers if they are published in Journal X, why should I publish in that journal? The scientific community still has to deal with not only the commercial publishers, but also with some of its professional organizations that still retain the same kind of protective positions that we decry in commercial publishers. However, scientists seem to be less willing to criticize when professional organizations maintain those publications and adhere to restrictive practices.

The next stage of this discussion, one that was not discussed here, is one that is the responsibility for all of us. We have to ask ourselves, who can we reach that can make a difference, and how can we put together the persuasive case for the legislators and the lobbyists that we have heard here? We can talk among ourselves, but it is in the rooms where the database legislation is being discussed, where there seems to be a little rising against the European Database Directive, where in the committee rooms the Digital Millennium Copyright Act might be reexamined. At this stage, we have collected a good deal of evidence that we need to put in a form to allow us to talk to people who can really change things.

The net result of this symposium is that this is a step forward. We must thank the organizing committee and Paul Uhlir and Jerry Reichman for putting together a very exciting and productive symposium.

APPENDIXES

Appendix A

Final Symposium Agenda

THURSDAY, SEPTEMBER 5

7:45 a.m. *Registration and continental breakfast*

8:30 Welcoming Remarks, *William A. Wulf, President, National Academy of Engineering*

8:40 Symposium Overview, *R. Stephen Berry, University of Chicago*

8:50 **Session 1: The Role, Value, and Limits of Scientific and Technical Data and Information in the Public Domain**

Discussion Framework, *Paul Uhlir, National Research Council*

9:10 —*in Society*, James Boyle, Duke University School of Law

9:30 —*for Innovation and the Economy*
- Intellectual Property—When is it the Best Incentive Mechanism for Scientific and Technical Data and Information? *Suzanne Scotchmer, University of California, Berkeley*
- "Open Science" Economics and the Logic of the Public Domain in Research: A Primer, *Paul David, Stanford University*

10:15 BREAK

10:35 —*for Innovation and the Economy (continued)*
- Scientific Knowledge as a Global Public Good: Contributions to Innovation and the Economy, *Dana Dalrymple, U.S. Agency for International Development*
- Opportunities for Commercial Exploitation of Networked Science and Technology Public-Domain Information Resources, *Rudolph Potenzone, LION Bioscience*

11:20 Discussion of Issues from Presentations

12:00 LUNCH

1:00	***—for Education and Research***
	• Education, *Bertram Bruce, University of Illinois, Urbana-Champaign*
	• Earth and Environmental Sciences, *Francis Bretherton, University of Wisconsin*
	• Biomedical Research, *Sherry Brandt-Rauf, Columbia University*
2:00	Discussion of Issues from Presentations
2:30	**Session 2: Pressures on the Public Domain**
	Discussion Framework, *Jerome Reichman, Duke University School of Law*
2:50	The Urge to Commercialize: Interactions between Public and Private Research and Development, *Robert Cook-Deegan, Duke University*
3:10	BREAK
3:30	Legal Pressures
	• in Intellectual Property Law, *Justin Hughes, Cardozo School of Law*
	• in Licensing, *Susan Poulter, University of Utah School of Law*
	• in National Security Restrictions, *David Heyman, Center for Strategic and International Studies*
4:45	The Challenge of Digital Rights Management Technologies, *Julie Cohen, Georgetown University School of Law*
5:10	Discussion of Issues from Session 2
5:55	ADJOURN
6:00–7:30	RECEPTION

FRIDAY, SEPTEMBER 6

8:00 a.m.	CONTINENTAL BREAKFAST
8:30	**Session 3: Potential Effects of a Diminishing Public Domain**
	Discussion Framework, *Paul Uhlir, National Research Council*
8:50	***—on Fundamental Research and Education,*** R. Stephen Berry, University of Chicago
9:20	***—in Two Specific Areas of Research***
	• Environmental Information, *Peter Weiss, National Weather Service*
	• Biomedical Research Data, *Stephen Hilgartner, Cornell University*
10:00	Discussion of Issues from Session 3
10:30	BREAK
11:00	**Session 4: Responses by the Research and Education Communities in Preserving the Public Domain and Promoting Open Access**
	Discussion Framework, *Jerome Reichman, Duke University School of Law*
11:20	Strengthening Public-Domain Mechanisms in the Federal Government: A Perspective from Biological and Environmental Research, *Daniel Drell, U.S. Department of Energy*
11:45	Academics as a Natural Haven for Open Science and Public-Domain Resources: How Far Can We Stray? *Tracy Lewis, University of Florida*
12:10	LUNCH

1:00   New Legal Approaches in the Private Sector, *Jonathan Zittrain, Harvard University School of Law*

1:20   Designing Public–Private Transactions that Foster Innovation, *Stephen Maurer, Esq.*

1:50   **New Paradigms**

   —in Academia
   - The Role of the Research University in Strengthening the Intellectual Commons: The OpenCourseWare and DSpace Initiatives at MIT, *Ann Wolpert, MIT*
   - Emerging Models for Maintaining Scientific Data in the Public Domain, *Harlan Onsrud, University of Maine*

2:20   *—in Industry*
   - Open-Source Software in Commerce, *Bruce Perens, Hewlett Packard*
   - Corporate Donations of Geophysical Data, *Shirley Dutton, University of Texas at Austin*
   - The Single Nucleotide Polymorphism Consortium, *Michael Morgan, Wellcome Trust*

3:20   BREAK

3:30   Discussion of Issues from Session 4

4:00   Closing Remarks, *R. Stephen Berry, University of Chicago*

4:15   ADJOURN

Appendix B

Biographical Information on Speakers and Steering Committee Members

SPEAKERS

R. Stephen Berry (*symposium chair*) is the James Franck Distinguished Service Professor Emeritus, Department of Chemistry, at the James Franck Institute at The University of Chicago. He received his A.B., A.M., and Ph.D. from Harvard University After an 18-month instructorship at Harvard, in 1957 he became an instructor in the Chemistry Department of the University of Michigan, and in 1960 he moved on to Yale as an assistant professor. In 1964 he joined the University of Chicago's Chemistry Department and James Franck Institute (then the Institute for the Study of Metals) as an associate professor. He became a professor in 1967 and James Franck Distinguished Service Professor in 1989. In 1983 he received a MacArthur Prize Fellowship, and he has spent extended periods at the University of Copenhagen, Oxford, Université de Paris-Sud, and the Frei Universität Berlin. Dr. Berry's research interests include electronic structure of atoms and molecules; photo-collisional detachment of negative ions; photochemistry of reactive organic molecules; vibronic coupling processes such as autoionization, predissociation, and internal vibrational relaxation; thermodynamics of finite-time processes; dynamics and structure of atomic and molecular clusters; phase changes in very small systems; chaos and ergodicity in few-body systems; and, most recently, as an outgrowth of the cluster studies, dynamics on many-dimensional potential surfaces and the origins of protein folding. He has also worked extensively with the efficient use of environmental energy and other resources. Dr. Berry is also interested with issues of science and law and with management of scientific data, activities that have brought him into the arena of electronic media for scientific information and issues of intellectual property in that context. He is a member and home secretary of the National Academy of Sciences (NAS). He has been involved in many activities of the National Academies, including chairing the National Resource Council's (NRC) study on the *Bits of Power: Issues in the Global Access to Scientific Data*.

James Boyle is William Neal Reynolds Professor of Law at the Duke University School of Law and author of *Shamans, Software and Spleens: Law and the Construction of the Information Society* (Harvard University Press, 1996) as well as many articles and essays about intellectual property and social and legal theory. He is a member of the Board of Creative Commons and of the academic advisory board of the Electronic Privacy and Information Center. He is working on a book called *The Public Domain*. His Web site is http://james-boyle.com.

Sherry Brandt-Rauf is associate research scholar at the Center for the Study of Society and Medicine of the Columbia College of Physicians and Surgeons. Trained at Columbia in law and sociology, she teaches and does research on areas in which law and medicine overlap. Particular areas of interest include the ownership of scientific data, occupational health, genetic testing, conflicts of interest, and the ethics of research on vulnerable populations. She recently completed an individual project fellowship at the Open Society Institute, researching the nature of the pharmaceutical industry's interactions with medical students and residents and the effects of such interactions on the practice of medicine. In addition, under a grant from the Jewish Women's Foundation of New York, she recently prepared an online information booklet for Ashkenazi Jewish women dealing with genetic testing for BRCA mutations. She sits on the Columbia-Presbyterian Medical Center Institutional Review Board and Pediatric Ethics Committee.

Francis Bretherton obtained his Ph.D. in applied mathematics at the University of Cambridge, England. His research areas include atmospheric dynamics and ocean currents. He has been a member of the faculty at Cambridge, at the Johns Hopkins University, and at the University of Wisconsin-Madison, where he is now professor emeritus in the Department of Atmospheric and Oceanic Sciences. From 1974 to 1981, he was director of the National Center for Atmospheric Research in Boulder, Colorado, and president of the University Corporation for Atmospheric Research. Since then, he has been deeply involved in planning national and international research programs on climate and changes in our global environment. From 1982 to 1987, he was chair of the NASA Earth System Sciences Committee, which formulated the strategy for the U.S. Global Change Research Program. Dr. Bretherton chaired the Global Observing System Space Panel under the auspices of the World Meteorological Organization from 1998 to 2000. He has served on many committees of the NRC, most recently as chair of the Committee on Geophysical and Environmental Data, and is also a member of advisory panels and boards to the National Oceanic and Atmospheric Administration (NOAA) on Climate and Global Change Research and on Climate System Modeling.

Bertram Bruce is a professor of library and information science at the University of Illinois at Urbana-Champaign, where he has been a member of the faculty since 1990. Before moving to Illinois, he taught computer science at Rutgers (1971-1974) and was a principal scientist at Bolt, Beranek and Newman (1974-1990). His research and teaching focus on new literacies, inquiry-based learning, and technology studies. A major focus of his work is with the Distributed Knowledge Research Collaborative studying new practices in scientific research. Other studies include research on education enhancements to Biology Workbench (a computational environment that facilitates bioinformatics research, teaching, and learning); Plants, Pathogens and People; Physics Outreach; and SEARCH. His analytical work has focused on changes in the nature of knowledge, community, and literacy. He serves on the editorial boards of *Educational Theory, Computers and Composition, Discourse Processes, Computer, International Journal of Educational Technology,* and *Interactive Learning Environments.*

Julie Cohen is professor of law at the Georgetown University Law Center. She teaches and writes about intellectual property and information privacy issues, with particular focus on computer software and digital works and on the intersection of copyright, privacy, and the First Amendment in cyberspace. She is a member of the Advisory Board of the Electronic Privacy Information Center, the Advisory Board of Public Knowledge, and the Board of Academic Advisors to the American Committee for Interoperable Systems. From 1995 to 1999, Professor Cohen taught at the University of Pittsburgh School of Law. From 1992 to 1995, she practiced with the San Francisco firm of McCutchen, Doyle, Brown & Enersen, where she specialized in intellectual property litigation. Professor Cohen received her A.B. from Harvard and her J.D. from Harvard School of Law. She was a former law clerk to the Honorable Stephen Reinhardt of the U.S. Court of Appeals for the Ninth Circuit.

Robert Cook-Deegan is director of the Center for Genome Ethics, Law, and Policy at Duke University. He is also a Robert Wood Johnson Health Policy Investigator at the Kennedy Institute of Ethics, Georgetown University, where he is completing a primer on how national policy decisions are made about health research. Until July 2002, he directed the Robert Wood Johnson Foundation Health Policy Fellowship program at the Institute of Medicine (IOM), the National Academies. In 1996 Dr. Cook-Deegan was a Cecil and Ida Green Fellow at the University of Texas, Dallas, following his work on the National Academies' report on *Allocating Federal Funds for Science and Technology*. From 1991 through 1994, he directed IOM's Division of Biobehavioral Sciences and Mental Disorders (since renamed Neuroscience and Behavioral Health). He worked for the National Center for Human Genome Research from 1989 to 1990, after serving as acting executive director of the Biomedical Ethics Advisory Committee of the U.S. Congress, 1988-1989. An Alfred P. Sloan Foundation grant culminated in *The Gene Wars: Science, Politics, and the Human Genome*. He continues as a consultant to the DNA Patent Database at Georgetown University. Dr. Cook-Deegan came to Washington, D.C. in 1982 as a congressional science fellow and stayed five more years at the congressional Office of Technology Assessment, ultimately becoming a senior associate. He received his B.S. degree in chemistry, magna cum laude, from Harvard, and his M.D. degree from the University of Colorado. Dr. Cook-Deegan chairs the Royalty Fund Advisory Committee for the Alzheimer's Association, is secretary and trustee of the Foundation for Genetic Medicine, and former chair of Section X (Social Impacts of Science and Engineering) for the American Association for the Advancement of Science (AAAS), where he is also a fellow.

Dana Dalrymple is senior research advisor in the Office of Agriculture and Food Security, U.S. Agency for International Development and agricultural economist in the Foreign Agricultural Service of the U.S. Department of Agriculture. He has helped administer U.S. government involvement in, and support of, the Consultative Group on International Agricultural Research and its network of 16 centers for 30 years. Dr. Dalrymple is a long-time student of the development and adoption of agricultural technology in agriculture, both in the United States and internationally, and has published widely on this subject. He has also served as an analyst on studies of agricultural research conducted by the NAS (as part of a larger project) and the Office of Technology Assessment of the U.S. Congress. He has recently reviewed the role of international agricultural research as a global public good. He received his B.S. and M.S. degrees from Cornell University, his Ph.D. in agricultural economics from Michigan State University, and was a member of the Department of Agricultural Economics at the University of Connecticut.

Paul David is professor of economics at Stanford University and senior fellow of the Stanford Institute for Economic Policy Research, where he leads the Program on Knowledge Networks and Institutions for Innovation. Since 1994 he has held a joint appointment as senior research fellow of All Souls College, Oxford. He is an elected fellow of the International Econometrics Society, of the American Academy of Arts and Sciences, and of the British Academy; he was president elect and president of the Economic History Association and served until 2001 as an elected member of the Council of the Royal Economics Society. In 1997 the University of Oxford conferred on Professor David the title professor of economics and economic history "in recognition of distinction." He is known internationally for his contributions in American economic history, economic and historical demography, and the economics of science and technology. Much of Professor David's research has been directed toward characterizing the conditions under which microeconomic and macroeconomic processes exhibit "path-dependent dynamics." He has published more than 130 journal articles and chapters in edited books, in addition to authoring and editing a number of books under his name.

Daniel Drell is a biologist with the Human Genome Program, Office of Biological and Environmental Research (OBER), of the U.S. Department of Energy. His responsibilities include oversight of the component of the OBER Genome Program devoted to Ethical, Legal, and Social Implications. Prior to joining the OBER in April 1991, Dr. Drell worked as a visiting scientist in the HLA Laboratory of the American Red Cross Holland Laboratory where he participated in ongoing research and service activities on HLA typing of donor blood samples for the bone marrow matching program. From 1986 to 1990, he was at the George Washington University Medical Center, ending up as an assistant professor in the Department of Medicine and associate director of the Immunogenetics and Immunochemistry Laboratories. From 1983 to 1986, he was a staff fellow in the Laboratory of Oral Medicine at the National Institute of Dental Research, National Institutes of Health, conducting research in the areas of the cell-mediated immunology of type 1 diabetes mellitus and autoimmune myositis. He has also held positions at the

Baylor College of Medicine, the Population Council's Center for Biomedical Research, and the Laboratory of Developmental Genetics at Sloan-Kettering. Dr. Drell received his Ph.D. in immunology from the University of Alberta, Edmonton, and his B.A. in biology from Harvard College. He has published nearly 30 scientific articles and abstracts.

Shirley P. Dutton is a senior research scientist and senior technical advisor to the director, Bureau of Economic Geology, the University of Texas at Austin. She received a B.A. in geological sciences from the University of Rochester, graduating with highest honors and election to Phi Beta Kappa. She earned M.A. and Ph.D. degrees, also in geological sciences, from the University of Texas at Austin. She has worked at the Bureau of Economic Geology since 1977, and her technical expertise is in clastic sedimentology and diagenesis. Dr. Dutton was a distinguished lecturer for the American Association of Petroleum Geologists in 1987. She has authored or coauthored 172 published papers and abstracts.

David Heyman joined the Center for Strategic and International Studies in November 2001 as a senior fellow for science and security initiatives—studies that explore the increasingly interconnected world of science, technology, and national security policy. Previously he served as a senior advisor to the Secretary of Energy at the U.S. Department of Energy from 1998 to 2001 and in the White House in the Office of Science and Technology Policy in the National Security and International Affairs division from 1995 to 1998. Prior to joining the Clinton administration, Mr. Heyman briefly worked as a consultant with Ernst & Young in their International Privatization and Economics Group in London, and was the director of International Operations for a New York software company developing supply-chain management systems for Fortune 100 firms. He has worked in Europe, Russia, and the Middle East. Mr. Heyman carried out his undergraduate work at Brandeis University, with a concentration in biology; his graduate work at the Johns Hopkins University School of Advanced International Studies was in technology policy and international economics.

Stephen Hilgartner is associate professor in the Department of Science and Technology Studies at Cornell University. Professor Hilgartner's research focuses on social studies of science and technology, especially biology, biotechnology, and medicine; biology, ethics, and politics; science as property; ethnography of science; and risk. His recent book, *Science on Stage: Expert Advice as Public Drama* (Stanford, 2000), explores the processes through which the expertise of science advisors is established, contested, and maintained. He is the author of many articles, book chapters, and reviews, including a series of works on data access and ownership. His work has appeared in such journals as the *Journal of Molecular Biology*; the *Journal of the American Medical Association*; *Social Studies of Science*; *Science, Technology, and Human Values*; *Science Communication*; and the *American Journal of Sociology*. Professor Hilgartner is chair of the Ethical, Legal, and Social Issues committee of the Cornell Genomics Initiative. He is a member of the Council of the Society for Social Studies of Science. He is also a member of the Steering Group of the Section on Societal Impacts of Science and Engineering of the AAAS. He is currently completing a book on genome mapping and sequencing in the 1990s.

Justin Hughes is assistant professor at the Benjamin N. Cardozo School of Law and was an attorney advisor in the U.S. Patent and Trademark Office from 1997 to 2001, focusing on initiatives in Internet-related intellectual property issues, Eleventh Amendment immunity issues, intellectual property law in developing economies, and on copyright appellate filings for the United States (including the Napster litigation). Professor Hughes practiced law in Paris and Los Angeles and clerked for the Lord President of the Malaysian Supreme Court in Kuala Lumpur. He is a former Henry Luce Scholar, Mellon Fellow in the Humanities, and American Bar Association (ABA) Baxter Scholar at the Hague Court. He was a visiting professor at the University of California at Los Angeles School of Law. He received his B.A. from Oberlin and his J.D. from Harvard.

Tracy Lewis was the James W. Walter Eminent Scholar of Entrepreneurship at the University of Florida and became the newly appointed director of the Duke University Innovation Center and professor of Economics at the Fuqua School of Business in January 2003. His research and teaching interests are in the areas of innovation processes, industrial organization, and financial and incentive contracting. Professor Lewis has published numerous articles on the management of open-access resources and has been a consultant for several federal regulatory agencies and private corporations on issues of patent policy and protection of intellectual property.

Stephen M. Maurer has practiced intellectual property litigation since 1982. He currently co-teaches a course on Internet law and economics at the School of Public Policy at the University of California, Berkeley. Mr. Maurer has written extensively on patent reform, proposed database legislation, and science policy. Over the past four years, he has worked with academic scientists trying to build a variety of commercially self-supporting databases. In 2000 he helped a worldwide community of academic biologists (The Mutations Database Initiative or MDI) negotiate a $3.2 million Memorandum of Understanding with Incyte Genomics. Under the proposed agreement, Incyte would have helped MDI scientists build a unified, computationally advanced archive for worldwide human mutations data. In return, MDI would have granted Incyte the exclusive right to host the database on a commercial web site. Academic and commercial users would have remained free to download and use the database for all other purposes. Mr. Maurer has also helped Lawrence Berkeley National Laboratory physicists design a commercial venture for archiving high-energy accelerator data. He is a graduate of Yale University and Harvard School of Law. He has published in many journals, including *Nature*, *Science*, *Economica*, *Sky & Telescope*, and *Beam Line*.

Michael Morgan is director of Research Partnerships & Ventures and chief executive of the Wellcome Trust Genome Campus in Cambridge, England. Dr. Morgan joined the Wellcome Trust in 1983 and now is responsible for development of new enterprises such as the Synchrotron Project (a joint project with the U.K. and French governments) and the Single Nucleotide Polymorphism Consortium. He plays a major role in the international coordination of the Human Genome Project and is also responsible for scientific establishments such as the Wellcome Trust Genome Campus. Three independent institutions are located on the campus: the Wellcome Trust Sanger Institute, the European Bioinformatics Institute, and the MRC Human Genome Mapping Resource Centre. Dr. Morgan is a graduate of Trinity College, Dublin, and obtained his Ph.D. from Leicester University.

Harlan Onsrud is professor of spatial information science and engineering at the University of Maine and a research scientist with the National Center for Geographic Information and Analysis. He is immediate past president of the University Consortium for Geographic Information Science, a nonprofit organization of over 70 universities and other research institutions dedicated to advancing understanding of geographic processes, relationships, and patterns through improved theory, methods, technology, and data. Professor Onsrud is chair of the NRC's U.S. National Committee on Data for Science and Technology (CODATA) and also serves on the NRC's Mapping Science Committee. A licensed attorney and engineer, his research focuses on the analysis of legal, ethical, and institutional issues affecting the creation and use of digital spatial databases and the assessment of the social impacts of spatial technologies. He teaches a range of legal courses to students in engineering, information systems, and computer science, including a graduate course in information systems law.

Bruce Perens is one of the founding Linux developers and is best known as the creator of the Open Source Definition, the canonical guidelines for open-source licensing. Mr. Perens is cofounder of the Open Source Initiative, the Linux Standard Base, and Software in the Public Interest. His first Free Software program, Electric Fence, is widely used in both commercial and free software development. His Busybox software became a standard tool kit in the embedded systems field and is thus included in many commercial print servers, firewalls, and storage devices. Mr. Perens has 19 years experience in the feature film business and is credited on the films "A Bug's Life" and "Toy Story II." As both a leader of open-source software and a long-time participant in the film industry, he is uniquely qualified to comment on the oft-painful intersection of copyright protection and the public domain. Mr. Perens was formerly employed as senior strategist for Linux and Open Source at Hewlett Packard. He is the primary author of that corporation's open-source policy manual.

Susan R. Poulter is professor of law at the S.J. Quinney College of Law, University of Utah, in Salt Lake City. She holds B.S. and Ph.D. degrees in chemistry, both from the University of California, Berkeley. After a period during which she taught chemistry at the University of Utah, she received a J.D. from the University of Utah College of Law in 1983, where she was executive editor of the *Utah Law Review* and was inducted into the Order of the Coif. She joined the faculty of the College of Law in 1990, after seven years in private law practice. She teaches in the areas of environmental law, intellectual property, and torts. Professor Poulter has written and lectured on scientific evidence and environmental law. She is a coauthor of the Reference Guide on Medical Testimony in the second edition of the Federal Judicial Center's *Reference Manual on Scientific Evidence*. She is a member of the Council of the Section of Science and Technology Law of the ABA and has served as a section representative to the National Conference of Lawyers and Scientists (NCLS), a joint committee of the ABA and the AAAS. Currently, she is the ABA cochair of the NCLS. She is a member of the Advisory Board of the AAAS project on Court-Appointed Scientific Experts and participated in the AAAS Workshop on Intellectual Property and Scientific Publishing in the Electronic Age.

Rudolph Potenzone became president and CEO of LION Bioscience, Inc. in March 2001. With over 20 years of cheminformatics experience, he most recently served as senior vice president for marketing and development at MDL, where he managed the design, development, and marketing of MDL's software and database products. Previously, Dr. Potenzone was director of research and new product development at Chemical Abstracts Service, as well as holding senior positions at Polygen/Molecular Simulations, Inc. Dr. Potenzone holds a Ph.D. in macromolecular science from Case Western Reserve University in Cleveland, Ohio, and a B.S. in biophysics and microbiology from the University of Pittsburgh.

Jerome H. Reichman became Bunyan A. Womble Professor of Law at Duke University in July 2000, where he teaches in the field of contracts and intellectual property. Before coming to Duke, he taught at Vanderbilt, Michigan, Florida and Ohio State universities and at the University of Rome, Italy. He graduated from the University of Chicago (B.A.) and attended Yale School of Law, where he received his J.D. degree. Professor Reichman has written and lectured widely on diverse aspects of intellectual property law, including comparative and international intellectual property law and the connections between intellectual property and international trade law. Other recent writings have focused on intellectual property rights in data, the appropriate contractual regime for online delivery of computer programs and other information goods, and new ways to stimulate investment in subpatentable innovation without impoverishing the public domain. Professor Reichman has served as consultant to the U.S. National Committee for CODATA at the National Academies on the subject of legal protection for databases. He also is an academic advisor to the American Committee for Interoperable Systems; a consultant to the Technology Program of the UN Conference on Trade and Development; and was a consultant on the UN Development Program's flagship project on Innovation, Culture, Biogenetic Resources, and Traditional Knowledge.

Suzanne Scotchmer is professor of economics and public policy at the University of California, Berkeley. Her graduate degrees are a Ph.D. in economics and an M.A. in statistics, both from the University of California, Berkeley. She has held visiting appointments or fellowships at the University of Cergy-Pontoise, Tel Aviv University, University of Paris I (Sorbonne), Boalt School of Law, the University of Toronto School of Law, Yale University School of Law, Stanford University, the Hoover Institution, and the New School of Economics, Moscow. She has published extensively on the economics of intellectual property, rules of evidence, tax enforcement, cooperative game theory, club theory, and evolutionary game theory. Her work has appeared in *Econometrica, American Economic Review, Quarterly Journal of Economics*, the *RAND Journal of Economics*, the *Journal of Public Economics, Journal of Economic Theory*, the *New Palgrave Dictionary of Law and Economics*, and as commentary in *Science*. She was previously on the editorial board of the *American Economic Review*, the *Journal of Economic Perspectives*, and the *Journal of Public Economics*, and is currently on the editorial board of the *Journal of Economic Literature* and *Regional Science and Urban Economics*. In 1998-1999 she served on the NRC Committee on Promoting Access to Scientific and Technical Data for the Public Interest and has appeared before several other committees of the NRC, mostly regarding intellectual property. She has served as a consultant for the U.S. Department of Justice, Antitrust Division, and for private clients in disputes regarding intellectual property.

Paul F. Uhlir is director of the Office of International Scientific and Technical Information Programs at the National Academies in Washington, D.C. His current area of emphasis is on scientific and technical data management and policy and on the relationship of intellectual property law in digital data and information to research and development policy. In 1997 he received the NAS Special Achievement Award for his work in this area. Mr. Uhlir has been employed at the National Academies since 1985, first as a senior staff officer for the Space Studies Board, where he worked on solar system exploration and environmental remote sensing studies for NASA, and then as associate executive director of the Commission on Physical Sciences, Mathematics, and Applications. Before joining the National Academies, he worked in the general counsel's office and as a foreign affairs officer at NOAA. He has directed and published over 20 NRC studies and written or edited over 40 articles and books. Mr. Uhlir has a B.A. in history from the University of Oregon, and a J.D. and M.A. degrees in international relations, with a focus on space law and arms control, from the University of San Diego.

Peter N. Weiss began work with the Strategic Planning and Policy Office of the NOAA National Weather Service, in March 2000. His responsibilities include domestic and international data policy issues, with a view toward fostering a healthy public-private partnership. Mr. Weiss was a senior policy analyst and attorney in the Office of Information and Regulatory Affairs, Office of Management and Budget (OMB), since 1991. Mr. Weiss analyzed policy and legal issues involving information resources and information technology management, with particular emphasis on electronic data interchange and electronic commerce. He is primary author of the information policy sections of OMB Circular A-130, "Management of Federal Information Resources," and was a member of the administration's Electronic Commerce Working Group (See "A Framework for Global Electronic Commerce"). From 1990 to 1991, Mr. Weiss was deputy associate administrator for procurement law, Office of Federal Procurement Policy. In this position, he analyzed legal and policy issues affecting the procurement process. Major projects included examination of legal and regulatory issues involving procurement automation, policies and Federal Acquisition Regulation revisions to facilitate electronic data interchange, as well as automatic data processing procurement legal and policy issues. From 1985 to 1990, Mr. Weiss was the assistant chief counsel for procurement and regulatory policy, Office of Advocacy, U.S. Small Business Administration. From 1981 to 1985, Mr. Weiss was in private practice in Washington, D.C. Mr. Weiss holds a B.A. from Columbia University and a J.D. from the Catholic University of America, Columbus School of Law. A recent publication is "International Information Policy in Conflict: Open and Unrestricted Access versus Government Commercialization," in *Borders in Cyberspace*, Kahin and Nesson, eds. (MIT Press, 1997).

Ann J. Wolpert is director of libraries for the Massachusetts Institute of Technology (MIT), and has reporting responsibility for the MIT Press. Her responsibilities include membership on the Committee on Copyright and Patents, the Council on Educational Technology, the Dean's Committee, and the President's Academic Council. She chairs the Management Board of the MIT Press and the Board of Trustees of Technology Review, Inc. Prior to joining MIT, Ms. Wolpert was executive director of library and information services at the Harvard Business School. Her experience previous to Harvard included management of the Information Center of Arthur D. Little, Inc., an international management and technology consulting firm, where she was also engaged in consulting assignments. She is active in the professional library community, currently serving on the Board of the Boston Library Consortium, on the Board of the Association of Research Libraries; and as a member of the editorial boards of *Library & Information Science Research* and *The Journal of Library Administration.* She is a trustee of Simmons College in Boston, Massachusetts, and presently serves as an advisor to the Publications Committee of the Massachusetts Medical Society. Recent consulting assignments have taken her to the campuses of INCAE in Costa Rica and Nicaragua and to the Malaysia University of Science and Technology, Selangor, Malaysia. A frequent speaker and writer, Ms. Wolpert has recently contributed papers on such topics as library service to remote library users, intellectual property management in the electronic environment, and the future of research libraries in the digital age. She received a B.A. from Boston University and an M.L.S. from Simmons College.

Wm. A. Wulf is president of the National Academy of Engineering and vice chair of the NRC, the principal operating arm of the National Academies of Sciences and Engineering. He is on leave from the University of

Virginia, Charlottesville, where he is AT&T Professor of Engineering and Applied Sciences. Among his activities at the university are a complete revision of the undergraduate computer science curriculum, research on computer architecture and computer security, and an effort to assist humanities scholars exploit information technology. Dr. Wulf has had a distinguished professional career that includes serving as assistant director of the National Science Foundation; chair and chief executive officer of Tartan Laboratories Inc., Pittsburgh; and professor of computer science at Carnegie Mellon University, Pittsburgh. He is the author of more than 80 papers and technical reports, has written three books, and holds two U.S. patents.

Jonathan Zittrain is the Jack N. and Lillian R. Berkman Assistant Professor of Entrepreneurial Legal Studies at Harvard School of Law, and a faculty director of its Berkman Center for Internet & Society. His research includes digital property, privacy, and speech, and the role played by private "middle people" in Internet architecture. He currently teaches "Internet & Society: The Technologies and Politics of Control" and has a strong interest in creative, useful, and unobtrusive ways to deploy technology in the classroom. He holds a J.D. from Harvard School of Law, an M.P.A. from the J.F.K. School of Government, and a B.S. in cognitive science and artificial intelligence from Yale. He is also a 15-year veteran sysop of CompuServe's online forums.

STEERING COMMITTEE MEMBERS

David R. Lide, Jr. (*chair*) is a consultant and the former director of the National Institute of Standards and Technology's Standard Reference Data Division. His expertise is in physical chemistry and scientific information. He has served as president of the international CODATA, chair of CODATA's publication committee, and chair of the U.S. National Committee for CODATA. He is currently a member of the International Council for Science/CODATA Ad Hoc Group on Data and Information, which focuses on problems of intellectual property rights and access to data, and a fellow of the International Union of Pure and Applied Chemistry. Dr. Lide is the editor-in-chief of CRC's *Handbook of Chemistry and Physics*.

Hal Abelson is the Class of 1922 Professor of Computer Science and Engineering with MIT's Department of Electrical Engineering and Computer Science. He holds an A.B. from Princeton, and a Ph.D. in mathematics from MIT. In 1992, Professor Abelson was designated as one of MIT's six inaugural MacVicar Faculty Fellows in recognition of his significant and sustained contributions to teaching and undergraduate education. He was 1992 recipient of the Bose Award (MIT's School of Engineering teaching award) and the winner of the 1995 Taylor L. Booth Education Award given by the Institute of Electrical and Electronic Engineers (IEEE) Computer Society, cited for his continued contributions to the pedagogy and teaching of introductory computer science. Professor Abelson has a longstanding interest in using computation as a conceptual framework in teaching, and he is currently teaching a class on Ethics and Law on the Electronic Frontier. He is a fellow of IEEE.

Mostafa El-Sayed is the Julius Brown Chair, Regents' Professor and Director of the Laser Dynamics Laboratory at the School of Chemistry and Biochemistry of the Georgia Institute of Technology, where he joined the faculty in 1994. Professor El-Sayed received his B.Sc. from Ain Shams University, Cairo, Egypt, and his Ph.D. from Florida State University. After being a research associate at Harvard, Yale, and the California Institute of Technology, he was appointed to the faculty of the University of California at Los Angeles in 1961. He was an Alfred P. Sloan, John Simon Guggenheim fellow, a visiting professor at the University of Paris, a Sherman Fairchild distinguished scholar at California Institute of Technology, and a Senior Alexander von Humboldt fellow at the Technical University of Munich. Professor El-Sayed is the editor-in-chief of the *Journal of Physical Chemistry A and B* and was an editor of the *International Reviews of Physical Chemistry*. Professor El-Sayed is a member of the NAS and the Third World Academy of Science and is a fellow of the American Academy of Arts and Sciences, the American Physical Society, and the AAAS.

Mark Frankel, Ph.D., directs the Scientific Freedom, Responsibility and Law Program at the AAAS. He is responsible for developing and managing AAAS activities related to science, ethics, and law. He serves as staff officer to two AAAS committees—the Committee on Scientific Freedom and Responsibility and the AAAS-ABA National Conference of Lawyers and Scientists. He is editor of *Professional Ethics Report*, the program's quarterly newsletter, and is a fellow of AAAS.

Maureen Kelly recently served as vice president for planning at BIOSIS, which is the largest abstracting and indexing service for the life sciences community, and is now a consultant. Ms. Kelly worked in different capacities for BIOSIS since 1969. Previously she had production responsibility for the bibliographic and scientific content of BIOSIS products. While in that position, she led the team that developed the system for capturing and managing indexing data in support of BIOSIS's new relational indexing. Ms. Kelly has authored a number of papers on managing and accessing biological information. She is currently secretary of the AAAS Section on Information, Computing, and Communication. She has served on various professional society research and publishing committees, including participating in several NAS E-Journal Summit meetings. Ms. Kelly is a member of the U.S. National Committee for CODATA and served on the NRC Committee on Promoting Access to Scientific and Technical Data in the Public Interest: An Assessment of Policy Options.

Pamela Samuelson is a professor of law and information management at the University of California at Berkeley and codirector of the Berkeley Center for Law and Technology. Her expertise is in intellectual property law, and she has written and spoken extensively about the challenges that new information technologies are posing for public policy and traditional legal regimes. Prior to joining the faculty at Berkeley, Professor Samuelson was at the University of Pittsburgh School of Law, where she had taught since 1981. A graduate of Yale School of Law, she has also practiced with the New York firm of Willkie Farr & Gallagher and served as the principal investigator for the Software Licensing Project at Carnegie Mellon University. In 1997 Professor Samuelson was named a fellow of the John D. and Catherine T. MacArthur Foundation. In 1998 she was recognized by the *National Law Journal* as being among the 50 most influential female lawyers in the country and among the eight most influential in Northern California. She was recently elected to membership in the American Law Institute and named a fellow of the Association of Computing Machinery. In 2001 she was appointed to a University of California, Berkeley, Chancellor's Professorship for distinguished research, teaching, and service for her contributions to both Boalt Hall and the School of Information Management and Systems. Professor Samuelson was a member of the U.S. National Committee for CODATA and the NRC's Computer Science and Telecommunications Board's Committee on Intellectual Property Rights and the Emerging Information Infrastructure.

Martha E. Williams is director of the Information Retrieval Research Lab and a professor of Information Science at the University of Illinois at Urbana-Champaign. Her research interests include digital database management, online retrieval systems, systems analysis and design, chemical information systems, and electronic publishing. She has published widely on these topics and has been editor of the *Annual Review of Information Sciences and Technology* (since 1975), *Computer Readable Databases: A Directory & Data Sourcebook* (1976-1987), and *Online Review* (since 1977). Professor Williams was chair of the Board of Engineering Information, Inc. from 1980 to 1988, was appointed to the National Library of Medicine's Board of Regents from 1978 to 1981, and served as chair of the board in 1981. In addition, she served on several NRC committees, including the Numerical Data Advisory Board (1979-1982). She has an A.B. from Barat College and an M.A. from Loyola University. Ms. Williams was a member of the NRC's Committee for a Study on Promoting Access to Scientific and Technical Data for the Public Interest.

Appendix C

Symposium Attendees

Hal Abelson
Massachusetts Institute of Technology
hal@mit.edu

Allan Adler
Association of American Publishers
adler@publishers.org

Prudence Adler
Association of Research Libraries
prue@arl.org

Kerri Allen
SPARC
kerri@arl.org

Debra Aronson
Federation of American Societies for Experimental Biology
daronson2op@faseb.org

Cynthia Banicki
U.S. Patent and Trademark Office
cbanicki@uspto.gov

Bruce Barkstrom
NASA Langley Research Center
b.r.barkstrom@larc.nasa.gov

Olga Barysheva
The National Library of Russia
barysh@nlr.ru

David Beckler
Consultant
d2beckler@aol.com

R. Stephen Berry
University of Chicago
berry@uchicago.edu

Gail Blaufarb
National Cancer Institute
blaufarg@mail.nih.gov

Kim Bonner
University of Maryland University College
kbonner@umuc.edu

Scott Boren
boren@mindspring.com

James Boyle
Duke University School of Law
boyle@law.duke.edu

Susan Bragdon
International Plant Genetic Resources Institute
s.bragdon@cgiar.org

Sherry Brandt-Rauf
Columbia University
spejo@aol.com

Francis Bretherton
University of Wisconsin
fbretherton@charter.net

Glenn Brown
Creative Commons
glenn@creativecommons.org

Michael Brown
The Institute for Genomic Research
mbrown@tigr.org

Bertram Bruce
University of Illinois, Urbana-Champaign
chip@uiuc.edu

Elizabeth Buffum
NASA/Center for Aerospace Information for ASRC Aerospace Corp.
ebuffum@sti.nasa.gov

Paul Bugg
U.S. Office of Management and Budget
pbugg@omb.eop.gov

Merry Bullock
American Psychological Association
mbullock@apa.org

San Cannon
Federal Reserve Board
scannon@frb.gov

Bonnie Carroll
Information International Associates, Inc.
bcarroll@infointl.com

Amy Centanni
U.S. Department of Veterans Affairs
amy.centanni@mail.va.gov

Joan Cheverie
Georgetown University Library
cheverij@georgetown.edu

Sangdai Choi
University of Maryland
sdchoi777@hotmail.com

Bonnie Chojnaki
White Memorial (Chemistry) Library
bc128@umail.umd.edu

Charles Clark
Defense Intelligence Agency
Afclacx@dia.osis.gov

Jeff Clark
James Madison University
clarkjc@jmu.edu

Julie Cohen
Georgetown University School of Law
jec@law.georgetown.edu

Sarah Comley
International Observers
scomley@mail.com

Robert Cook-Deegan
Duke University
bob.cd@duke.edu

Charlotte Cottrill
U.S. Environmental Protection Agency
cottrill.charlotte@epa.gov

Christian Cupp
Defense Technical Information Center
ccupp@dtic.mil

Paul Cutler
The National Academies
pcutler@nas.edu

Lisa DaCosta
The Institute for Genomic Research
ldacosta@tigr.org

APPENDIX C

Dana Dalrymple
U.S. Agency for International Development
ddalrymple@usaid.gov

Paul David
Stanford University
pad@stanford.edu

Rudolph Dichtl
University of Colorado Cooperative Institute for Research in Environmental Sciences
dichtl@kryos.colorado.edu

Julie Dietzel-Glair
University of Maryland
jdietzel@wam.umd.edu

Dr. Sidney Draggan
U.S. Environmental Protection Agency
draggan.sidney@epamail.epa.gov

Daniel Drell
U.S. Department of Energy
daniel.drell@science.doe.gov

Beatrice Droke
U.S. Food and Drug Administration
bdroke@oc.fda.gov

Nhat-Hang Duong
Defense Technical Information Center
cduong@dtic.mil

Shirley Dutton
University of Texas at Austin
shirley.dutton@beg.utexas.edu

Alice Eccles
University of Maryland
aeccles@wam.umd.edu

Anita Eisenstadt
National Science Foundation
aeisenst@nsf.gov

Julie Esanu
The National Academies
jesanu@nas.edu

Lloyd Etheredge
Policy Sciences Center
lloyd.etheredge@yale.edu

Cynthia Etkin
U.S. Government Printing Office
cetkin@gpo.gov

V. Jeffrey Evans
National Institutes of Health
Jeff_Evans@nih.gov

Phoebe Fagan
National Institute of Standards and Technology
phoebe.fagan@nist.gov

Susan Fallon
Community of Science
skf@cos.com

Cindy Fang
University of Maryland, College Park
xfang@wam.umd.edu

Suzanne Fedunok
New York University Bobst Library
suzanne.fedunok@nyu.edu

Martha Feldman
National Agricultural Library
mfeldman@nal.usda.gov

E.T. Fennessy
The Copyright Group
efennessy@att.net

Walter Finch
National Technical Information Service
wfinch@ntis.gov

Nigel Fletcher-Jones
Nature America, Inc.
n.fletcher-jones@natureny.com

Carolyn Floyd
NASA Langley Research Center
c.e.floyd@larc.nasa.gov

Peter Folger
American Geological Union
pfolger@agu.org

Olga Francois
University of Maryland University College
ofrancois@umuc.edu

Amy Franklin
The National Academies
afranklin@nas.edu

Ken Fulton
The National Academies
kfulton@nas.edu

John Gardenier
Center for Disease Control and Prevention, National Center for Health Statistics
drgarden@starpower.net

Lorrin Garson
American Chemical Society
lgarson@acs.org

Ruggero Giliarevsky
All-Russian Institute for Scientific and Technical Information
giliarevski@viniti.ru

Karl Glasener
Agronomy, Soil Science Societies
Karlglasener@cs.com

Jerry Glenn
Stars of the Future
jglenn@igc

Kyrille Goldbeck
University of Maryland, College Park
kkg4@georgetown.edu

Karen Goldman
National Institutes of Health
goldmank@mail.nih.gov

Dov Greenbaum
Yale University
dov.greenbaum@yale.edu

Jane Griffith
National Library of Medicine
jbgriffith@nlm.nih.gov

Robert Hardy
Council on Governmental Relations
rhardy@cogr.edu

Sam Hawala
U.S. Census Bureau
sam.hawala@census.gov

Janet Heck
University of Maryland, College of Information Studies
jheck@wam.umd.edu

Stephen Heinig
Association of American Medical Colleges
sheinig@aamc.org

Stephen Heller
National Institute of Standards and Technolgy
srheller@nist.gov

Jennifer Hendrix
American Library Association
jhendrix@alawash.org

Robert Hershey
Engineering and Management Consulting
hershey@cpcug.org

David Heyman
Center for Strategic and International Studies
dheyman@csis.org

Stephen Hilgartner
Cornell University
shh6@cornell.edu

Derek Hill
National Science Foundation
dhill@nsf.gov

Will Hires
Johns Hopkins University
will.hires@jhuapl.edu

APPENDIX C

Cynthia Holt
George Washington University
holt@gwu.edu

Justin Hughes
Cardozo School of Law
hughes@ymail.yu.edu

RuthAnn Humphreys
Johns Hopkins University Applied Physics Lab
R.Humphreys@jhuapl.edu

Joe Ingram
John Wiley & Sons, Inc.
jingram@wiley.com

Vsevolod Isayev
Russian Embassy
vsevis@rategier.ru

Kenan Jarboe
Athena Alliance
kpjarboe@athenaalliance.org

Jennifer Jenkins
Duke University School of Law
jenkins@law.duke.edu

Scott Jenkins
FDC Reports, Inc.
s.jenkins@elsevier.com

Laura Jennings
National Imagery and Mapping Agency
jenningl@nima.mil

Douglas Jones
University of Arizona
jonesd@u.library.arizona.edu

Heather Joseph
BioOne, Inc.
heather@arl.org

Janet Joy
The National Academies
jjoy@nas.edu

Brian Kahin
University of Maryland
bk90@umail.umd.edu

Subhash Kuvelker
Kuvelker Law Firm
kuvelker@mindspring.com

Myra Karstadt
U.S. Environmental Protection Agency
karstadt.myra@epa.gov

Atsushi Kato
University of Tokyo
akato@ip.rcast.u-tokyo.ac.jp

Maureen Kelly
Consultant
mckelly@ix.netcom.com

Christopher Kelty
Rice University
ckelty@rice.edu

Miriam Kelty
National Institute on Aging
mk46u@nih.gov

Bonnie Klein
U.S. Department of Defense
bklein@dtic.mil

Jeremiah Knight
University of Maryland
jeremiahknight@yahoo.com

David Korn
Association of American Medical Colleges
dkorn@aamc.org

Thomas Krause
U.S. Patent and Trademark Office
thomas.krause@uspto.gov

Keith Kupferschmid
Software and Information Industry Association
keithk@siia.net

Catherine Langrehr
Association of American Universities
catherine_langrehr@aau.edu

Carol Lee
International Finance Corporation
clee@ifc.org

Ron Lee
Arnold & Porter
ronald_lee@aporter.com

Rolf Lehming
National Science Foundation
rlehming@nsf.gov

Dave Levy
U.S. Environmental Protection Agency
levy.dave@epa.gov

David Lewis
Blackmask Online
llewis@blackmask.org

Tracy Lewis
University of Florida
lewistr@dale.cba.ufl.edu

Rose Li
Analytical Sciences, Inc.
rli@asciences.com

Jean Liddell
Auburn University
liddeje@auburn.edu

David Lide, Jr.
National Institute of Standards and Technology
drlide@post.harvard.edu

Young Lim
Electronics and Telecommunications Research Institute
yylim@etri.re.kr

David Lipman
National Institutes of Health
lipman@ncbi.nlm.nih.gov

Jacqueline Lipton
Case Western Reserve University
jdl14@po.cwru.edu

Richard Llewellyn
Iowa State University
rllew@iastate.edu

Karin Lohman
House Science Committee
karin.lohman@mail.house.gov

Jun Luo
University of Maryland, College of Information Studies
jun@wam.umd.edu

Benjamin Lum
FDC Reports
b.lum@elsevier.com

Patrice Lyons
palyons@bellatlantic.net

Mary Lyons
University of Maryland, College of Information Studies
Lyons_Mary@yahoo.com

Neil MacDonald
Federal Technology Report
Namacdee@aol.com

Constance Malpas
New York Academy of Medicine
cmalpas@nyam.org

Cheryl Marks
Division of Cancer Biology, National Cancer Institute
marksc@mail.nih.gov

David Martinsen
American Chemical Society
d_martinsen@acs.org

Carol Mason
University of Maryland, College of Information Studies
masoncl@wam.umd.edu

APPENDIX C

Eric Massant
Reed Elsevier Inc.
eric.massant@lexisnexis.com

Paul Massell
U.S. Census Bureau
paul.b.massell@census.gov

Yukiko Masuda
University of Washington School of Law
ymasuda@u.washington.edu

Stephen Maurer
Consultant
maurer@econ.berkeley.edu

Anne-Marie Mazza
The National Academies
amazza@nas.edu

Patrice McDermott
OMB Watch
patricem@rtk.net

Stephen Merrill
The National Academies
smerrill@nas.edu

Peggy Merryman
U.S. Geological Survey Library
mmerryma@usgs.gov

Henry Metzger
National Institutes of Health
metzgerh@exchange.nih.gov

Linda Miller
University Corporation for Atmospheric Research/ Unidata
lmiller@unidata.ucar.edu

John Mitchell
Public Knowledge
jmitchell@publicknowledge.org

Kurt Molholm
Defense Technical Information Center
kmolholm@dtic.mil

Richard Monastersky
Chronicle of Higher Education
richm@chronicle.csm

Michael Morgan
Wellcome Trust
m.morgan@wellcome.ac.uk

Rebecca Moser
U.S. Environmental Protection Agency
moser.rebecca@epa.gov

Suzy Mouchet
INSERM, Paris
mouchet@tolbiac.inserm.fr

Teresa Mullins
National Snow and Ice Data Center
tmullins@kryos.colorado.edu

Vincent Munch
Audrey Cohen College
munchv@audreycohen.edu

G. Craig Murray
University of Maryland
gcraigm@glue.umd.edu

Michael Nelson
IBM Corporation
mrn@us.ibm.com

Richard Nelson
Columbia University
ldb3@columbia.edu

Miriam Nisbet
American Library Association
mnisbet@alawash.org

Vivian Nolan
U.S. Geological Survey
vpnolan@usgs.gov

Carlo Nuss
National Institutes of Health
nussc@mail.nlm.nih.gov

Terry O'Bryan
U.S. Enviromental Protection Agency
Obryan.terry@epa.gov

Ann Okerson
Yale University
ann.okerson@yale.edu

Jay Olin
University of Maryland

Jill O'Neill
National Federation of Abstracting and Information Services
jilloneill@nfais.org

Nahoko Ono
University of Tokyo
nahoko_ono@ip.rcast.u-tokyo.ac.jp

Harlan Onsrud
University of Maine
onsrud@spatial.maine.edu

Dianne Ostrow
National Cancer Institute
ostrowd@mail.nih.gov

Pierre Oudet
INSERM
pierre.oudet@chru-strasbourg.fr

Aristides A. Patrinos
U.S. Department of Energy
ari.patrinos@science.doe.gov

Bruce Perens
Hewlett Packard
bruce@perens.com

Virginia Peterman
Howard County, Maryland
vpeterman@co.ho.md.us

Troy Petersen
Duke University
troy.petersen@law.duke.edu

Katharina Phillips
Council on Governmental Relations
kphillips@cogr.edu

Nicole Pinhas
INSERM, Paris
nicole.pinhas@dicdoc.inserm.fr

Neal Pomea
Univeristy of Maryland University College
npomea@umuc.edu

Rudolph Potenzone
LION Bioscience
rudolph.potenzone@lionbioscience.com

Robert Potts
U.S. Department of Veterans Affairs
bob.potts@mail.va.gov

Susan Poulter
University of Utah School of Law
poulters@law.utah.edu

Matt Powell
National Science Foundation
mpowell@nsf.gov

Matthew Quint
Embassy of Australia
matthew.quint@auestemb.org

Roberta Rand
University of Miami
rrand@rsmas.miami.edu

Alan Rapoport
National Science Foundation
arapoport@nsf.gov

Jerome Reichman
Duke University School of Law
reichman@law.duke.edu

Elspeth Revere
The John D. and Catherine T. MacArthur Foundation
erevere@macfound.org

APPENDIX C

Cody Rice
U.S. Environmental Protection Agency
rice.cody@epa.gov

Alison Roeske
IOS Press Inc.
iosbooks@iospress.com

John Rumble
National Institute of Standards and Technology
john.rumble@nist.gov

Carrie Russell
American Library Association
crussell@alawash.org

Joshua Sarnoff
American University
jsarnoff@wcl.american.edu

Suzanne Scotchmer
University of California, Berkeley
scotch@socrates.berkeley.edu

Joan Shigekawa
The Rockefeller Foundation
jshigekawa@rockfound.org

Ituki Shimbo
RIKEN, GSC
shimbo@gsc.go.jp

Joan Sieber
National Science Foundation
jsieber@nsf.gov

Elliot Siegel
National Library of Medicine
siegel@nlm.nih.gov

Pamela Sieving
National Institutes of Health Library
sievingp@ors.od.nih.gov

Mary Silverman
America's Bulletin

William Sittig
The Library of Congress
wsit@loc.gov

F. Hill Slowinski
Worthington International
hslowinski@worthingtoninternational.com

Abby Smith
Council on Library and Information Resources
asmith@clir.org

Kent Smith
National Library of Medicine
kent_smith@nlm.nih.gov

Lowell Smith
U.S. Environmental Protection Agency
smith.lowell@epa.gov

Larry Snavley
Rensselaer Polytechnic Institute
snavll@rpi.edu

Gregory Snyder
U.S. Geological Survey
gsnyder@usgs.gov

Jack Snyder
National Library of Medicine
snyderj@mail.nlm.nih.gov

Anthony So
The Rockefeller Foundation
aso@rockfound.org

Gigi Sohn
Public Knowledge
gbsohn@publicknowledge.org

Sohyun Son
Electronics and Telecommunications Research Institute
shson@etri.re.kr

Shelley Sperry
Coalition for Networked Information
shelley@cni.org

Miron Straf
The National Academies
mstraf@nas.edu

Marti Szczur
National Library of Medicine
szczurm@mail.nlm.nih.gov

Glenn Tallia
National Oceanic and Atmospheric Administration
Glenn.E.Tallia@noaa.gov

Valerie Theberge
The National Academies
vtheberge@nas.edu

Henry Tomburi
California State University at Hayward

Samuel Trosow
University of Western Ontario
strosow@uwo.ca

Kelly Turner
National Oceanic and Atmospheric Administration
kelly.turner@noaa.gov

Paul Uhlir
The National Academies
puhlir@nas.edu

Justin VanFleet
American Association for the Advancement of Science
jvanflee@aaas.org

Walter Warnick
U.S. Department of Energy
warnickw@osti.gov

Jennifer Washburn
jaw529@aol.com

Linda Washington
Centers for Disease Control and Prevention
lrw1@cdc.gov

Rebecca Wason
The American Society for Cell Biology
rwason@ascb.org

Lee Watkins
Johns Hopkins University
lwatkins@jhu.edu

Peter Weiss
National Weather Service
peter.weiss@noaa.gov

Stephen Weitzman
Datapharm Foundation
saw519x@aol.com

Eric Wieser
Nuclear Waste News
ewieser@bpinews.com

Martha Williams
University of Illinois at Urbana-Champaign
m-will13@staff.uiuc.edu

Ann Wolpert
Massachusetts Institute of Technology Libraries
awolpert@mit.edu

Tamae Wong
The National Academies
twong@nas.edu

Stephen Woo
The National Academies
swoo@nas.edu

Julia Woodward
Anne Arundel Community College
jbwoodward@aacc.edu

William Wulf
National Academy of Engineering
wwulf@nae.edu

Young Ye Lin
Electronics and Telecommunications Research Institute
yylim@etri.re.kr

John Young
American Council for the U.N. University Millennium Project
vistasjy@md.prestige.net

Jonathan Zittrain
Harvard School of Law
zittrain@law.harvard.edu

Appendix D

Acronyms and Initialisms

ABA	American Bar Association
BP	British Petroleum
CAS	Chemical Abstracts Service
CGIAR	Consultative Group on International Agricultural Research
CODATA	Committee on Data for Science and Technology
CP	Challenge Program
CSS	content scrambling system
DMCA	Digital Millennium Copyright Act
DOE	Department of Energy
DRM	digital rights management
ELSI	Ethical, Legal and Social Implications
EOSAT	Earth Observation Satellite (Company)
EST	expressed sequence tag
E.U.	European Union
GCC	GNU C compiler
GUI	graphical user interface
GNU	a recursive name for GNU not Unix; refers to Stallman's free software movement
GPL	general public license
HGS	Human Genome Sciences
HP	Hewlett Packard
HR	House Resolution

IEEE	Institute of Electrical and Electronic Engineers
IOM	Institute of Medicine
IP	intellectual property
IPR	intellectual property rights
IT	information technology
LASL	Los Alamos Sequence Library
MDI	Mutations Database Initiative
MIT	Massachusetts Institute of Technology
MTA	Material Transfer Agreement
NAS	National Academy of Sciences
NASA	National Aeronautics and Space Administration
NCDC	National Climate Data Center
NCLS	National Conference of Lawyers and Scientists
NIH	National Institutes of Health
NOAA	National Oceanic and Atmospheric Administration
NRC	National Research Council
NSF	National Science Foundation
OBER	Office of Biological and Environmental Research (DOE)
OMB	Office of Management and Budget
PTO	Patent and Trademark Office
R&D	research and development
RIAA	Recording Industry Association of America
SNP	single nucleotide polymorphism
S&T	Scientific and Technical
STI	scientific and technical data and information
TAC	Technical Advisory Committee
UCITA	Uniform Computer Information Transactions Act

OHIO UNIVERSITY LIBRARY

Please return this book as soon as you he
finished with it. In order to
be returned by the lat
low. All books a